Quantitative Modeling of Earth Surface Processes

Geomorphology is undergoing a renaissance made possible by new techniques in numerical modeling, geochronology and remote sensing. Earth surface processes are complex and richly varied, but analytical and numerical modeling techniques are powerful tools for interpreting these systems and the landforms they create.

This textbook describes some of the most effective and straightforward quantitative techniques for modeling earth surface processes. By emphasizing a core set of equations and solution techniques, the book presents state-of-the-art models currently employed in earth surface process research, as well as a set of simple but practical tools that can be used to tackle unsolved research problems. Detailed case studies demonstrate application of the methods to a wide variety of processes including hillslope, fluvial, eolian, glacial, tectonic, and climatic systems. The computer programming codes used in the case studies are also presented in a set of appendices so that readers can readily utilize these methods in their own work. Additional references are also provided for readers who wish to fine-tune their models or pursue more sophisticated techniques.

Assuming some knowledge of calculus and basic programming experience, this quantitative textbook is designed for advanced geomorphology courses and as a reference book for professional researchers in Earth and planetary sciences looking for a quantitative approach to earth surface processes. Exercises at the end of each chapter begin with simple calculations and then progress to more sophisticated problems that require computer programming. All the necessary computer codes are available online at www.cambridge.org/ 9780521855976.

Jon Pelletier was awarded a Ph.D. in geological sciences from Cornell University in 1997. Following two years at the California Institute of Technology as the O.K. Earl Prize Postdoctoral Scholar, he was made an associate professor of geosciences at the University of Arizona where he teaches geomorphology. Dr Pelletier's research involves mathematical modeling of a wide range of surface processes on Earth and other planets, including the evolution of mountain belts, the transport and deposition of dust in arid environments, and fluvial and glacial processes on Mars.

Quantitative Modeling of Earth Surface Processes

Jon D. Pelletier
University of Arizona

CAMBRIDGE UNIVERSITY PRESS
Cambridge, New York, Melbourne, Madrid, Cape Town, Singapore, São Paulo, Delhi

Cambridge University Press
The Edinburgh Building, Cambridge CB2 8RU, UK

Published in the United States of America by Cambridge University Press, New York

www.cambridge.org
Information on this title: www.cambridge.org/9780521855976

First published 2008

Printed in the United Kingdom at the University Press, Cambridge

A catalog record for this publication is available from the British Library

Library of Congress Cataloging-in-Publication data
Pelletier, Jon D.
Quantitative modeling of earth surface processes / Jon D. Pelletier.
    p.   cm.
Includes bibliographical references and index.
ISBN 978-0-521-85597-6 (hardback)
1. Geomorphology–Mathematical models.   2. Geomorphology–Data processing.
3. Earth sciences–Mathematical models.   4. Earth sciences–Data processing.
I. Title.
GB400.42.M33P45 2008
551.4101′5118–dc22
                       2008003225

ISBN 978-0-521-85597-6 hardback

Contents

Preface   *page* ix

Chapter 1 Introduction   1

1.1 A tour of the fluvial system   1
1.2 A tour of the eolian system   12
1.3 A tour of the glacial system   19
1.4 Conclusions   29

Chapter 2 The diffusion equation   30

2.1 Introduction   30
2.2 Analytic methods and applications   34
2.3 Numerical techniques and applications   57
     Exercises   63

Chapter 3 Flow routing   66

3.1 Introduction   66
3.2 Algorithms   66
3.3 "Cleaning up" US Geological Survey DEMs   70
3.4 Application of flow-routing algorithms to estimate
     flood hazards   72
3.5 Contaminant transport in channel bed sediments   74
     Exercises   85

Chapter 4 The advection/wave equation   87

4.1 Introduction   87
4.2 Analytic methods   88
4.3 Numerical methods   90
4.4 Modeling the fluvial-geomorphic response of the southern
     Sierra Nevada to uplift   93
4.5 The erosional decay of ancient orogens   101
     Exercises   107

Chapter 5 Flexural isostasy   109

5.1 Introduction   109
5.2 Methods for 1D problems   111
5.3 Methods for 2D problems   113
5.4 Modeling of foreland basin geometry   116
5.5 Flexural-isostatic response to glacial erosion in the
     western US   120
     Exercises   123

Chapter 6 Non-Newtonian flow equations 125

6.1 Introduction 125
6.2 Modeling non-Newtonian and perfectly plastic flows 125
6.3 Modeling flows with temperature-dependent viscosity 130
6.4 Modeling of threshold-sliding ice sheets and glaciers over complex 3D topography 132
6.5 Thrust sheet mechanics 147
6.6 Glacial erosion beneath ice sheets 149
Exercises 160

Chapter 7 Instabilities 161

7.1 Introduction 161
7.2 An introductory example: the Rayleigh–Taylor instability 162
7.3 A simple model for river meandering 164
7.4 Werner's model for eolian dunes 166
7.5 Oscillations in arid alluvial channels 169
7.6 How are drumlins formed? 174
7.7 Spiral troughs on the Martian polar ice caps 183
Exercise 187

Chapter 8 Stochastic processes 188

8.1 Introduction 188
8.2 Time series analysis and fractional Gaussian noises 188
8.3 Langevin equations 191
8.4 Random walks 193
8.5 Unsteady erosion and deposition in eolian environments 194
8.6 Stochastic trees and diffusion-limited aggregation 196
8.7 Estimating total flux based on a statistical distribution of events: dust emission from playas 199
8.8 The frequency-size distribution of landslides 205
8.9 Coherence resonance and the timing of ice ages 210
Exercises 221

Appendix 1 Codes for solving the diffusion equation 222

Appendix 2 Codes for flow routing 235

Appendix 3 Codes for solving the advection equation 242

Appendix 4 Codes for solving the flexure equation 256

Appendix 5 Codes for modeling non-Newtonian
flows   263

Appendix 6 Codes for modeling instabilities   267

Appendix 7 Codes for modeling stochastic
processes   274

References   278
Index   290

The colour plates are to be found between pages 108 and 109.

Preface

Geomorphology is undergoing a renaissance made possible by new techniques in numerical modeling, geochronology, and remote sensing. Advances in numerical modeling make it possible to model surface processes and their feedbacks with climate and tectonics over a wide range of spatial and temporal scales. The Shuttle Radar Topography Mission (SRTM) has mapped most of Earth's topography at much higher spatial resolution and accuracy than ever before. Cosmogenic dating and other geochronologic techniques have provided vast new data on surface-process rates and landform ages. Modeling, geochronology, and remote sensing are also revolutionizing natural-hazard assessment and mitigation, enabling society to assess the hazards posed by floods, landslides, windblown dust, soil erosion, and other geomorphic hazards.

The complexity of geomorphic systems poses several challenges, however. First, the relationship between process and form is often difficult to determine uniquely. Many geomorphic processes cannot be readily quantified, and it is often unclear which processes are most important in controlling a particular geomorphic system, and how those processes interact to form the geomorphic and sedimentary records we see today. Terraces and sedimentary deposits on alluvial fans, for example, are controlled by climate, tectonics, and internal drainage adjustments in a way that geomorphologists have not been able to fully unravel. Second, surface processes are strongly influenced by fluid motions, and most classic geomorphic techniques (e.g. field mapping) are not well suited to quantifying fluid dynamics and their interactions with the surface. Third, the geomorphic community must bridge the gap between process-based geomorphology and Quaternary geology. Process-based geomorphologists have made great strides in quantifying transport and erosion laws for geomorphic systems, but this approach has not yet led to major advances in big geologic questions, such as how quickly the Grand Canyon was carved, or how mountain belts respond to glacial erosion, for example. Quaternary geolo-

gists, on the other hand, are adept at reading the geomorphic and sedimentary records to address big questions, but those records often cannot be fully interpreted using field observations and geochronology alone. A fourth challenge is geomorphic prediction. In order for applied geomorphology to realize its full potential, geomorphologists must be able to predict where geomorphic hazards are most likely to occur, taking into account the full complexity of processes, feedbacks, and the multi-scale heterogeneity of Earth's surface.

Analytical and numerical modeling are powerful tools for addressing these challenges. First, modeling is useful for establishing relationships between process and form. Using quantitative models for different processes, modeling allows us to determine the signatures of those processes in the landscape. In some cases, the histories of external forcing mechanisms (e.g. climate and tectonics) can also be inferred. These linkages are important because there is generally no direct observation of landforms that enables us to understand how they evolved. Modeling has played a significant role in recent contributions to our understanding of many classic landform types. In this book, I will present analytic and numerical models for many different landform types, including drumlins, sand dunes, alluvial fans, and bedrock drainage networks, just to name a few. Modeling is particularly useful for exploring the feedbacks between different components of geomorphic systems (i.e. hillslopes and channel networks). Channel aggradation and incision, for example, controls the base level of hillslopes, and hillslopes supply primary sediment flux to channels. Yet many geomorphic textbooks treat these as essentially independent systems. In some cases, the boundary conditions posed by one aspect of a geomorphic system can be considered to be fixed. In many of the most interesting unsolved problems, however, they cannot. An understanding of fluvial-system response to climate change, for example, cannot be achieved without a quantitative understanding of the coupled evolution and feedbacks between hillslopes and channels.

Numerical modeling is also useful in geomorphology because of the central role that fluid dynamics play in landform evolution. Geomorphology can be roughly defined as the evolution of landforms by the fluid flow of wind, liquid water, and ice above the surface. Yet, fluid mechanics generally plays a minor role in many aspects of geomorphic research. Many landform types evolve primarily by a dynamic feedback mechanism in which the topography influences the fluid flow (and hence the shear stresses) above the topography, which, in turn, controls how the topography evolves by erosion and deposition. Numerical modeling is one of the most successful ways to quantify fluid flow in a complex environment, and hence it can and should play a central role in nearly all geomorphic research. In this book, numerical models of fluid flow serve as the basis for many of the book's numerical landform evolution models. Despite the power of numerical models, they cannot be used in a vacuum. All numerical modeling studies must integrate field observations, digital geospatial data, geochronology, and small-scale experiments to motivate, constrain, and validate the numerical work.

Earth surface processes are complex and richly varied. Most books approach the inherent variety of geomorphic systems by serially cataloging the processes and landforms characteristic of each environment (hillslope, fluvial, eolian, glacial, etc.). The disadvantage of this approach is that each geomorphic environment is presented as being essentially unique, and common dynamical behaviors are not emphasized. This book follows a different path, taking advantage of a common mathematical framework to emphasize universal concepts. This framework focuses on linear and nonlinear diffusion, advection, and boundary-value problems that quantify the stresses in the atmosphere and lithosphere. The diffusion equation, for example, can be used to describe hillslope evolution, channel-bed evolution, delta progradation, hydrodynamic dispersion in groundwater aquifers, turbulent dispersion in the atmosphere, and heat conduction in soils and the Earth's crust. By first providing the reader with a solid foundation in the behavior and solution methods for diffusion, the reader will be poised to understand diffusive phenomena as they arise in different contexts. Similarly, boundary-layer and non-Newtonian fluid flows are also examples of equation sets that have broad applicability in Earth surface processes. Non-Newtonian fluid flows are the basis for understanding and modeling lava flows, debris flows, and alpine glaciers. In the course of developing my quantitative skills and applying them to a wide range of research problems, I have been continually amazed at how often a concept or technique from one area can provide the missing piece required to solve a long-standing problem in another field. It is this cross-fertilization of ideas and methods that I want to foster and share with readers. By emphasizing a core set of equations and solution techniques, readers will come away with powerful tools that can be used to tackle unsolved problems, as well as specific knowledge of state-of-the-art models currently used in Earth surface process research. The book is designed for use as a textbook for an advanced geomorphology course and as a reference book for professional geomorphologists. Mathematically, I assume that readers have mastered multivariable calculus and have had some experience with partial differential equations. The exercises at the end of each chapter begin with simple problems that require only the main concepts and a few calculations, then progress to more sophisticated problems that require computer programming.

The purpose of this book is not to provide an exhaustive survey of all analytic or numerical methods for a given problem, but rather to focus on the most powerful and straightforward methods. For example, advective equations can be solved using the upwind-differencing, Lax–Wendroff method, staggered-leapfrog, pseudospectral, and semi-Lagrangian methods, just to name a few. Rather than cover all available methods, this book focuses primarily on the simpler methods for readers who want to get started quickly or who need to solve problems of modest computational size. In this sense, the book will follow the model of *Numerical Recipes* (Press *et al.*, 1992) by providing tools that work for most applications, with additional references for readers who want to fine-tune their models. The appendices provide the reader with computer code to illustrate technique application in real-world research problems. Hence, many of the applications I cover in this book necessarily come from my own research. In focusing on my own work, I don't mean to imply that my work is the only or best approach to a given problem. The book is not intended to be a complete survey of geomorphology, and I knowingly have left out many important contributions in favor of a more focused, case-study approach.

Modelers are not always consistent in the way that they use the terms one- (1D), two- (2D), and three-dimensional (3D) when referring to a model. Usually when physicists and mathematicians use the term 1D they are referring to a model that has one independent spatial variable. This can be confusing when applied to geomorphic problems, however, because a model for the evolution of a topographic profile, $h(x, t)$, for example, would be called 1D even though it represents a 2D profile. Similarly, a model for the evolution of a topographic surface $h(x, y, t)$ would be described as 2D even though the surface itself is three-dimensional. In this book we will use the convention that the dimensionality of the problem refers to the number of independent spatial variables. Therefore, when we model 2D topographic profiles we will classify the model as 1D and when we model the evolution of surfaces we will refer to the model as 2D. This convention may seem strange at first, but it makes the book more consistent overall. For example, if we model heat flow in a thin layer using the diffusion equation, everyone would agree that we are solving a 2D problem for $T(x, y, t)$. If we use the diffusion equation to model the evolution of a hillslope described by $h(x, y, t)$, that, too, is *mathematically* a 2D problem even though it represents a 3D landform.

I wish to acknowledge my colleagues at the University of Arizona for taking a chance on an unconventional geomorphologist and for creating such a collegial working environment. I also wish to thank my graduate students Leslie Hsu, Jason Barnes, James Morrison, Steve DeLong, Michael Cline, Joe Cook, Maria Banks, Joan Blainey, Jennifer Boerner, and Amy Rice for helpful conversations and collaborations. I hope they have learned as much from me as I have from them. Funding for my research has come from the National Science Foundation, the Army Research Office, the US Geological Survey, the NASA Office of Space Science, the Department of Energy, and the State of Arizona's Water Sustainability Program. I gratefully acknowledge support from these agencies and institutions. Finally, I wish to thank my wife Pamela for her patience and support.

Chapter 1

Introduction

1.1 | A tour of the fluvial system

The fluvial system is classically divided into erosional, transportational, and depositional regimes (Schumm, 1977). In the Basin and Range province of the western US, Cenozoic tectonic extension has produced a semi-periodic topography with high ranges (dominated by erosion) and low valleys (dominated by deposition). In this region, all three fluvial-system regimes can be found within distances of 10–20 km. As an introduction to the scientific questions addressed in this book, we start with a tour of the process zones of the fluvial system, using Hanaupah Canyon, Death Valley, California, as a type example (Figure 1.1).

1.1.1 Large-scale topography of the basin and range province

The large-scale geomorphology of the Basin and Range is a consequence of the geometry of faults that develop during extension and the subsequent isostatic adjustment of each crustal block. In the late Cretaceous, the Basin and Range was an extensive high-elevation plateau, broadly similar to the Altiplano-Eastern Cordillera of the central Andes today. When subduction of the Farallon plate ceased beneath the western US, this region became a predominantly strike-slip plate boundary and the horizontal compressive force that supported the high topography and over-thickened crust of the region could no longer withstand the weight of the overlying topography. The result was regional extension (Sonder

and Jones, 1999). The structural style of extensional faulting varies greatly from place to place in the western US, but Figure 1.2 illustrates one simple model of extension that can help us understand the topography of the modern Basin and Range. The weight of the high topography in the region created a vertical compressive force on the lower crust. That compressive force was accompanied by an extensional force in the horizontal direction. Mechanically, all rocks respond to a compressive force in one direction with an extensional force in the other direction in order to preserve volume. This combination of compressive vertical forces and extensional horizontal forces created faults at angles of $\approx$ 5–30° to the horizontal (with smaller angles at greater depths).

Crustal blocks float on the mantle, with the weight of the crust block supported by the buoyancy force that results from the bottom of the crustal block displacing higher-density mantle rock. The shape of each crustal block partially determines how high above Earth's surface the block will stand. Normal faulting creates a series of trapezoidal crustal blocks that, while mechanically coupled by faults, can also rise and fall according to their shape. A crustal block with a wide bottom displaces a relatively large volume of high-density mantle. The resulting buoyancy requires that the block stand higher above Earth's geoid in order to be in isostatic equilibrium. Conversely, a crustal block with a relatively narrow bottom will displace a smaller volume of mantle and will hence stand lower relative to the geoid. This kind of adjustment is the fundamental

Fig 1.1 Hanaupah Canyon and its alluvial fan, Death Valley, California.

(a)

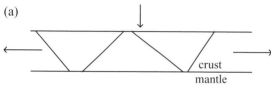

(b)

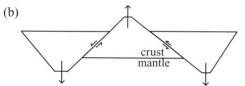

Fig 1.2 Schematic diagram of a simple model for Basin and Range extension and isostatic adjustment. In (a), horizontal extension and vertical compression cause the creation of trapezoidal crustal blocks bounded by a series of normal faults. Crustal blocks with wide bottoms (ranges) rise to maintain isostatic balance, while blocks with narrow bottoms (basins) fall relative to Earth's geoid.

Fig 1.3 Oblique perspective image of the high peaks of Hanaupah Canyon. Reproduced with permission of DigitalGlobe.

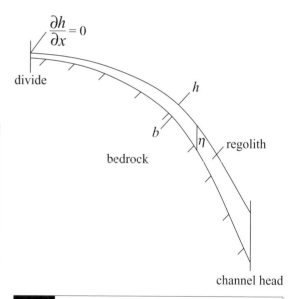

Fig 1.4 Schematic diagram of a hillslope profile from divide to channel head.

reason why extension causes the semi-periodic topography of the Basin and Range Province.

1.1.2 The hillslope system

Hillslopes in the high elevations of Hanaupah Canyon are characterized by steep, planar topographic profiles and abrupt, knife-edge divides (Figure 1.3). West of the divide, water and sediment drains towards Panamint Valley, while on the east side they drain towards Death Valley. If we quantify the flux of water and sediment along a profile that includes the drainage divide, the flux of water and sediment must be negative on one side of the divide and positive on the other

because of this change in drainage direction. At the divide itself, the water and sediment flux is zero, since this is the transition point from western drainage to eastern drainage. Sediment flux on hillslopes is directly related to the topographic gradient. Therefore, since the sediment flux is zero, the topographic gradient, $\partial h / \partial x$, must also be zero (Figure 1.4).

Hillslopes in bedrock-dominated landscapes are composed of a system of two interacting surfaces: the topographic surface, with elevations given by $h(x, y)$, and the underlying weathering front, given by $b(x, y)$ (Figure 1.4). The difference between these two surfaces is the regolith depth, $\eta(x, y)$. The topographic and weathering-front surfaces are strongly coupled because the shape of the topography controls erosion and deposition, which, in turn, changes the values of $\eta(x, y)$ (Furbish and Fagherazzi, 2001). The values of $\eta(x, y)$, in turn, control weathering rates. The simplest system of equations that describes this feedback relationship is given by:

$$\frac{\partial \eta}{\partial t} = -\frac{\rho_b}{\rho_s}\frac{\partial b}{\partial t} - \kappa\frac{\partial^2 h}{\partial x^2} \qquad (1.1)$$

and

$$\frac{\partial b}{\partial t} = -P_0 e^{-\eta/\eta_0} \qquad (1.2)$$

where t is time, ρ_b is the bedrock density, ρ_s is the sediment density, κ is the hillslope diffusivity, x is the distance along the hillslope profile, P_0 is the regolith-production rate for bare bedrock, and η_0 is a characteristic regolith depth. Equation (1.1) states that the rate of change of regolith thickness with time is the difference between a "source" term equal to the rate of bedrock lowering multiplied by the ratio of bedrock to sediment density, and a "sink" term equal to the curvature of the topographic profile. This curvature-based erosion model is the classic diffusion model of hillslopes, first proposed by Culling (1960). The diffusion model of hillslope evolution, discussed in Chapter 2, is a consequence of conservation of mass along hillslope profiles, and the fact that sediment flux is proportional to topographic gradient if certain conditions are met. The diffusion model of hillslope evolution does not apply to the steep hillslopes of Hanaupah Canyon, however. In steep landscapes, sediment flux increases nonlinearly with the topographic gradient as the angle of stability is approached and landsliding becomes the predominant mode of sediment transport. The effects of mass movements on hillslope evolution can be captured in a nonlinear hillslope diffusion model that will also be discussed in Chapter 2. The steep, planar hillslopes and abrupt, knife-edge drainage divides of Hanaupah Canyon are a signature of landslide-dominated, nonlinear transport on hillslopes (Roering et al., 1999).

Equation (1.2) states that bedrock lowering is a maximum for bare bedrock slopes and decreases exponentially with increasing regolith thickness. This relationship has been inferred from cosmogenic isotope analyses on hillslopes (Heimsath et al., 1997). Conceptually, the inverse relationship between weathering/regolith-production rate and regolith thickness is a consequence of the fact that regolith acts as a buffer for the underlying bedrock, protecting it from diurnal temperature changes and infiltrating runoff that act as drivers for physical and chemical weathering in the subsurface. In Figure 1.4, the weathering front is shown as an abrupt transition from bedrock to regolith, but in nature this boundary is usually gradual.

Equations (1.1) and (1.2) can be solved for the steady-state case in which regolith thickness is independent of time at all points along the hillslope profile:

$$\eta = \eta_0 \ln\left(\frac{\rho_b}{\rho_s}\frac{P_0}{\kappa}\frac{1}{-\frac{\partial^2 h}{\partial x^2}}\right) \qquad (1.3)$$

Equation (1.3) implies that, in steady state, regolith thickness decreases as the negative (downward) curvature increases, becoming zero (i.e. bare bedrock) where the curvature reaches a critical value of

$$\left.\frac{\partial^2 h}{\partial x^2}\right|_{bare} = -\frac{\rho_b}{\rho_s}\frac{P_0}{\kappa} \qquad (1.4)$$

Equation (1.2) is not universally applicable. As regolith thickness decreases below a critical value in arid regions, for example, the landscape is unable to store enough water to support significant plant life. Plants act as weathering agents (e.g. root growth can fracture rock, canopy cover can decrease evaporation, etc.). Therefore, in some arid environments weathering rates actually *increase* as regolith thickness increases, rather than decreasing with thickness as the exponential term in Eq. (1.2) predicts (Figure 1.5). Using a vegetation-limited model of weathering in landscape evolution models results in landscapes with a bimodal distribution of slopes (i.e. cliffs and talus slopes) similar to many

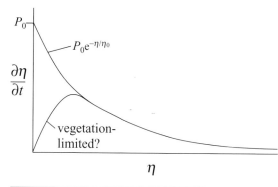

Fig 1.5 Models for the relationship between regolith production and regolith thickness, illustrating the exponential model of Heimsath et al. (1997) and the alternative vegetation-limited model of Anderson and Humphrey (1990). The density contrast between the bedrock and sediment is assumed to be zero for simplicity.

arid-region hillslopes (Anderson and Humphrey, 1990; Strudley et al., 2006). It should also be noted that the parameters P_0 and κ may be approximately uniform along some hillslope profiles, but in general the hillslope and regolith-thickness profiles will influence the hillslope hydrology (e.g. the ratio of surface to subsurface flow), which will, in turn, modify the values of P_0 and κ through time. Thinner regolith, for example, can be expected to increase surface runoff, thereby increasing κ and decreasing P_0 in a positive feedback.

Equations (1.1) and (1.2) suggest that even in well-studied geomorphic systems such as hillslopes, quantifying the behavior of even the most basic elements is a significant challenge. This challenge provides an opportunity for mathematical modeling to play an essential role in geomorphic research, however, because most geomorphic systems are too complex to be fully understood and interpreted with field observations, measurements, and geochronologic techniques alone. The coupled hillslope evolution model of Eqs. (1.1) and (1.2) also illustrates a larger point: many geomorphic systems of greatest interest involve the coupling of different process domains (in this case the hillslope weathering and sediment transport regimes). Other key process-domain linkages occur at the juncture between the hillslope and the channel head, the

channel bed and channel bank, and the channel bed and shoreline, just to name a few.

The transition between the hillslope and fluvial channel system occurs where the shear stress of overland flow is sufficient to entrain hillslope material. In addition, the rate of sediment excavation from the channel head by overland flow must be greater than the rate of sediment infilling by creep and other hillslope processes. Empirically, this transition occurs where the product of the topographic gradient and the square root of contributing area is greater than a threshold value (Montgomery and Dietrich, 1999):

$$S A^{\frac{1}{2}} = X^{-1} \tag{1.5}$$

where S is the topographic gradient or slope, A is the contributing area, and X is the drainage density (i.e. equal to the ratio of the total length of all the channels in the basin to the basin area). The value of X depends on the texture and permeability of the regolith, hillslope vegetation type and density, and on the relative importance of different hillslope processes.

1.1.3 Bedrock channels

Channels are divided into alluvial and bedrock channel types depending on whether or not alluvium is stored on the channel bed. In bedrock channels, the transport capacity of the channel is greater than the upstream sediment flux. Sediment storage is rare or nonexistent in these cases. In alluvial channels, upstream sediment flux is greater than the transport capacity of the channel and alluvium fills the channel bed as a result. The distinction between bedrock and alluvial channels is important for understanding how they evolve. In order for bedrock channels to erode their beds they must pluck or abrade rock from the bed. Once rock material is eroded from a bedrock river it is usually transported far from the site of erosion because the flow velocity needed to transport material is typically much lower than the velocity required for plucking or abrasion. Bedrock channels erode their beds by a combination of plucking, abrasion, and cavitation. Plucking occurs when the pressure of fast-flowing water over the top of a jointed rock causes sufficient Bernoulli lift to dislodge rock from the bed. Abrasion occurs as saltating bedload impacts the bed, chipping away small pieces of rock. Cavitation occurs when water boils under conditions of very high pressure and shear stress. Imploding bubbles in the water create pressures sufficient to pulverize the rock. Conditions conducive to cavitation are most likely to occur in rare large floods. The relative importance of plucking and abrasion depend primarily on lithology. Abrasion is generally considered to be the dominant process in massive bedrock lithologies such as granite (Whipple *et al.*, 2000). In sedimentary rocks, abrasion and plucking are both likely to be important.

Several mathematical models exist for modeling bedrock channel evolution. In the stream-power model, first proposed by Howard and Kerby (1983), the rate of increase or decrease in channel bed elevation is equal to the difference between the uplift rate and the erosion rate. The erosion rate is a power function of drainage area and channel-bed slope:

$$\frac{\partial h}{\partial t} = U - K_w A^m \left| \frac{\partial h}{\partial x} \right|^n \tag{1.6}$$

where h is the local elevation, t is time, U is uplift rate, K_w is a constant that depends on bedrock erodibility and climate, A is the drainage area, m and n are empirical constants, and x is the distance along the channel. The general form of the stream-power model also includes an additional constant term that represents a threshold shear stress for erosion (Whipple and Tucker, 1999). The stream-power model is empirically based: Howard and Kerby (1983) found that Eq. (1.6) successfully reproduced observed erosion rates at a site in Perth Amboy, New Jersey, measured by repeat survey. Their analysis cannot, however, rule out whether other variables correlated with drainage area and slope are the fundamental controlling variables of bedrock erosion. Nevertheless, the stream power is based on the physically-reasonable assumption that the erosive power of floods increases as a function of drainage area (which controls the volume of water routed through the channel) and channel-bed slope (which controls the velocity of that water for otherwise similar drainage areas and precipitation intensities). Calibrations of the stream-power model in natural systems suggest that

erosion is usually linear with slope (i.e. $n \approx 1$ and $m \approx 1/2$ (Kirby and Whipple, 2001)).

In steady state, uplift and erosion are in balance and $\partial h/\partial t = 0$. Equation (1.6) predicts that channel slope, $S = |\partial h/\partial x|$, is inversely proportional to a power-law function of drainage area, A, for drainage basins in topographic steady state:

$$S = \frac{U}{K_w} A^{-m/n} \tag{1.7}$$

Equation (1.7) predicts that, as drainage area increases downstream, the channel slope must decrease in order to maintain uniform erosion rates across the landscape. This inverse relationship between slope and area is responsible for the concave form of most bedrock channels.

Equation (1.7) can be used to derive an analytic expression for the steady-state bedrock channel longitudinal profile. To do this, we must relate the drainage area, A, to the distance along the channel from the channel head, x. The relationship between area and distance depends on the planform (i.e. map-view) shape of the basin. For a semi-circular basin, the square root of drainage area is proportional to the distance from the channel head. Adopting that assumption together with the empirical observation $m/n = 0.5$, Eq. (1.7) becomes

$$\frac{\partial h}{\partial x} = -\frac{C}{x} \tag{1.8}$$

where $\partial h/\partial x$ has been substituted for S and C is a constant that combines the effects of uplift rate, bedrock erodibility, climate, and basin shape. Integrating Eq. (1.8) gives

$$h - h_0 = -C \ln\left(\frac{x}{L}\right) \tag{1.9}$$

where h_0 is the elevation of the base of the channel at $x = L$. Figure 1.6 shows the longitudinal profile of the main channel of Hanaupah Canyon together with a plot of Eq. (1.9). The model fits the observed profile quite well using $C = 0.9$ km.

The evolution of a hypothetical bedrock channel governed by the stream-power model is illustrated qualitatively in Figure 1.7a. A small reach of the channel's longitudinal profile is shown at times t_1, t_2, and t_3. If the section is relatively small and has no major incoming tributaries, the drainage area can be considered uniform

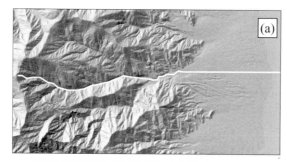

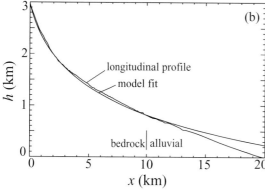

Fig 1.6 (a) Shaded-relief image of the Hanaupah Canyon drainage network and alluvial fan. Location of longitudinal profile shown as white curve. (b) Longitudinal profile of main Hanaupah Canyon channel, together with best-fit to Eq. (1.8).

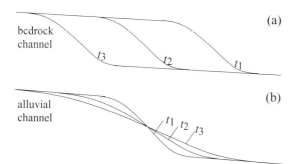

Fig 1.7 Schematic diagrams of the evolution of (a) bedrock and (b) alluvial channels through time, illustrating the advective behavior of bedrock channels and the diffusive behavior of alluvial channels.

throughout the reach. If drainage area is uniform and $n = 1$ is assumed, the erosion rate is proportional to the channel slope according to the stream-power model. Accordingly, steeper portions of the bed will erode faster, result-

ing in the propagation of knickpoints upstream as topographic waves. The rate of knickpoint propagation is equal to $K_w A^m$, i.e. knickpoints propagate faster in larger channels, wetter climates, and areas of more easily-eroded bedrock. The stream-power model will serve as the type example of the advection/wave equation studied in Chapter 4.

The Kern River (Figures 1.8 and 1.9) provides a nice example of knickpoint propagation in action. Two distinct topographic surfaces have long been recognized in the landscape of the southern Sierra Nevada (Webb, 1946) (Figure 1.8). The Boreal surface is a high-elevation, low-relief plateau that dips to the west at 1° (Figure 1.8b). The Chagoopa Plateau is an intermediate topographic "bench" that is restricted to the major river canyons and inset into the Boreal Plateau (Webb, 1946; Jones, 1987). Figure 1.8b maps the maximum extents of the Chagoopa and Boreal Plateaux based on elevation ranges of 1750–2250 m (Chagoopa) and 2250–3500 m a.s.l. (Boreal). Associated with each surface are prominent knickpoints along major rivers. Knickpoints along the North Fork Kern River, for example (Figure 1.9b), occur at elevations of 1600–2100 m and 2500–3300 m a.s.l. The stepped nature of the Sierra Nevada topography is generally considered to be the result of two pulses of Cenozoic and/or late Cretaceous uplift (Clark et al., 2005; Pelletier, 2007c). According to this model, two major knickpoints were created during uplift, each initiating a wave of incision that is still propagating headward towards the range crest.

Recent work has highlighted the importance of abrasion in controlling bedrock channel evolution. In the abrasion process it is sediment, not water, that acts as the primary erosional agent. In the stream-power model, the erosive power is assumed to be a power function of drainage area. Although sediment flux increases with drainage area, upstream relief also plays an important role in controlling sediment flux. As such, the stream-power model does not adequately represent the abrasion process. Sklar and Dietrich (2001, 2004) developed a saltation-abrasion model to quantify this process of bedrock channel erosion. Insights into their model can be gained by replacing drainage area with sediment flux in the

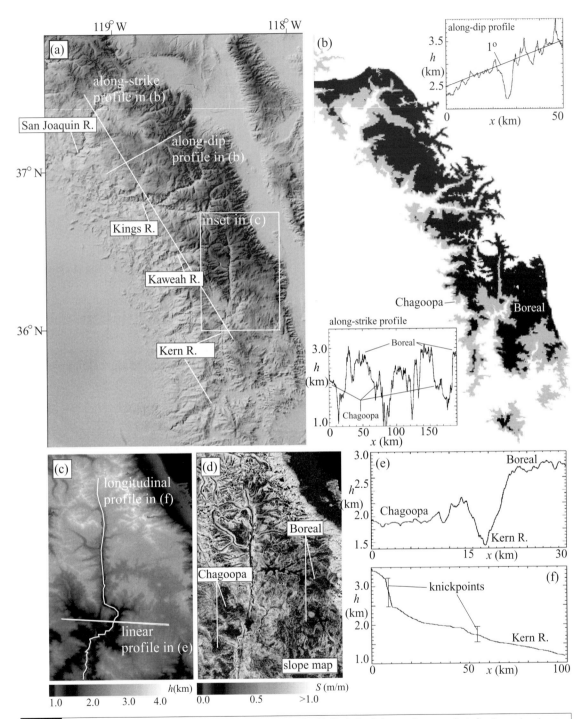

Fig 1.8 Major geomorphic features of the southern Sierra Nevada. (a) Shaded relief map of topography indicating major rivers and locations of transects plotted in b. (b) Maximum extents of the Chagoopa and Boreal Plateaux based on elevation ranges of 1750–2250 m and 2250–3500 m a.s.l. Also shown are along-strike and along-dip topographic transects illustrating the three levels of the range in the along-strike profile (i.e. incised gorges, Chagoopa and Boreal Plateaux) and the westward tilt of the Boreal Plateau in the along-strike transect. (c) and (d) Grayscale map of topography (c) and slope (d) of the North Fork Kern River, illustrating the plateau surfaces (e) and their associated river knickpoints (f). For color version, see plate section. Modified from Pelletier (2007c). Reproduced with permission of Elsevier Limited.

Fig 1.9 Virtual oblique aerial photographs of portions of the Kern River basin. (a) The upper portion of the basin is characterized by the low-relief Boreal Plateau. The mainstem Kern River is incised 1–2 km into the Boreal Plateau. (b) Relief between the Boreal Plateau and Kern River is accommodated by a series of channel knickpoints along the Kern River and its tributaries.

stream-power model to obtain a sediment-flux-driven model:

$$\frac{\partial h}{\partial t} = U - K_s Q_s^m \left| \frac{\partial h}{\partial x} \right|^n \qquad (1.10)$$

where Q_s is the sediment flux and K_s is a new coefficient of erodibility.

The predictions of the stream-power and sediment-flux-driven models are similar for cases of uniformly uplifted, steady-state mountain belts. In such cases, erosion is spatially uniform (everywhere balancing uplift) and therefore sediment flux Q_s is proportional to drainage area A. The predictions of the two models are very different, however, following the uplift of an initially low-relief plateau. In such cases, the stream-power model predicts a relatively rapid response to uplift because large rivers with sub-

stantial stream power drain the plateau. In the sediment-flux-driven model, however, low erosion rates on the plateau limit the supply of sediment that acts as cutting tools to channels draining the plateau edge. As a result, the sediment-flux-driven model leads to much slower rates of knick-point retreat than those predicted by the stream-power model in a plateau-type landscape (Gasparini et al., 2006).

The sediment-flux-driven model also implies that the evolution of hillslopes and channels is more intimately linked than the stream-power model would suggest. In the stream-power model, hillslope evolution plays no explicit role in bedrock channel evolution. In the sediment-flux-driven model, however, increased hillslope erosion will supply more cutting tools to bedrock channels downstream. Downstream channels will respond to this increased supply with faster incision, further lowering the base level for hillslopes in a positive feedback. These ideas will be further explored in Chapter 4 where the stream-power and sediment-flux-driven models are compared in detail.

As one travels through the fluvial system, the transport capacity of the main channel generally increases with downstream distance as a result of increasing contributing area. The effect of declining channel slope, however, partially offsets that effect. As a result, transport capacity increases downstream, but not as rapidly as the sediment supply that the channel is required to transport. The bedrock-alluvial transition occurs where the sediment supply exceeds the transport capacity. In most drainage basins of the western US, the bedrock-alluvial transition occurs upstream of the mountain front (e.g. Figure 1.10). Only in regions that are tectonically most active does the bedrock-alluvial transition occur at the mountain front itself.

1.1.4 Alluvial channels
Alluvial channels evolve in a fundamentally different style than that of bedrock channels (Figure 1.7). Alluvial channel evolution is governed by a conservation of mass relationship which states that the change in channel-bed elevation is equal to the gradient in bedload sediment flux along

Fig 1.10 Oblique perspective image of channels of
Hanaupah Canyon. Smaller channels carve directly into
bedrock, while the largest channels have wider beds filled
with alluvium. Reproduced with permission of DigitalGlobe.

the channel profile:

$$\frac{\partial h}{\partial t} = -\frac{1}{c_0}\frac{\partial(wq_s)}{\partial x} \tag{1.11}$$

where h is the elevation of the channel bed, t is time, c_0 is the volumetric concentration of bed sediment, w is the channel width, q_s is the sediment discharge per unit channel width, and x is the distance downstream. Equation (1.11) is simply a statement of conservation of mass, i.e. the channel bed must aggrade if the sediment-flux gradient is negative (if more sediment enters the reach from upstream than leaves it downstream) and incise if the sediment-flux gradient is positive (if more sediment leaves the reach than enters it). A number of different relationships exist for quantifying bedload sediment flux in alluvial channels, but one common approach quantifies sediment discharge as a linear function of channel gradient and a nonlinear function of the discharge per unit channel width:

$$q_s = -B\left(\frac{Q}{w}\right)^b \frac{\partial h}{\partial x} \tag{1.12}$$

where B is a mobility parameter related to grain size, Q is water discharge, and b is a constant. The value of b is constrained by sediment rating curves and is generally between 2 and 3 for bed-

load transport. If channel width is assumed to be uniform along the longitudinal profile, the combination of Eqs. (1.11) and (1.12) gives the classic diffusion equation (Begin et al., 1981):

$$\frac{\partial h}{\partial t} = \kappa \frac{\partial^2 h}{\partial x^2} \tag{1.13}$$

where the diffusivity is given by $\kappa = (BQ^b)/(c_0 w_0)$, and w_0 is the uniform channel width. The diffusive evolution of alluvial channels of uniform width is illustrated schematically in Figure 1.7b.

1.1.5 Alluvial fans
Deposition occurs on alluvial fans primarily because of channel widening near the mountain front. Abrupt widening decreases the transport capacity with distance downfan. Under such conditions, Eq. (1.11) predicts aggradation and fan development.

Hanaupah Canyon has one of the most spectacular alluvial fans in the western US. The size of the Hanaupah Canyon fan reflects the large sediment flux draining from Hanaupah Canyon, which, in turn, is a result of the high relief and semi-arid climate of this drainage basin. The semi-arid climate maximizes runoff intensity while minimizing erosion-suppressing vegetation cover. Many alluvial fans in the western US, including Hanaupah Canyon fan, have a series of distinct terraces (Figure 1.11) that rise like a flight of stairs from the active channel, with older terraces occurring at higher topographic positions relative to the active channel. Alluvial-fan terraces in many areas of the western US can be correlated on the basis of their elevation above the active channel, their degree of desert pavement and varnish development, and their extent of degradation. Terraces form as a result of changes in the ratio of sediment supply to transport capacity through time. During times when the ratio of sediment supply to transport capacity is high, aggradation and channel widening operate in a positive feedback that grows the fan vertically and radially. During times when the ratio of sediment supply to transport capacity is low, incision and channel narrowing operate in a positive feedback that causes the channel to entrench into older fan deposits, leaving an abandoned terrace.

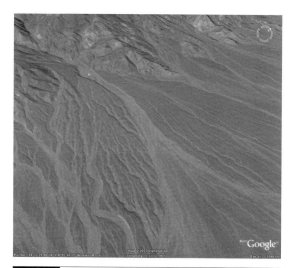

Fig 1.11 Oblique perspective image of the terraces of Hanaupah Canyon fan. Reproduced with permission of DigitalGlobe. Younger terraces are lighter in color in this image, representing the relatively limited time available for desert pavement and varnish formation on younger surfaces.

Over time, episodes of aggradation, incision, and lateral reworking produce a nested sequence of terraces.

Multiple cut and fill cycles on alluvial fans create a spatially-complex, distributary channel network that presents a challenge to floodplain managers in the western US. The Tortolita Mountains fan northwest of Tucson, Arizona is a classic example. Figure 1.12 presents four views of this topographic complexity using a shaded relief image, aerial photo, grayscale map of a numerical model of flow depth during a recent extreme flood, and a surficial geologic map. The surficial geologic map (Figure 1.12d) was constructed by integrating soil development and other indicators of terrace age to group the terraces into distinct age ranges (Gile *et al.*, 1981; McFadden *et al.*, 1989). The surface age represents the approximate time since deep flooding occurred on the terrace because soils would be stripped from a terrace subjected to deep scour and buried on a surface subjected to significant fluvial deposition. Surficial geologic mapping indicates that flood risk (which is inversely correlated with surface age) varies greatly even at scales less than 1 km. The modeled flow depths also illustrate the spatial complexity of flooding. As Figure 1.12c, the

main channels on the fan branch into dozens of distributary channels separated by horizontal distances of only a few hundred meters or less.

The triggering mechanisms for alluvial-fan terrace formation have long been debated, but climatic changes most likely play a significant role in controlling the changes in sediment supply that trigger fan aggradation and incision. A growing database of surface and stratigraphic age estimates suggests that Quaternary geomorphic surfaces and underlying deposits of the western US can be correlated regionally (Christensen and Purcell, 1983; Bull, 1991; Reheis *et al.*, 1996; Bull, 1996). Several studies have documented fill events and/or surface exposure dates between 700–500 ka, 150–120 ka, 70–50 ka, and 15–10 ka corresponding to the Q2a, Q2b, Q2c, and Q3 geomorphic surfaces identified by Bull (1991) (surface exposure ages correspond to youngest ages in these ranges). Similarity of ages regionally has provided preliminary support for the hypothesis that Quaternary alluvial-fan terraces are generated by climatic changes.

Changes in sediment supply due to climatic changes can result from several factors, but variations in drainage density are likely to play a significant role in controlling the temporal variations in sediment flux from drainage basins in the western US. Terrace formation during the Pleistocene–Holocene transition is associated with a ten-fold increase in sediment supply (e.g. Weldon, 1980). It is unlikely that a change in precipitation alone could account for such large increases in sediment supply. During Pleistocene climates, vegetation densities were higher at most elevations across the western US. Higher vegetation density results in a lower drainage density. During times of lower drainage density, accommodation space is created in hollows for the deposition of sediment eroded from higher up on the hillslope, thereby lowering sediment fluxes from the basin relative to long-term geologic averages. During humid-to-arid transitions, drainage densities increase, removing sediment stored as colluvium in hollows during the previous humid interval. Sediment fluxes decrease when the drainage density reaches a new maximum in equilibrium with the drier climate. According to this model, it is the *change* in climate,

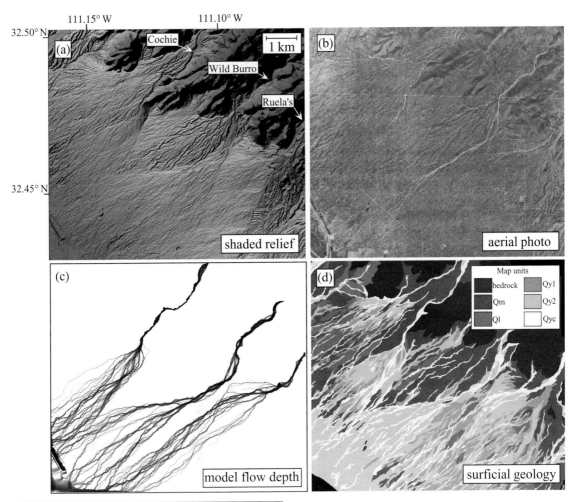

Fig 1.12 (a) Shaded-relief image, (b) aerial photo, (c) flow depth model prediction, and (d) surficial-geologic map of the Tortolita Mountains fan northwest of Tucson. Multiple cycles of cutting and filling on this fan have resulted in a complex distributary channel network in which flooding from confined channels upstream branches into dozens of active channels downstream on the fan.

not the absolute state of climate (wet or dry), that most strongly controls sediment fluxes from drainage basins.

Figure 1.13 schematically illustrates the relationships between precipitation, vegetation density, and sediment flux in the western US during fluctuating climates according to the fluctuating drainage density model. Precipitation and vegetation changes oscillate in phase over geologic time scales. Sediment flux, however, oscillates out of

phase with climate because it is responding to the rate of change of climate, not its absolute state. The duration of the climate oscillation also plays a role in landscape response. Short-term climatic oscillations may not allow sufficient time for sediment to be stored on the landscape during times of lower drainage density. Long-term climate changes, however, allow time for more sediment to be stored during low-drainage-density humid intervals, resulting in a larger response when the climate shifts to more arid conditions.

In the western US, the elevation zone of greatest vegetation change is well constrained through the Last Glacial Maximum (LGM) by packrat middens. Today, the modern tree line (i.e. the location where shrub vegetation transitions to mature woody species) occurs at approximately 2000 m a.s.l. (with variations of a few hundred

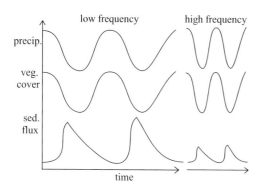

Fig 1.13 Schematic diagram of the relationships between precipitation, vegetation density, and sediment flux for low-frequency (long-duration) and high-frequency (short duration) climate changes.

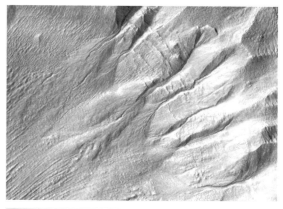

Fig 1.14 A HiRISE Camera image of erosional gullies on Mars and their adjacent alluvial fans.

meters due to slope aspect). At the LGM, the tree line was approximately 500 m a.s.l. In the western US, therefore, the elevational range from 500–2000 m has undergone significant changes in vegetation type and density many times within the Plio–Quaternary period (Spaulding, 1990). It is within this elevation range that we can expect to see the most significant impact of climate change on fluvial drainage basin processes. Of course, at much higher elevations some drainage basins in the western US were also glaciated. Glaciation has a huge impact on fluvial processes downstream by delivering large volumes of sediment due to the erosional efficiency of alpine glaciers.

The floor of Death Valley is a saline playa that has been an ephemeral shallow lake many times in the past. Pluvial lakes occupied large areas in the western US during much of the Pleistocene. During pluvial lake highstands, prominent shorelines were created by wave-cut action in many areas of Utah (Bonneville shoreline) and Nevada (Lahontan shoreline) (Reheis, 1999). Degradation of these shorelines over time provides natural experiments in hillslope evolution that have been exploited by hillslope geomorphologists. Fluctuations in lake levels also store and release vast amounts of windblown dust in arid regions, thereby influencing soil formation and hydrology on nearby alluvial-fan terraces (Reheis *et al.*, 1995).

One of the most exciting developments in surface process research is the wealth of new data emerging on the surface of Mars. Planetary scientists are currently grappling with many of the same science questions that geomorphologists puzzle over on Earth. Figure 1.14 presents one of many images of the Martian surface that is tantalizingly similar to fluvial landforms on Earth. Figure 1.14 shows a series of erosional valleys in the upper right-hand side of the image, transitioning into entrenched depositional fans with multiple terrace levels. Whether these terraces represent the effects of deposition during dry or wet flow events is not currently known. Either way, the similarity between many of the fluvial landforms on Mars and those on Earth is striking, especially given that liquid water is unstable on Mars under the temperature and pressure conditions of today. The wealth of imagery and elevation data from Mars provides an excellent opportunity to compare and contrast Martian landforms to those on Earth in order to learn more about how landforms evolved on both planets.

1.2 | A tour of the eolian system

The eolian system can be divided into two types: systems dominated by the transport of silt and clay, and systems dominated by the transport of sand. Both types of systems involve particle entrainment from the ground, atmospheric transport, and redeposition by gravitational settling. The primary difference between these two types

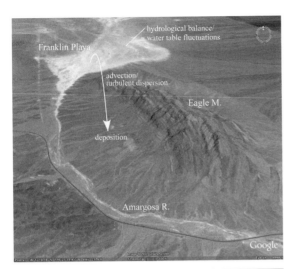

Fig 1.15 Oblique perspective image of Franklin Lake Playa and adjacent Eagle Mountain piedmont. Franklin Lake Playa acts as a strong regional source of dust. Terraces on piedmonts such as that of Eagle Mountain act as net sinks for dust in arid environments. Reproduced with permission of DigitalGlobe.

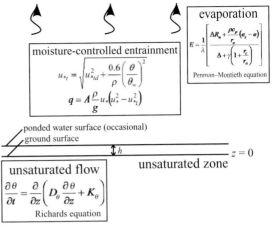

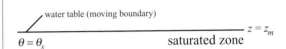

Fig 1.16 Schematic figure illustrating the coupling between the hydrologic and eolian systems on playa surfaces.

of systems is the length and time scales involved in particle motion. Silt and clay transport involves travel distances on the order of 1 km or greater and residence times on the order of 1 hour or more in the atmosphere. Sand, in contrast, moves primarily in saltation. Saltation involves trajectories on the order of 10–1000 cm and residence times on the order of 1 s. In this section, we will focus first on the transport of dust in arid environments (i.e. silt and clay transport), using Franklin Lake Playa in Amargosa Valley, California, as a type example. Second, we will explore the transport of sand across an alluvial fan using Whiskey Petes' alluvial fan as an example. Finally, we will explore the formation mechanisms of eolian dunes.

1.2.1 The dust cycle in arid environments

The dust cycle in arid environments refers to the transport of dust as it moves from sources (e.g. playas and dry channel beds) to sinks (e.g. alluvial fan terraces) and ultimately back to sources, primarily by the action of floods. Franklin Lake Playa and the Amargosa River drainage basin (Figure 1.15) provide a classic example of the dust cycle in action. Most of the dust that is deposited on the

alluvial fan terraces of Eagle Mountain piedmont are sourced from Franklin Lake Playa. The dust of Franklin Lake Playa is, in turn, sourced from the Amargosa Valley drainage system, of which Eagle Mountain piedmont is a part. As the Amargosa River drains into Franklin Lake Playa, its low gradient and distributary channel geometry causes most of the sediments it carries to settle out and be deposited on the playa. This dust is then susceptible to windblown transport when the playa becomes dessicated. Some of this windblown dust is deposited immediately downwind (e.g. on the alluvial fan terraces of Eagle Mountain alluvial fan), while some is transported thousands of kilometers away to the ocean or to a river that ultimately drains into the ocean. As such, the dust cycle is not a closed system on the continents: atmospheric transport results in a net transport of dust from continents to the ocean. Rock weathering continually generates new dust for this cycle.

The amount of dust emitted from playas varies greatly in space and time. The hydrologic state of a playa plays a significant role in controlling this spatial and temporal variation in dust-emitting potential (Fig. 1.16) (Reynolds *et al.*,

2007). Generally speaking, wet playas (those with shallow groundwater tables and hence significant surface moisture) can be expected to emit less dust than dry playas under similar wind conditions, because in wet playas the surface moisture creates a cohesive force between grains that is not present in dry playas. Franklin Lake Playa, however, is one of the most active playas in the western US despite its very shallow water table (less than 3 m below the surface in most portions of the playa). The shallow water table beneath Franklin Lake Playa causes a vapor discharge that disrupts the formation of surface crusts that would otherwise serve to protect the surface from erosion. The result is a soft puffy surface on Franklin Lake Playa that promotes dust emission (Czarnecki, 1997). In general, dust production on playas is relatively low for very wet or very dry playas, and relatively high for playas in an intermediate state.

Temporal variations in dust emissions from a single playa are also quite complex. In many playas, dust emissions are observed to be negatively correlated with antecedent precipitation. In such cases, flood events cause a pulse of recharge that raises the water table, wetting the surface and increasing the threshold wind velocity. In other playas, however, dust emissions are positively correlated with antecedent rainfall (Reheis, 2006). In these playas, dust emissions are limited by the supply of fine-grained material to the playa. In such playas, significant dust storms are produced only when antecedent floods bring sufficient sediment to the playa to be mobilized when the playa becomes dessicated.

Particles are entrained by the wind when the Bernoulli lift created by wind flowing over the top of the particle exceeds the weight of the particle. The Bernoulli force, in turn, is directly related to the shear velocity exerted by the flow, given by (Bagnold, 1941)

$$u_* = \frac{\kappa u_m}{\ln\left(\frac{z_m}{z_0}\right)} \tag{1.14}$$

where κ is the von Karman constant (equal to 0.4), u_m is the wind velocity measured at a height z_m above the ground, and z_0 is the aerodynamic roughness length of the surface. The roughness length z_0, in turn, is related to the size and shape of the surface elements (e.g. sand ripples, plants, rocks, etc.). Particles can be entrained by the wind when the shear velocity is greater than a threshold value, u_{*td}, given by

$$u_{*td} = A\sqrt{\frac{\rho_s - \rho_a}{\rho_s}gD} \tag{1.15}$$

where A is a constant of proportionality (equal to 0.1 for air), ρ_s is the density of sediment, ρ_a is the density of air, g is the acceleration due to gravity, and D is grain size diameter.

If the soil is wet, Eq. (1.15) must be modified to include the effects of cohesion. Chepil (1956) developed an empirical equation to represent the effects of moisture on the threshold shear velocity:

$$u_{*t} = \sqrt{u_{*td}^2 + \frac{0.6}{\rho_a}\left(\frac{\theta}{\theta_{1.5}}\right)^2} \tag{1.16}$$

where θ is the volumetric soil moisture and $\theta_{1.5}$ is the soil moisture at a pressure of -1.5 MPa (i.e. the wilting point).

Quantifying the dust cycle is important for geomorphic, pedologic, and hydrologic reasons. Dust deposition, for example, controls the permeability of alluvial fan terraces and their rates of soil development. Windblown dust is also a very significant human health hazard. Windblown particulate matter includes both natural (e.g. mineral dust from playas) and anthropogenic (e.g. smoke from fires), but natural sources represent a significant portion of the total. Studies have shown a correlation between *daily* mortality and high levels of particulate matter (PM) in the atmosphere (Samet *et al.*, 2000). This means that, in addition to causing long-term respiratory ailments, high dust concentrations also cause sudden death in some individuals. Figure 1.17 is a photograph of a dust storm that hit Las Vegas, Nevada on April 15, 2002. Although such dust storms have long been a hazard, rapid population growth in the western US may be contributing to more dust storms through groundwater withdrawal and land disturbance.

Atmospheric transport of particulate matter can be modeled as a combination of downwind advection, turbulent diffusion, and gravitational settling. The steady-state equation describing

Fig 1.17 Photograph of a dust storm in Las Vegas, Nevada, on April 15, 2002, when particulate matter (PM) concentrations reached above 1400 μg/m^3.

these processes is given by

$$K \left(\frac{\partial^2 c}{\partial x^2} + \frac{\partial^2 c}{\partial y^2} + \frac{\partial^2 c}{\partial z^2} \right) - u \frac{\partial c}{\partial x} + q \frac{\partial x}{\partial z} = 0$$

$$(1.17)$$

where K is the turbulent diffusivity, c is the particle concentration, x is the downwind distance, y is the crosswind distance, z is the vertical distance from the ground, u is the mean wind velocity, and q is the settling velocity (model geometry shown in Figure 1.18a). Solutions to Eq. (1.17) are known as Gaussian plumes. This version of the advection-diffusion-settling equation assumes that K and u are uniform. More complex models are also available in which K and u vary with height to better represent transport processes close to the ground (e.g. Huang, 1999).

Deposition from a Gaussian plume is modeled by treating the ground as a sink for particles. Deposition in this model is characterized by a deposition velocity, p, defined as the fraction of the particle concentration just above the ground that undergoes deposition per unit time. In this model framework, deposition at the ground surface is equal to the downward flux due to turbulent diffusion and particle settling. This balance provides a flux boundary condition at $z = 0$:

$$K \left. \frac{\partial c}{\partial z} \right|_{z=0} + qc(x, y, 0) = pc(x, y, 0) \qquad (1.18)$$

Physically, the deposition velocity represents the trapping ability of the surface, independent of whether or not the particles are falling under gravity. Silt, for example, has a negligible settling velocity but a finite depositional velocity because silt particles are deposited as they fall or lodge between the crags of a rough surface (e.g. clasts in a desert pavement).

Dust deposition plays an important role in creating desert pavement, one of the most enigmatic landforms of the arid environment. Desert pavement is a type of surface landform formed on alluvial fan terraces in low-elevation parts of the western US and other arid regions around the world. Desert pavement is characterized by a monolayer of stones at the surface, each stone tightly sutured to the next like pieces in a jigsaw puzzle. Digging into the pavement reveals that, in most places, the desert pavement sits atop a silt-rich layer comprised predominantly of wind-blown material (Figure 1.19) (McFadden et al., 1987; Wells et al., 1995). While desert pavement was originally thought to represent a lag of coarse material left behind as finer material was eroded out, the observation of a layer of fine material underneath the stony monolayer is most consistent with a model for pavement formation in which windblown silt and clay is deposited in the interstitial spaces between surface stones. Once an initial layer of silt and clay is deposited on and immediately underneath the surface, the reorganization of stony clasts can occur during wetting and drying events, freeze–thaw cycles, and bioturbation (all of which require, or are enhanced by, fine-grained sediments). The reorganization of clasts, in turn, helps protect the underlying fine-grained layer from erosion. In this way, the desert pavement and underlying eolian deposit coevolve. This coevolution is thought to take several thousand years to initiate, but may occur faster in areas of higher dust deposition rates (Pelletier et al., 2008a).

The texture of the alluvial-fan parent material plays an important role in determining the specific mode of pavement formation. In gravel-rich parent material (i.e. those found near the proximal part of an alluvial fan close to the mountain front), clast motion on the surface is unlikely to be significant until the underlying eolian layer is sufficiently thick to cause expansion/contraction of the near-surface layer

(a)

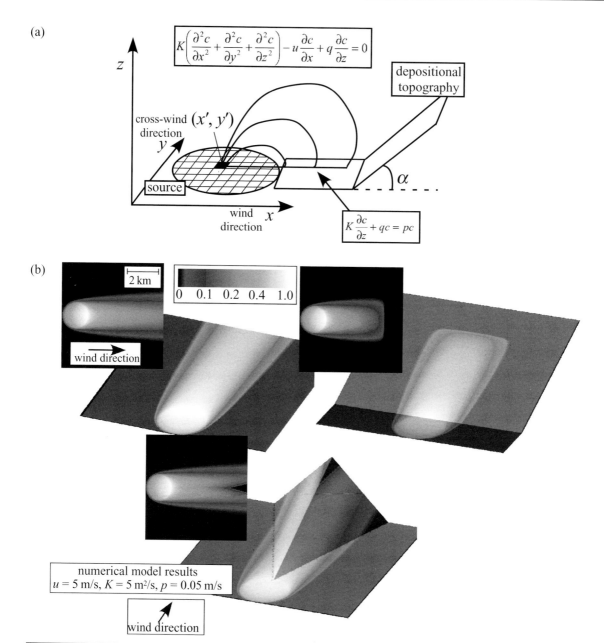

$$K\left(\frac{\partial^2 c}{\partial x^2} + \frac{\partial^2 c}{\partial y^2} + \frac{\partial^2 c}{\partial z^2}\right) - u\frac{\partial c}{\partial x} + q\frac{\partial c}{\partial z} = 0$$

z

depositional topography

cross-wind (x', y') direction

y

source

wind x direction

α

$$K\frac{\partial c}{\partial z} + qc = pc$$

(b)

2 km

0   0.1   0.2   0.4   1.0

wind direction

numerical model results
$u = 5$ m/s, $K = 5$ m²/s, $p = 0.05$ m/s

wind direction

Fig 1.18 (a) Schematic diagram of model geometry. Depositional topography shown in this example is an inclined plane located downwind of source (but model can accept any downwind topography). (b) Grayscale maps of 3D model results illustrating the role of variable downwind topography. In each case, model parameters are $u = 5$ m/s, $K = 5$ m²/s, and $p = 0.05$ m/s. Downwind topography is, from left to right, a flat plane, an inclined plane, and a triangular ridge. Modified from Pelletier and Cook (2005).

through wetting–drying and freeze–thaw cycles. In cases of sand-dominated parent material (e.g. the distal portions of many alluvial fans), the initial surface has relatively few clasts with which to form a pavement. In these cases, pavement clasts must first be pushed to the surface by freeze–thaw or wetting–drying cycles, or possibly made available through progressive fracturing of larger clasts (McFadden *et al.*, 2005). As the surface gains clast-material coverage, lateral migration serves to interlock and suture the clasts as in the

Fig 1.19 Layered stratigraphy of typical alluvial fan terrace (Eagle Mountain piedmont), illustrating stony monolayer comprising the pavement, underlain by silt-rich eolian deposits sitting atop parent alluvium.

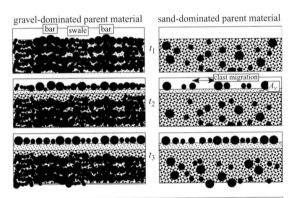

Fig 1.20 Schematic illustration of two modes of parent-material formation associated with gravel-dominated and sand-dominated parent material.

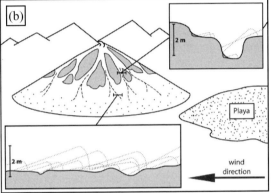

Fig 1.21 (a) Virtual oblique aerial photograph of the margin of Roach Playa and the adjacent Whiskey Pete's alluvial fan along the southern Nevada/California border. (b) Schematic diagram showing how significant along-strike relief prevents continuous saltation in the proximal portion of the fan, resulting in local storage and accumulation of sand. On the distal portion of the fan, conversely, continuous saltation enables an eolian corridor to form.

gravel-dominated case. These two distinct modes of pavement formation are illustrated schematically in Figure 1.20.

1.2.2 Sand-dominated eolian systems

Many playas are dominated by sand rather than silt or clay. Roach Playa, located along the southern Nevada–California border, is one example of a sand-dominated playa. Sand from Roach Playa is transported off the playa and onto the nearby Whiskey Pete's alluvial fan (Figure 1.21). Because transport takes place by saltation rather than large-scale atmospheric mixing, the spatial pattern of sand deposition downwind from Roach

Playa is distinctly different from that of silt-dominated playas such as Franklin Lake Playa. In particular, the topographic relief of the alluvial fan plays a significant role in controlling the spatial pattern of deposition.

The distal portion of the Whiskey Pete's alluvial fan appears lighter in color than the proximal portion of the fan. In the field, this difference in surface reflectivity corresponds to an abrupt increase in the sand content of soils below a threshold elevation on the fan. Above this elevation, sand sourced from the playa and from channels draining the mountain accumulates locally and is not transported across the fan. Below this elevation, both coarse and fine eolian sand from nearby playa and channel sources is readily transported across the fan. These distal-fan eolian corridors are controlled by a threshold fan-surface relief (Cook and Pelletier, 2007). When along-strike relief falls below a threshold value, an eolian transportational surface develops. When the along-strike relief rises above the threshold value, sand is trapped locally in low spots and a continuous surface of transportation is prevented from developing. Along-strike relief promotes the storage of windblown sand in two ways. First, topographic obstacles create aerodynamic recirculation zones on their lee sides. These recirculation zones are characterized by very low bed shear stresses. Sand that is deposited behind those obstacles may be stored indefinitely. Second, the presence of topographic obstacles increases the force necessary to move the sand. It is not only sufficient for sand particles to be entrained from the surface, they must be picked up with sufficient force to be transported over large steps downwind. The transport of sand over a complex surface, therefore, involves topographic controls on both the shear stress exerted on the particle and on the value of threshold shear stress necessary to move the particle past downwind topographic obstacles.

Eolian bedforms can be classified into three basic types depending on their spatial scale and position within the bedform hierarchy: ripples, dunes, and megadunes (Wilson, 1972). Ripples are the smallest of the three bedforms, and are typically spaced by 0.1–1 m and have heights of 1–10 cm with higher, more widely spaced ripples forming in areas with stronger winds and/or coarser sand. Dunes are the next level in the bedform hierarchy and form only when ripples are present. Dunes commonly have spacings of 10–100 m and heights of 1–10 m. Megadunes form when dunes are present, and may attain spacings of several kilometers and heights of several hundred meters. All three bedform types exhibit a correlation between spacing and grain size in the Sahara Desert (Wilson, 1972) (Figure 1.22). This correlation is not observed everywhere, however, partly because geographic variations in wind speed also control bedform spacing. The precise mechanisms responsible for bedform genesis are still a subject of active research. All three bedform types, however, likely involve a positive feedback between the bedform shape, which con-

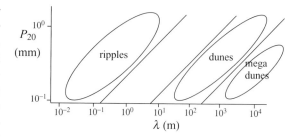

Fig 1.22 Relationship between coarsest fraction (20th percentile) of sand grains and bedform wavelength, λ, based on measurements of ripples, dunes, and megadunes in the Sahara Desert (after Wilson, 1972).

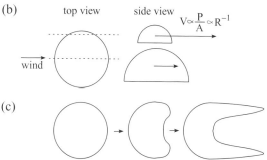

Fig 1.23 (a) Barchan dunes of the Salton Sea area, California. (b) Schematic diagram illustrating how a barchan gets its shape. The migration velocity of a dune is proportional to the ratio of the perimeter to the area of the cross section perimeter. Therefore, the central core of the dune will move slower than the sides, causing arms to form over time (c).

trols the air flow above it, and the pattern of erosion and deposition and subsequent modification of the bedform size and shape through time.

Eolian dunes come in a variety of types depending primarily on sand supply and wind-direction variability. Barchan dunes (Figure 1.23) provide a nice example of both of these controls. Barchan dunes have a central core flanked by two arms oriented downwind. Barchan dunes form in a steady wind direction when the sand supply is relatively low (leaving some areas of bare ground exposed). Dune formation under such conditions requires that sand coalesce and accumulate in patches separated by bare ground. How does this "coalescence" happen? The reason why sand tends to collect in patches rather than spread out uniformly on the valley floor has to do with the higher coefficient of restitution of soft sandy surfaces relative to hard bare ground. Sand in saltation will tend to bounce off of bare ground and land more softly on patches of sand. As a result, sand saltating across a valley floor has a higher probability of deposition in sand-rich areas than sand-free areas.

Next consider how different portions of an incipient sand dune will migrate downwind. Figure 1.23b illustrates an idealized incipient dune with a semi-circular cross-sectional shape. Dunes migrate by moving sand along their surfaces, and the rate of sand movement is proportional to the perimeter, P, of the dune cross section. In order to move the dune a given distance downwind, however, all of the sand in the cross sectional area, A, has to be moved. The rate of dune migration, therefore, is proportional to the ratio of the perimeter to area. In the center of an incipient dune, the ratio of perimeter to area is lower than it is on the sides of the dune. As such, the sides will migrate faster, causing the development of "arms" oriented downwind (Figure 1.23c). The wind direction must be steady for barchan dunes to form because a slight change in wind direction is sufficient for dunes of this type to become unstable. A shift in wind direction can cause sand from one arm of the barchan to be transferred to the other arm, lengthening one arm at the expense of the other over time. Eventually, a longitudinal dune is formed with a crestline oriented parallel to the wind.

Werner (1995) developed a numerical model for the formation of eolian dunes using simple geometric rules. Discrete units of sand in Werner's model are picked up from a sand bed at random and transported a characteristic distance downwind. Sand units are deposited back down on the bed with a probability that is relatively low on bedrock surfaces compared to sandy surfaces (reflecting the higher coefficient of restitution of a hard versus a soft bed). The probability

of deposition also depends on whether the sand had been transported into the lee side of an incipient dune. In the model, lee sides are mapped by defining a shadow zone (i.e. all areas that were located in shadow given a sun angle of $15°$ oriented parallel to the wind direction) with a high probability of deposition. Finally, sand deposited on the bed rolls down the direction of steepest descent if deposition causes the surface to be steeper than the angle of repose.

Werner's model is capable of reproducing the four principal dune types (transverse, barchan, star, and longitudinal) by varying the sand supply and wind direction variability (e.g. Figure 1.24b). This success is remarkable considering that his model does not include any details on the microscopic physics of grain-to-grain interactions that had long been assumed to be essential for understanding the formation of eolian bedforms. Although Werner's model does not successfully reproduce all aspects of dunes (e.g. topographic profiles), his model includes the essential feedbacks of dune formation necessary for understanding how different types of dunes are formed. First, Werner's model illustrates how sand blown from a river bed tends to coalesce into discrete mounds rather than spreading out. Because the probability of sand deposition is higher on an area already covered by sand, a steady supply of sand particles saltating across an area will be preferentially deposited on areas already covered by sand. In this way, discrete patches of sand will form and increase their heights relative to the surrounding topography. Second, as the heights of incipient bedforms increase, the length of their shadow zones (which mimics the recirculation zone downwind of the dune crest) will also lengthen. This lengthening allows dunes to grow higher since their heights are limited only by the angle of repose and the lengths of their shadow zones. Werner's model is further discussed in Chapter 7.

1.3 | A tour of the glacial system

Today, ice sheets and glaciers are restricted to a relatively small portion of the Earth's surface. Large ice sheets are found only on Greenland and

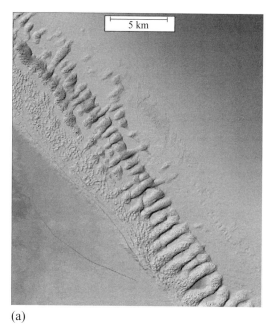

(a)

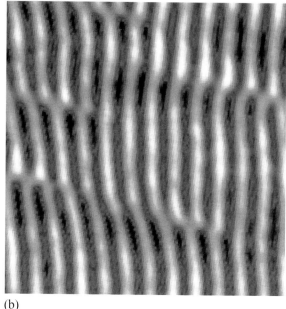

(b)

Fig 1.24 (a) Shaded-relief image of the Algodones Dunes located along the California–Arizona border near Yuma, Arizona. (b) Example output of Werner's model for eolian dunes for the transverse-dune case with large sand supply and constant wind direction.

Antarctica, and alpine glaciers are restricted to elevations above approximately 5 km a.s.l. in temperate latitudes. In contrast, ice covered nearly all of Canada, the northernmost United States, and most of northern Europe during the LGM. Alpine glaciers expanded to cover areas with elevations approximately 1 km lower than today (Porter *et al.*, 1983). While geomorphology in the southernmost United States means wind and water, geomorphology in Canada and the northern United States is primarily a story of ice (Flint, 1971).

Ice sheets and glaciers can be thought of as conveyor belts of ice. Some of the snow that falls at high elevations sticks around through the following summer. Successive years produce more snow that acts as a weight on underlying snow, turning the snow into firn (a state intermediate between snow and ice) and eventually ice as more snow is deposited above it. As snow and ice accumulate and the ground surface increases in elevation, more snow and ice accumulate because of the cooler temperatures at higher elevations. This positive feedback towards thicker and thicker ice is limited by the shear strength of the ice and friction at the base of the ice. Eventually, the thickness and slope of the ice become great enough to overcome the shear strength of ice or the basal friction (whichever is lower), causing the ice to surge outward by internal deformation or basal sliding. Zones of higher elevation have positive mass balance (i.e. more snow is precipitated on the landscape than melts away) and areas of lower elevation have negative mass balance. Ice flows from areas of positive mass balance (i.e. accumulation) to areas of negative mass balance (i.e. ablation) when sufficient thickness and slope have been built up. Although the mechanics of ice sheets and glaciers is very complex in detail, at large scales their shapes are often well adjusted to the threshold condition for shear strength or basal friction. Increases in snow accumulation cause shear stresses to increase above the threshold condition, causing rapid surging of the ice sheet or glacier in response. Therefore, the *shapes* of ice sheets and glaciers on Earth are primarily a function of their shear strength or basal friction. The *speed* of these ice bodies is primarily a function of their accumulation and ablation rates.

Ice sheets and glaciers move by a complex combination of internal ice flow, sliding of ice

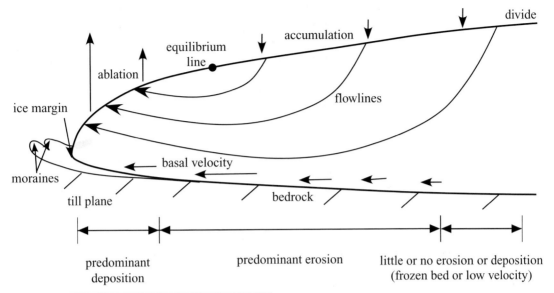

Fig 1.25 The subglacial environment can be broadly divided into three zones: (1) a zone of little or no erosion or deposition, located near the divide, where the ice sheet is frozen to the bed or where velocities are too small to significantly modify the surface, (2) a zone of predominant erosion, where basal velocities and meltwater concentrations increase toward the margin, transporting all of the bedrock debris removed from the bed, and (3) a zone of predominant deposition near the ice margin, where basal velocities wane and moraines and till plains are commonly formed. Also shown are the zones of accumulation, ablation, internal flow lines, and their relationship to the major subglacial zones.

on bedrock, and sliding of ice on glacial till. Whether an ice sheet or glacier deforms internally or slides along its base depends primarily on the thermal conditions at the base. In "cold-based" ice sheets and glaciers the ice is frozen to its bed, which precludes basal sliding. In "warm-based" ice sheets and glaciers, motion occurs by a combination of basal sliding and internal ice flow. The frictional force between an ice body and its bed depends on many factors, but sliding of ice over glacial till is generally thought to require lower basal shear stresses than sliding of ice over bedrock. The distinction between cold and warm-based ice sheets is very significant for geomorphology because only warm-based ice sheets perform significant glacial erosion. Millions of years of glacial activity have produced little measurable glacial erosion in areas dominated by cold-based glaciers (e.g. Dry Valleys of Antarctica).

The erosional potential of ice sheets and glaciers is not fully understood, but most researchers agree with Hallet (1979; 1996) that the erosion rate is proportional to a power-law function of the basal sliding velocity V:

$$\frac{\partial h}{\partial t} = -aV^b \tag{1.19}$$

where a is a coefficient that depends primarily on the bedrock erodibility, and b is an empirical exponent equal to ≈ 2 (Figure 1.25).

The ability of a warm-based ice sheet or glacier to erode its bed varies as a function of distance along the longitudinal profile. Ice flow velocities increase with distance downslope in a manner very similar to that of fluvial drainage networks. In a fluvial drainage network, water flows in the direction of the hillslope or channel orientation or aspect. Fluvial channels accomodate this continually increasing discharge primarily by increasing in width and depth downslope. Channel flow velocities also increase downslope, but at rates typically much lower than changes in width and depth. Flow in an ice sheet or glacier also occurs in the direction of the ice-surface gradient. As discharge accumulates with distance downslope, the ice velocity is the variable that changes most greatly to accommodate the increasing flow. The shape of the ice sheet or glacier does not vary as greatly as the velocity because a small change in thickness or slope causes a large change in discharge due to the threshold

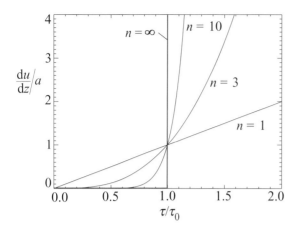

Fig 1.26 Plot of normalized velocity gradient versus normalized shear stress for a Newtonian fluid ($n = 1$), a non-Newtonian fluid (e.g. $n = 3$ and $n = 10$), and a perfectly plastic material ($n = \infty$).

nature of ice flow. Close to an ice divide, therefore, ice-flow velocities are quite low because the contributing area is small. Ice flow velocities are a combination of internal flow and basal sliding, so not all of the increased flow downstream occurs as basal sliding. Nevertheless, increasing ice discharge is usually associated with increased basal sliding velocities. Erosion rates close to an ice divide are, therefore, relatively low according to Eq. (1.19). Further downslope, flow velocities and erosion rates increase to accommodate ice delivered from upslope. Basal sliding velocities increase with distance downice, usually reaching a maximum beneath the equilibrium line. Past the equilibrium line, ice-flow velocities decrease due to ablation. This decrease, combined with the accumulation of erosional debris from further upice, usually causes a zone of sediment deposition beneath the ablation zone of an ice sheet or glacier.

1.3.1 How ice deforms

The flow of liquid water and air over the Earth's surface are both examples of *Newtonian* flows. Understanding glacial erosion, however, requires quantifying the behavior of a non-Newtonian fluid. The difference between Newtonian and non-Newtonian fluids is illustrated in Figure 1.26, where graphs of velocity gradient versus shear stress are illustrated for three types of fluids. In

a Newtonian fluid, strain rate is proportional to shear stress. The coefficient of proportionality is the viscosity, a measure of stiffness of the fluid. In a non-Newtonian fluid, strain rate and shear stress are nonlinearly related:

$$\frac{du}{dz} = a \left(\frac{\tau}{\tau_0} \right)^n \tag{1.20}$$

where u is the fluid velocity, z is the distance along the profile, τ is the shear stress, τ_0 is a reference or yield stress, and a and n are rheological parameters. The values of a and n can be considered to be constant in some cases, but more often they depend on the temperature or density of the flow. Experiments indicate that ice deforms according to Eq. (1.20) with $n \approx 3$ (i.e. Glen's Flow Law (Glen, 1955)). Some materials, including rock in Earth's upper crust, have highly nonlinear rheologies characterized by Eq. (1.20) with $n = 10$ or greater. Such materials deform very slowly until brittle fracture occurs, after which deformation can be accommodated relatively easily by slippage along fractures. Equation (1.20) does not resolve the detailed structure of fracture zones in such materials, but it can be used to study the large-scale behavior of deformation in such materials.

In the limit $n = \infty$, the material is called perfectly plastic. Perfectly plastic materials are completely rigid below their yield stress, τ_0, but flow entirely without resistance above it. The perfectly plastic model is an idealization, but in some cases it is very useful within certain limitations. Glaciers, ice sheets, lava flows, and debris flows, are all examples of flows driven by gravity. Flow in these systems occurs when the material thickens locally, increasing the shear stress at the base, causing the flow to surge forward, thereby lowering the shear stress back below a threshold value. The dynamics of this type of surge behavior depends sensitively on the value of n in Eq. (1.20). If one is not interested in resolving the detailed dynamics of individual surge events, however, then the perfectly plastic model can be very useful. In nature, there is always some finite time lag between changes in accumulation and the associated glacier response. In the perfectly plastic model, however, glacier surges start and stop instantly, because any increase in flow

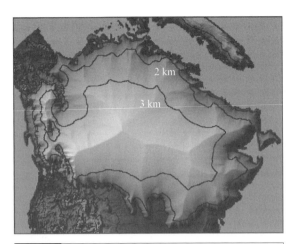

Fig 1.27 Model reconstruction of the Laurentide Ice Sheet at 18 ka constructed using a perfectly plastic ice sheet model and assuming isostatic balance.

thickness that causes the shear stress to increase above a threshold value instantaneously triggers flow to correct the imbalance. Perfectly plastic glaciers and ice sheets, therefore, are constantly in balance with the pattern of accumulation and ablation.

Figure 1.27 illustrates a model reconstruction of the Laurentide Ice Sheet over North America at 18 ka. This reconstruction was developed using the perfectly plastic model with basal shear stresses similar to modern ice sheets (e.g. Greenland, East Antarctica) and assuming isostatic balance. Isostatic balance in this context means that the weight of the ice sheet causes the crust to subside into the mantle until the point where the additional load is balanced by the buoyancy associated with the displaced mantle material. Despite decades of research, the shape and thickness of the Laurentide Ice Sheet (LIS) is still a subject of active research and debate. Sophisticated mass-balance thermomechanical models are used to reconstruct the ice sheet, but uncertainty in paleoclimatic, rheological, and basal-flow parameters results in uncertainty in the resulting reconstructions (Marshall *et al.*, 2000). Interpretation of past sea-level changes is another method by which the shape of the former ice sheet can be inferred. In this method, the past ≈100 yr of sea-level records are used to determine the present pattern of mantle flow beneath for-

mer ice sheets. These mantle-flow patterns are difficult to interpret uniquely in terms of ice-sheet size and shape, however, because the response is a function of lithospheric rigidity and the mantle-viscosity profile in addition to the size and shape of the ice sheet. Therefore, many different ice-sheet sizes and shapes may be equally consistent with the observed data pattern of sea-level change. Landforms such as drumlins, flutes, and mega-scale lineations can also be used to reconstruct ice-flow directions and provide information on the locations of ice domes (Clarke *et al.*, 2000). While these geomorphic markers provide solid evidence that ice once flowed in a certain direction, it is often difficult to reconstruct a spatially complete flow field or to determine the time that the observed flow directions were imprinted on the landscape.

1.3.2 Glacial landforms

Upstate New York provides some of the most striking examples of erosional and depositional glacial landforms on Earth. Figure 1.28a illustrates the topography of a portion of upstate New York immediately south of Lake Ontario. This Ontario Lowlands region has the largest drumlin field in North America, with nearly 10 000 drumlins. Drumlins are subglacial bedforms elongated parallel to the ice-flow direction and composed primarily of subglacial till or sediment. Drumlins in the New York State drumlin field come in a wide variety of shapes and sizes. The controls on drumlin size and shape are not well understood, but we will explore a possible model for drumlin formation in Chapter 7. Subsurface bedding in drumlins often parallels the topographic surface, suggesting that they form by localized, upward flow of sediment into low pressure zones beneath the ice sheet.

Along the southern portion of Figure 1.28a, the elevation rises up to the Allegheny Plateau and the Finger Lakes Region. The Finger Lakes Region is characterized by five major elongated lakes that fan out like the fingers of a hand. More generally, finger lakes are any kind of elongated glacial lake formed near the margin of a former ice sheet. Figure 1.28b illustrates the topography in the vicinity of the town of Ithaca, located near the southern tip of Cayuga Lake (the

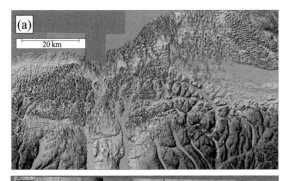

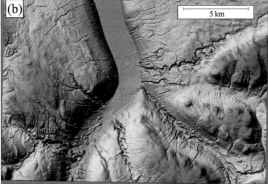

Fig 1.28 Shaded relief images of two portions of upstate New York. (a) Drumlin field of the Ontario Lowlands. (b) Southern Cayuga Lake (one of the five Finger Lakes) in the vicinity of the town of Ithaca.

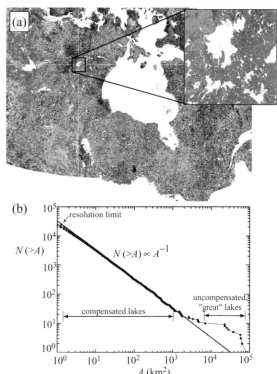

Fig 1.29 Analysis of lakes in central Canada. (a) Radarsat mosaic image of central Canada (resolution 250 m), including a closeup of Dubawnt Lake region to illustrate the full resolution of the image. (b) Cumulative frequency-size distribution of lakes extracted from (a). Also included in this plot are the Great Lakes, even though they are not included in (a). A power-law relationship $N(>A) \propto A^{-1}$ is observed from $A = 10^0$ to $A \approx 10^4$ km². Above $A \approx 10^4$ km², the largest lakes do not fit the power-law relationship. The larger, more uniform size of the great lakes is consistent with the influence of lithospheric flexure on their formation.

largest of the Finger Lakes). The topography of the Ithaca area is characterized by wide (>1 km), deep (>100 m), glacially scoured valleys cut into the broad, low-relief Allegheny Plateau (von Engeln, 1961). Fluvial channels in the region are said to be underfit because channels occupy only a small portion of the broad valley floors that were once occupied by ice up to 1 km in thickness. Fluvial drainage density is low, reflecting the short period of geologic time since small-scale landforms were last smoothed by glacial processes. The deeply scoured valleys and smooth uplands illustrate that glacial processes smooth the landscape at small (<5 km) scales, while increasing relief through trough formation at larger scales.

In addition to elongated finger lakes, the northern US and Canada contain innumerable glacially carved lakes that come in a wide range of sizes (Figure 1.29a). The cumulative frequency-size distribution is one tool for quantifying the population of lakes in a given region. The cumulative frequency-size distribution of lakes is the number of lakes larger than a given area. Consider Figure 1.29b, which illustrates the cumulative frequency-size distribution of all the lakes in Canada resolvable in a 250 m-resolution Radarsat mosaic of the country. Certainly not all Canadian lakes are erosional in origin; some of the smaller lakes occur in kettles and hummocky moraine, for example. Nonetheless, the majority of Canadian lakes, particularly the largest lakes, are erosional in origin. Figure 1.29a shows a subset of the Radarsat mosaic, illustrating the mosaic image data at both small and large scales.

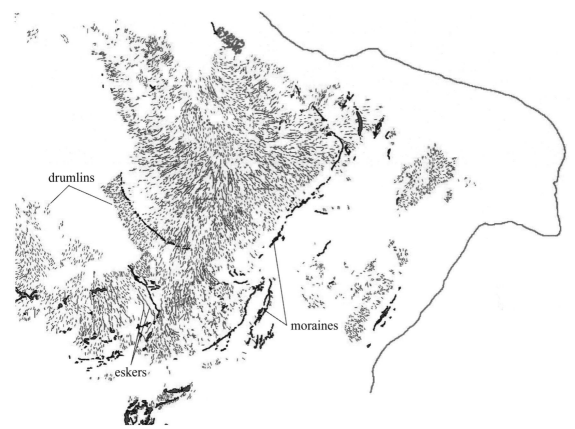

drumlins

moraines

eskers

Fig 1.30 Surficial geologic map of eastern Canada, illustrating moraines (thick curves oriented perpendicular to ice flow), eskers (long, thin curves oriented parallel to ice flow) and drumlins (short, thin curves parallel to ice flow).

Lakes were extracted from the image by scanning the grid for pixels with water coverage. Once a water-covered pixel was identified, all neighboring pixels were searched recursively, calculating the number of pixels in the connected lake "domain." In this way, the size of each lake in the image was determined. We have also included the five Great Lakes in this population, even though they are not included in the mosaic image. The distribution of Figure 1.29b includes two distinct lake types. Lakes smaller than $\approx 10^4$ km^2 in area follow a simple power-law distribution characterized by $N(> A) \propto A^{-1}$. Above 10^4 km^2, the largest lakes do not follow the trend of smaller lakes. Instead, these lakes are larger than the size predicted by an extrapolation of the trend for smaller lakes. This distribution suggests that the largest lakes in North America may be controlled by some additional process not involved in the erosion of smaller lakes. In Chapter 6, we will propose that the bending of the lithosphere beneath the ice is that additional process that controls the formation of the largest North American glacial lakes.

Depositional landforms beneath ice sheets include moraines, eskers, drumlins, kettles, and kames. Figure 1.30 includes a map of depositional landforms in eastern Canada. As ice sheets have waxed and waned, nearly all glaciated areas have experienced a combination of erosion and deposition. As such, glacial landscapes are classic palimpsests. This poses a problem for the interpretation of formerly glaciated topography. For example, predominantly erosional topography may be covered with till of varying thickness, obscuring the bedrock erosional surface.

Alpine glaciers differ from ice sheets primarily in their size and the extent to which ice flow is controlled by subglacial topography. In an ice sheet, the thickness of the ice is often several

Fig 1.31 U-shaped and hanging valley of the Beartooth Mountains, Montana.

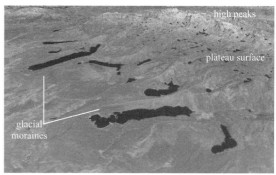

Fig 1.32 Oblique virtual aerial photograph of the Wind River Range, illustrating the three distinct topographic levels characteristic of many glaciated mountain ranges of the western US.

times greater than the relief of the bed topography (except near the margins). As a result, the pattern of ice flow is controlled principally by the planform shape of the ice sheet with valleys and ridges beneath the ice sheet playing a relatively minor role. In contrast, ice thickness in a glacier is typically much less than the relief of the underlying bed topography (e.g. ≈ 100 m for a glacier spanning ≈ 1 km in bed elevation). Alpine glaciers are strongly controlled by bed topography, with ice thickness greatest in areas of lowest bed slope (i.e. valley bottoms). The tendency of ice thickness and flow velocity to be greatest in valley bottoms suggests that glaciers may increase subglacial landscape relief more readily than ice sheets. Alpine glaciers are also strongly controlled by elevation. The relationship between the ELA (Equilibrium Line Altitude) and glacial erosion is much tighter in alpine glaciers than in ice sheets.

Figure 1.31 illustrates examples of two classic alpine glacial landforms: U-shaped valleys and hanging valleys. Fluvial valleys are classically V-shaped, reflecting the contrasting erosional power of fluvial hillslopes process regimes. In a V-shaped valley, hillslope erosion requires steep slopes in order to "keep up" with the concentrated power of rivers to incise narrowly into the valley bottom. In alpine glacial systems, in contrast, ice fills the valley floor, resulting in a "plug" of ice that erodes the valley floor and side slopes more uniformly than in the fluvial case. A hanging valley forms when glacial erosion of

a main valley takes place much faster than erosion of a side tributary, leaving the side tributary "hung" above the floor of the main valley. Hanging valleys are, therefore, a consequence of the more spatially discontinuous nature of glacial erosion compared with fluvial erosion. It should be noted, however, that U-shaped valleys and hanging valleys are not unique to glacial systems. Fluvial channels that incise very rapidly can oversteepen adjacent hillslopes to the point where water and sediment draining through side tributaries will separate from the channel bed (Wobus *et al.*, 2006). As such, fluvial systems in tectonically very active areas or in areas with strong structural control can also form hanging valleys.

Figure 1.32 illustrates the large-scale geomorphology of alpine glacial terrain using the Wind River Range as a type example. The Wind River Range has three distinct topographic levels. The high peaks of the Wind River Range (≈ 4 km a.s.l.) stand ≈ 1 km above the surrounding low-relief plateau surface. This large-scale pattern of narrow high peaks surrounded by a broad plateau is common in the glaciated mountain ranges of the western US. The formation of these plateau surfaces is not well understood. Did glaciers emanating from the high peaks "plane" the topography to a relatively smooth plateau, or was the plateau exhumed by isostatic rebound responding to erosion concentrated in the high peaks?

The most intense glacial erosion occurs near the equilibrium line in most alpine glacial

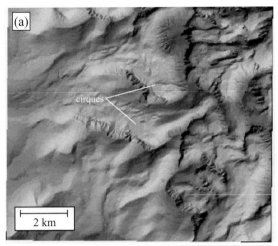

Fig 1.33 Alpine glacial landforms near the equilibrium line: (a) Shaded relief image of glacial topography in the Uinta Mountains, Utah, illustrating steep cirque walls and overdeepened cirque lakes. (b) Cirques of the Beartooth Mountains, Montana.

systems. Figure 1.33a illustrates the topography of a portion of the Uinta Mountains using a shaded-relief image of a US Geological Survey Digital Elevation Model (DEM). The topography of the Uinta Mountains is characterized by relatively flat divide surfaces, steep cirque walls, and overdeepened cirque floors. The divides are relatively flat because of the limited erosion that takes place at elevations far above the equilibrium line. Near the equilibrium line, glacial erosion is concentrated spatially, decreasing both up and down-slope. This local maximum in erosion causes localized glacial scour. In river systems, bed scour decreases the local slope, which acts as a negative feedback to limit scour. In glacial systems, however, scour continues as long as the *ice surface* points downslope. Alpine glaciers can maintain sloping ice surfaces even when the bed beneath the ice has been scoured to form a closed depression. When the ice retreats, a cirque lake is formed. Figure 1.33b illustrates the cirque form at the margin of the Beartooth Mountains in Montana.

The formation of overdeepenings beneath alpine glaciers is illustrated schematically in Figure 1.34. Figure 1.34 shows a longitudinal profile

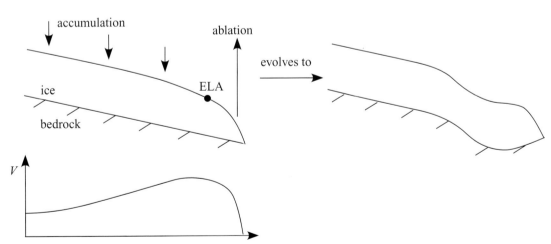

Fig 1.34 Conceptual model of flow in an alpine valley glacier. Basal sliding velocity and glacial erosion are concentrated beneath the equilibrium line. Over time, concentrated erosion leads to the formation of an overdeepened bed that forms a cirque lake when the ice retreats.

of an alpine valley glacier. The basal sliding velocity beneath the glacier is a maximum beneath the ELA. Over time, glacial erosion concentrated beneath the ELA will cause the formation of an overdeepened bed. Numerical models exist that form realistic cirques (MacGregor *et al.*, 2000), but

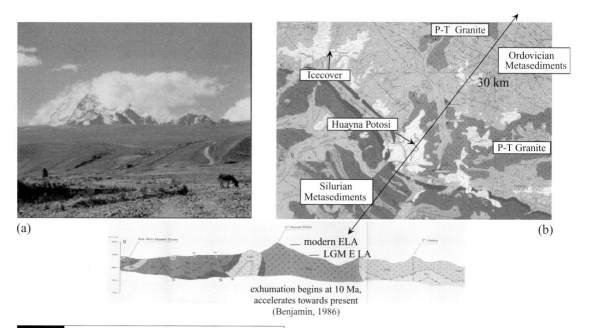

(a)

(b)

exhumation begins at 10 Ma,
accelerates towards present
(Benjamin, 1986)

Fig 1.35 Topography, ice cover, and structural geology of the Cordillera Real, central Andes. (a) View of the glaciated peak of Huayna Potosi, one of the high peaks of the Cordillera Real, with the Altiplano in the foreground. (b) Structural geology of a portion of the Cordillera Real, showing the intrusion of Permo-Triassic granite of the high peaks into Paleozoic metasediments of the Eastern Cordillera. Apatite fission track (AFT) data indicate that the high peaks of the Cordillera Real were exhumed rapidly beginning at 10 Ma, accelerating towards the present day.

many questions remain. For example, do most cirques primarily reflect stable ELAs during LGM conditions, or are their morphologies controlled by fluctuations in ELAs through time?

The tight coupling between glacial ice cover, glacial erosion, and elevation suggests that some interesting feedback relationships may exist in glaciated mountain belts. For example, the "glacial-buzzsaw" hypothesis states that glaciers limit the elevation of mountain belts because any uplift that raises the height of mountain belts above the ELA will quickly be met with enhanced ice cover and glacial erosion that will lower elevations back below the ELA (Whipple et al., 2004). This model predicts a tight correlation between the peak elevation of mountain ranges and the local ELA. On the other hand, the fact that glaciers erode most vigorously in valley bottoms and cirque floors suggests that glaciers

may actually *increase* the elevations of the high peaks of the range. In this "relief-production" hypothesis, glacial erosion causes isostatic imbalance. Because the erosion is concentrated at valley bottoms with minimal erosion of the high peaks, however, the presence of isostatic rebound together with the absence of significant erosion causes the peaks to grow even higher in elevation. The difference between the glacial-buzzsaw and relief-production hypotheses is partly an issue of scale. At scales of 100 km or larger, glacial erosion most likely does limit the relief of mountain belts. At scales of a single valley and peak, however, the relief-production model appears to be most consistent with the results of glacial landform evolution models. Feedbacks between glaciers and topography may also involve climate through enhanced orographic precipitation in areas of higher/steeper topography.

The Cordillera Real of the central Andes is one potential example of the role of glacial erosion in enhancing topographic relief. The Cordillera Real comprises the boundary between the Altiplano to the west and the Eastern Cordillera to the east (Figure 1.35). This region is topographically and structurally distinct from both of its surrounding regions. Topographically, the range is a very narrow (i.e. ≈ 15–20 km) series of peaks up to ≈ 2 km above the surrounding topography.

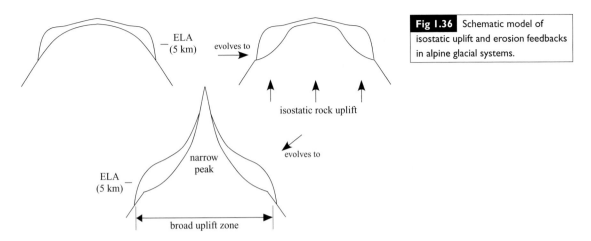

Fig 1.36 Schematic model of isostatic uplift and erosion feedbacks in alpine glacial systems.

Structurally, the Cordillera Real is comprised of granite exhumed between the Paleozoic sedimentary rocks that make up the Eastern Cordillera. If the Cordillera Real displayed the usual signs of active uplift (modern seismicity, active Quaternary faults), then its steep, high topography could simply be attributed to active tectonics. However, active tectonic uplift in the central Andes occurs only in the Subandean Zone today, which is the lowest part of the Andes (i.e. ≈ 1–2 km a.s.l.). Exhumation rates in the Cordillera Real measured thermochronologically (e.g. Benjamin et al., 1987; Gillis et al., 2006), however, show high rates from 10 Ma to the present.

Figure 1.36 illustrates a schematic model for the exhumation and peak uplift of the Cordillera Real. According to this model, the Cordillera Real was a broad anticline with peak elevations near 5 km before 10 Ma. Global cooling of ≈ 5° at 10 Ma lowered the ELA from above 6 km to approximately 5 km. This ELA lowering would have initiated glacial erosion in the highest portions of the Cordillera Real. Because glacial erosion increases as a function of distance from the divide, the high peaks of the Cordillera Real would not have been subject to significant erosion, however, while lower areas within the glaciated zone would have been more intensely eroded. As such, an initially broad anticline would have been sculpted to a narrow peak. Glacial erosion would also have triggered isostatic rebound.

This rebound would have uplifted the flanks of the Cordillera Real (in addition to its glaciated high peaks), increasing the area of the Cordillera Real that stands above the ELA. This increase in glaciated area would further increase glacial erosion and isostatic uplift in a positive feedback, causing localized exhumation of igneous and metamorphic rocks from the subsurface.

1.4 | Conclusions

The remainder of the book is organized according to mathematical themes. Starting with Chapter 2, each chapter focuses on a particular type of equation or algorithm and includes example applications that span Earth surface processes. This structure serves to emphasize the common mathematical language that underlies many of the disparate process zones of Earth's surface. Chapter 2, for example, focuses on techniques used to solve the classic diffusion equation and the general class of equations that are similar to it. Since the diffusion equation can be used to describe hillslope evolution, channel-bed evolution, delta progradation, hydrodynamic dispersion in groundwater aquifers, turbulent dispersion in the atmosphere, and heat conduction in soils and the Earth's crust, Chapter 2 will be relevant to many aspects of Earth surface processes.

Chapter 2

The diffusion equation

2.1 Introduction

The diffusion equation is perhaps the most widely used differential equation in science. It is the equation that describes heat transport in solids and in many turbulent fluids. In geomorphology, the diffusion equation quantifies how landforms, especially hillslopes, are smoothed over time. Diffusional processes act to smooth elevation in the same way that conduction smooths temperature in solids. The diffusion equation has played a particularly important role in the study of well-dated landforms, such as pluvial shoreline scarps and volcanic cones, in which the initial topography can be relatively well constrained at a known time in the past. Certain conditions must be met for the diffusion equation to apply to hillslope evolution, however. The diffusion equation is applicable to gently sloping hillslopes comprised of relatively uniform, unconsolidated alluvium or regolith subject to creep, rainsplash, and bioturbation. The accuracy of the model breaks down as hillslope gradients increase and mass-movement processes become significant.

The applicability of the diffusion equation has two requirements: slope-proportional transport and conservation of mass. First, the flux of sediment per unit length, q, must be proportional to the hillslope gradient:

$$q = -\rho \kappa \frac{\partial h}{\partial x} \qquad (2.1)$$

where ρ is the bulk density of sediment on the hillslope, h is the elevation, κ is the diffusivity in units of L^2/T (where L is length and T is time) and x is the distance from the divide. In Eq. (2.1), q is a mass flux with units of $M/L/T$ (where M is mass, L is length, and T is time). Mass must also be conserved. In this context, conservation of mass means that any change in sediment flux along a hillslope results in either an increase or decrease in the elevation. One way to think about conservation of mass is to consider a small segment of a hillslope profile (e.g. the section between x_3 and x_4 in Figure 2.1). If more sediment enters the segment from upslope than leaves the segment downslope, that hillslope segment must store the difference, resulting in an increase in the average elevation. Conversely, if more sediment leaves the segment downslope than enters the segment upslope (as in the section between x_1 and x_2), there is a net loss of sediment and the elevation must decrease. The diffusion equation applies this logic to infinitesimally-small hillslope segments in the same way that the first derivative of calculus calculates the average slope $\Delta h/\Delta x$ of a function $h(x)$ in the limit as Δx goes to zero. Mathematically, conservation of mass requires that the increase or decrease in the elevation be equal to the *change* in flux per unit length, divided by the bulk density ρ:

$$\frac{\partial h}{\partial t} = -\frac{1}{\rho} \frac{\partial q}{\partial x} \qquad (2.2)$$

where t is time. Substituting Eq. (2.1) into Eq. (2.2) gives the classic diffusion equation:

$$\frac{\partial h}{\partial t} = \kappa \frac{\partial^2 h}{\partial x^2} \qquad (2.3)$$

If, alternatively, q is expressed as a volumetric flux per unit length, the ρ terms are removed from both Eqs. (2.1) and (2.2). We will use both the mass and volumetric fluxes in this chapter.

Equation (2.3) is applicable to hillslopes with a constant cross-sectional topographic profile along strike (i.e. a fault scarp). More generally, hillslopes evolve according to the 2D diffusion equation, given by

$$\frac{\partial h}{\partial t} = \kappa \left(\frac{\partial^2 h}{\partial x^2} + \frac{\partial^2 h}{\partial y^2} \right) = \kappa \nabla^2 h \qquad (2.4)$$

Figure 2.1a illustrates a hypothetical fault scarp 2 m high and 10 m wide after some erosion has taken place. Figures 2.1b and 2.1c illustrate the gradient and curvature of the scarp, respectively. The diffusion equation states that the sediment flux q at any point along a hillslope is proportional to the hillslope gradient (i.e. Figure 2.1b). The magnitude of the flux is illustrated in Figure 2.1a at several points along the profile using arrows of different lengths. At the top of the scarp, the flux increases from left to right, indicating that more material is moving out of the section than is being transported into it from upslope. This results in erosion along the top of the scarp where the *change* in gradient along the profile (i.e. the curvature) is negative. Conversely, flux decreases from left to right at the bottom of the scarp, indicating that more material is moving into that segment than out of it. The result is an increase in surface elevation (i.e. deposition) along the base of the scarp where curvature is positive. The rate of erosion or deposition varies with time (Figures 2.1d–2.1f) according to the magnitude of curvature. Immediately after faulting, for example, the change in flux per unit length (i.e. the curvature in Figure 2.1c) is very large and concentrated right near the top and bottom of the scarp. The result is very rapid, concentrated erosion and deposition at the top and bottom of the scarp, respectively. Over

time, however, the rate of erosion and deposition decreases and the widths of the top and bottom of the scarps where erosion and deposition occurs increases (Figure 2.1f).

Many physical processes can be classified as either diffusive or advective. Therefore, one reason for studying the simple diffusion and advection equations in detail is that they provide us with tools for studying more complex processes that can be broadly classified as one of these two process types. Conceptually, diffusive processes result in the smoothing of some quantity (temperature, elevation, concentration of some chemical species, etc.) over time. For instance, if one injects a colored dye into a clear fluid being vigorously mixed, the average dye concentration smooths out over time (rapidly at first when concentration gradients are high, then more slowly) according to the diffusion equation. Similarly, if faulting offsets an alluvial fan, the scarp elevation smooths out over time as sediment is eroded from the top of the scarp and deposited at the base.

Advection, in contrast, involves the lateral translation of some quantity without spreading. If a colored dye is injected into a laminar fluid moving down a pipe, for example, the dye will be transported down the pipe with the same velocity as the fluid but with little spreading. Similarly, if a fault offsets a bedrock pediment instead of an alluvial fan, the fault will evolve primarily by weathering-limited slope retreat instead of smoothing. In such cases, slow bedrock weathering acts to maintain a steep fault scarp over time because transport of the weathered material is rapid enough that sediment does not accumulate at the scarp base and no deposition occurs. This type of slope evolution can be modeled by the advection equation, where the erosion rate is proportional to the hillslope gradient:

$$\frac{\partial h}{\partial t} = c \frac{\partial h}{\partial x} \qquad (2.5)$$

The variable c is the advection velocity in Eq. (2.5), and it equals the rate of horizontal retreat of the scarp. In general, transport-limited slopes and channels evolve diffusively while

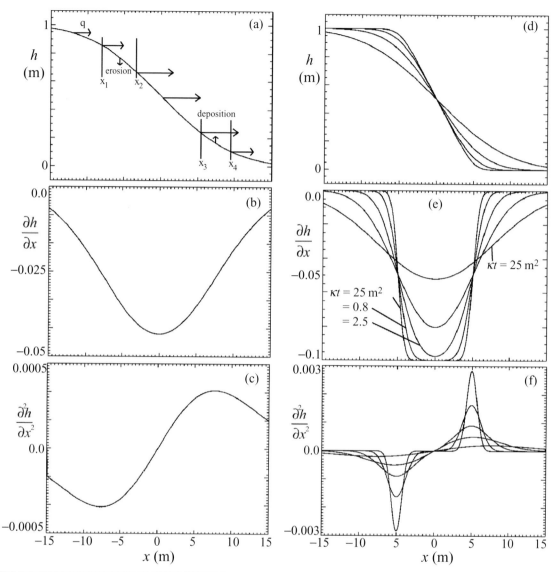

Fig 2.1 Evolution of a topographic scarp, illustrating (a) elevation, (b) slope, and (c) curvature. In (a), arrows of varying length represent the sediment flux at each point. In the diffusion model, the flux is proportional to the local slope, and the resulting raising or lowering rate of the surface is proportional to the *change* in flux per unit length, which, in turn, is proportional to the curvature. (d) and (e) Graphs of elevation, slope, and curvature for five times following scarp offset ($\kappa t = 0.01, 0.03, 0.1, 0.3, 1.0$.)

weathering-limited slopes and channels evolve advectively. Analytic and numerical techniques for solving the advection equation will be presented in Chapter 4.

In many applications, diffusion is one component of a more complex model. The transport of contaminants in groundwater flow, for example, can be quantified using a combination of advection (i.e. the translocation of contaminants as they are carried along with the fluid) and diffusion (i.e. the spreading of contaminant concentration from hydrodynamic dispersion as the contaminant travels along variable flow paths through a porous aquifer). Similarly, the concentration of dust downwind from a playa is the result of both advection and turbulent diffusion. Advection describes the motion of the dust as it is carried downwind, while diffusion describes the spreading of the plume as it is mixed by atmospheric turbulence. In many of the examples in this chapter, we will assume a fixed channel head at the base of a diffusing hillslope. Of course, channels have their own complex evolution and are rarely fixed at the same elevation over time. As such, a complete understanding of hillslopes or channels requires a fully coupled model of the fluvial system of which hillslope diffusion is one component. Although the techniques used in this chapter are framed around solving a single equation that has limited application in geomorphology, the techniques we will discuss form the basis for solving many real-world problems in which diffusion is one part of a more realistic model.

The applicability of the diffusion model to hillslope evolution depends on the processes acting to move sediment on the hillslope. Creep and rain splash are examples of hillslope processes that are accurately represented by the diffusion equation for gentle slope angles (Carson and Kirkby, 1972). In these processes, particle movement takes place by a series of small jumps triggered by freeze/thaw cycles, wetting/drying, or momentum imparted from rain drops. In these jumps, particles may move both upslope and downslope but the distance traveled by a particle moving downslope is slightly greater than the distance of those traveling upslope. Moreover, the increase in distance traveled by a particle moving downslope increases linearly with the hillslope angle or gradient for relatively small slopes. Bioturbation is another important hillslope process that can often be approximated as diffusive, particularly for gravelly hillslopes of arid environments where processes such as slope wash are relatively ineffective due to the coarse texture of the slope material. Slope wash, rilling, and mass movements are examples of hillslope processes that are not diffusive. Slope wash is not diffusive because the shear stress exerted by overland flow increases with distance from the divide as flow accumulates along the slope. While hillslopes dominated by slope wash are not governed by the diffusion equation, they can be described by a more general, spatially variable diffusion equation in which the flux is proportional to the hillslope gradient and the distance from

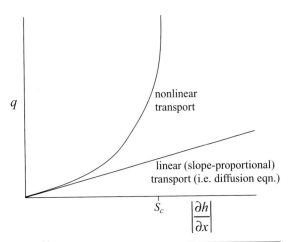

Fig 2.2 Schematic graph of flux versus slope. The diffusion equation assumes a linear relationship between flux and slope. More generally, sediment flux increases nonlinearly with slope, diverging at a critical slope value S_c. However, even the most general flux relationship can be approximated as linear for small slopes. As such, the predictions of linear and nonlinear transport models are the same for gently-sloping topography.

the divide (Carson and Kirkby, 1972):

$$q = -\kappa x \frac{\partial h}{\partial x} \tag{2.6}$$

The evolution of hillslopes governed by this flux relationship is considered in Section 2.3.2. Mass movements are not diffusive because the sediment flux increases rapidly and nonlinearly with small increases in hillslope gradient near the point of slope failure S_c, as illustrated in Figure 2.2.

The value of κ controls how fast diffusion takes place. In the Basin and Range Province of the western US, an approximate value of $\kappa = 1\,\mathrm{m}^2/\mathrm{kyr}$ has often been used based on studies of pluvial shorelines and other landforms of known age (Hanks, 2000). It is widely recognized, however, that the $1\,\mathrm{m}^2/\mathrm{kyr}$ value varies spatially according to climate, vegetation, soil texture, and other variables, even if that variation is not yet well characterized. One drawback of the diffusion approach in geomorphology is that we often do not know the value of κ for a given location with any precision. This is a valid criticism, but it is still important to study hillslope evolution even if the absolute rates of change cannot be precisely determined.

Three classic publications form the basis of many of the results in this chapter. Carslaw and Jaeger (1959) is the definitive text on analytic solutions to the diffusion equation. Although the book was written primarily for engineering applications to heat conduction in solids, the solutions presented in that book can be applied to diffusion problems that arise in any field of science. Culling (1963) was among the first authors to apply the diffusion equation to hillslope evolution. His paper is still the most complete reference on the subject. Culling's paper includes solutions appropriate for many different initial conditions as well as a stochastic treatment of the relationship between the random motion of sediment particles on the hillslope and the diffusive evolution of the hillslope. Carson and Kirkby (1972) is the most complete reference on the relationship between specific hillslope processes and their signature forms.

In the following sections we present solutions to the diffusion equation primarily within the context of hillslope evolution. As such, the diffusing variable h will represent elevation along a topographic profile. It should be emphasized, however, that the solutions we describe are general. Therefore, any solution that we obtain for h (for a particular initial condition and boundary conditions) can be applied to any other diffusion problem, whether the physical process is heat conduction, contaminant transport, etc. That is part of the power of applied mathematics: once a set of solutions or a solution method has been learned it can often be rapidly applied to many different physical systems.

2.2 | Analytic methods and applications

2.2.1 Steady-state hillslopes

Steady-state landscapes have received a great deal of attention in geomorphology in recent years. The topography of a steady-state landform is time-independent, such as when uplift and erosion are in perfect balance at every point in a rapidly uplifting mountain range. In such cases, the diffusion equation simplifies to the time-independent equation:

$$\kappa \frac{\partial^2 h}{\partial x^2} + U = 0 \tag{2.7}$$

where U is the uplift rate. Solving for the hillslope curvature, Eq. 2.7 gives:

$$\frac{\partial^2 h}{\partial x^2} = -\frac{U}{\kappa} \tag{2.8}$$

Integrating Eq. (2.8) gives an equation for the hillslope gradient:

$$\frac{\partial h}{\partial x} = -\frac{U}{\kappa} x + c_1 \tag{2.9}$$

where c_1 is an integration constant. In order to constrain the value of c_1, it is necessary to specify the value of $\partial h / \partial x$ at one end of the hillslope profile. The upslope end of the hillslope is chosen to coincide with a divide in this case. Divides are *defined* as locations across which no water or sediment passes. As such, a *no-flux* boundary condition is appropriate for the upslope end of this profile. The no-flux boundary condition is given by

$$\left. \frac{\partial h}{\partial x} \right|_{x=0} = 0 \tag{2.10}$$

which implies $c_1 = 0$. Integrating Eq. (2.9) gives

$$h(x) = -\frac{U}{2\kappa} x^2 + c_2 \tag{2.11}$$

where c_2 is another integration constant. At the downslope end of the profile, the elevation is assumed to be constant. This boundary condition is appropriate for a hillslope-channel boundary in which all of the sediment delivered to the channel is transported away from the slope base with no erosion or deposition. If the hillslope has a length L and an elevation of zero at the slope base:

$$h(L) = 0 \tag{2.12}$$

then $c_2 = U L / 2\kappa$ and the hillslope profile is given by

$$h(x) = \frac{U}{2\kappa} (L^2 - x^2) \tag{2.13}$$

Equation (2.13) shows that diffusive hillslope profiles are parabolas in cases of steady-state uplift. The hillslope relief in Eq. (2.13) is given by

$U L^2/2\kappa$, so the relief of diffusive hillslopes is proportional to uplift rate and slope length, and inversely proportional to diffusivity. Equations (2.9)–(2.13) illustrate how boundary conditions are used to solve differential equations.

Equation (2.8) suggests that in regions of similar climate and hillslope processes, hillslope curvature can be a relative measure of tectonic uplift rates. Equation (2.8) may be applicable to the early stage of uplift when hillslope gradients are still relatively low and weathering rates are sufficiently rapid for hillslopes to be regolith covered. In general, however, Eq. (2.8) is of limited use in many areas of rapid uplift because the requirements for the diffusion equation generally do not apply. Later in this chapter we will solve the solution to the nonlinear diffusion equation. That equation usually provides a more accurate representation of hillslope evolution in rapidly uplifting terrain.

2.2.2 Fourier series method

Two general classes of solutions exist for the diffusion equation: series solutions and similarity solutions. First consider the one-dimensional (1D) diffusion equation in a region bounded by $x = 0$ and $x = L$. By 1D, we mean a profile that varies with only one independent spatial variable, x. Let's assume that the initial condition at $t = 0$ is given by a function $f(x)$ and that the value of h is set to zero at both ends of the region for all time t (i.e. $h(0, t) = h(L, t) = 0$). If the initial condition $f(x)$ is given by

$$f(x) = a_n \sin\left(\frac{n\pi x}{L}\right) \tag{2.14}$$

then the solution to the diffusion equation is

$$h(x, t) = a_n \sin\left(\frac{n\pi x}{L}\right) e^{-\kappa n^2 \pi^2 t/L^2} \tag{2.15}$$

The correctness of this solution can be checked by differentiation and substitution of the results into the diffusion equation. This solution is only valid for the sinusoidal function given by Eq. (2.14), so it may not appear to be of much use. Equation (2.15), however, is actually a powerful building block that can be used to solve the diffusion equation for nearly any initial condition by using a Fourier series representation for $f(x)$.

Fourier showed that nearly any function could be represented as a series of sinusoidal functions:

$$f(x) = \sum_{n=1}^{\infty} a_n \sin\left(\frac{n\pi x}{L}\right) \tag{2.16}$$

where $f(x)$ is any function that goes to zero at $x = 0$ and $x = L$. Given a function $f(x)$ that obeys these boundary conditions, the values of the Fourier coefficients a_n can be calculated using

$$a_n = \frac{2}{L} \int_0^L f(x') \sin\left(\frac{n\pi x'}{L}\right) dx' \tag{2.17}$$

The solution to the diffusion equation with initial condition given by Eq. (2.16) is the summation of Eq. (2.15) from $n = 1$ to $n = \infty$:

$$h(x, t) = \sum_{n=1}^{\infty} a_n \sin\left(\frac{n\pi x}{L}\right) e^{-\kappa n^2 \pi^2 t/L^2} \tag{2.18}$$

Substituting Eq. (2.17) into Eq. (2.18) gives the general solution

$$h(x, t) = \frac{2}{L} \int_0^L f(x') \sum_{n=1}^{\infty} \left[\sin\left(\frac{n\pi x'}{L}\right) \right.$$
$$\left. \times \sin\left(\frac{n\pi x}{L}\right) e^{-\frac{\kappa n^2 \pi^2 t}{L^2}}\right] dx' \tag{2.19}$$

This approach suggests a powerful means of solving the diffusion equation in bounded regions: using the initial condition $f(x)$, solve for the coefficients a_n of the Fourier series representation of $f(x)$; then, plug the a_n values into Eq. (2.18) to obtain the solution. Other boundary conditions (e.g. values of $h(0, t)$ and $h(L, t)$ that are nonzero, or boundary conditions specified by hillslope gradients) can also be handled by using cosine, linear, and/or constant terms in the summation. For example, if we wish to place a divide at $x = 0$ then the series solution is

$$h(x, t) = \frac{1}{L} \int_{-L}^{L} f(x') \sum_{n=1}^{\infty} \left[\cos\left(\frac{n\pi x'}{2L}\right) \right.$$
$$\left. \times \cos\left(\frac{n\pi x}{2L}\right) e^{-\frac{\kappa n^2 \pi^2 t}{4L^2}}\right] dx' \tag{2.20}$$

Equation (2.20) is the same as Eq. (2.19) except that cosine terms are used and the hillslope domain is assumed to be from $x = -L$ to $x = L$. If $f(x)$ is chosen to be symmetric (i.e. $f(x) = f(-x)$) then Eq. (2.20) is consistent with the divide boundary condition given by Eq. (2.13).

The Fourier series approach is only applicable to bounded regions. Of course, all geomorphic applications are bounded, because the Earth is not infinite in extent! In some cases, however, it is useful to define mathematical models in infinite or semi-infinite domains. For example, if one is interested in modeling the first few years of radionuclide migration into an alluvial deposit that is hundreds of meters thick, a semi-infinite model is more convenient to work with and practically speaking more accurate than a Fourier series approach.

The Fourier series approach works because the diffusion equation is linear (i.e. it does not involve squared or higher-order terms of h or its derivatives). Only in linear equations can different solutions be superposed to obtain other solutions.

2.2.3 Similarity method

Series solutions with fixed-elevation boundary conditions are appropriate for hillslopes with a well-defined base-level control (such as a nearby channel that carries sediment away from the slope base to maintain a constant elevation). In some cases, however (such as when sediment is allowed to deposit and prograde at the slope base) it is more appropriate to assume that diffusion occurs over an infinite or semi-infinite region. In such cases, similarity solutions are a powerful approach. Let's consider the semi-infinite region given by $x = 0$ to $x = \infty$ with an initial condition $h(x, 0) = h_0$. Let's also assume that at $t = 0$ the elevation h is instantaneously lowered to zero at $x = 0$. In the context of hillslope evolution, this problem corresponds to instantaneous base level drop of an initially planar hillslope or terrace.

In order to solve this problem, it is useful to introduce a dimensionless elevation given by

$$\theta = \frac{h}{h_0} - 1 \tag{2.21}$$

The diffusion equation for θ is identical to the diffusion equation for h. The boundary conditions on θ are now given by $\theta(x, 0) = 0$, $\theta(0, t) = 1$, and $\theta(\infty, t) = 0$. Similarity solutions make use of a mathematical trick that reduces the diffusion equation (a partial differential equation) to an ordinary differential equation that is solved by

integration. The approach is to introduce a new variable that is a combination of x and t:

$$\eta = \frac{x}{2\sqrt{\kappa t}} \tag{2.22}$$

and rewrite the diffusion equation for θ and its boundary conditions in terms of η. This can be done (following the approach in Turcotte and Schubert (1992)) using the chain rule for differentiation:

$$\frac{\partial \theta}{\partial t} = \frac{\partial \theta}{\partial \eta} \frac{\partial \eta}{\partial t} = \frac{\partial \theta}{\partial \eta} \left(-\frac{1}{4} \frac{x}{\sqrt{\kappa t}} \frac{1}{t} \right) = \frac{d\theta}{d\eta} \left(-\frac{1}{2} \frac{\eta}{t} \right) \tag{2.23}$$

$$\frac{\partial \theta}{\partial x} = \frac{d\theta}{d\eta} \frac{\partial \eta}{\partial x} = \frac{d\theta}{d\eta} \frac{1}{2\sqrt{\kappa t}} \tag{2.24}$$

$$\frac{\partial^2 \theta}{\partial x^2} = \frac{1}{2\sqrt{\kappa t}} \frac{d^2 \theta}{d\eta^2} \frac{\partial \eta}{\partial x} = \frac{1}{4} \frac{1}{\kappa t} \frac{d^2 \theta}{d\eta^2} \tag{2.25}$$

The diffusion equation for θ becomes

$$-\eta \frac{d\theta}{d\eta} = \frac{1}{2} \frac{d^2 \theta}{d\eta^2} \tag{2.26}$$

and the boundary conditions become $\theta(\infty) = 0$ and $\theta(0) = 1$. Equation (2.26) is a second-order ordinary differential equation (ODE) that can be reduced to a first-order ODE by introducing the variable

$$\phi = \frac{d\theta}{d\eta} \tag{2.27}$$

Equation (2.26) then becomes

$$-\eta d\eta = \frac{1}{2} \frac{d\phi}{\phi} \tag{2.28}$$

Equation (2.28) can be integrated to obtain

$$-\eta^2 = \ln \phi - \ln c_1 \tag{2.29}$$

Taking the exponential of both sides of Eq. (2.29) gives

$$\phi = c_1 e^{-\eta^2} = \frac{d\theta}{d\eta} \tag{2.30}$$

Integrating Eq. (2.30) gives

$$\theta = c_1 \int_0^\eta e^{-\eta'^2} d\eta' + 1 \tag{2.31}$$

where η' is an integration variable and the condition $\theta(0) = 1$ was used to constrain the integration constant. Application of $\theta(\infty) = 0$ requires that

$$0 = c_1 \int_0^\infty e^{-\eta'^2} d\eta' + 1 \tag{2.32}$$

The integral in Eq. (2.36) is given by $\sqrt{\pi}/2$, so the constant c_1 has the value $-2/\sqrt{\pi}$. Equation (2.31) becomes

$$\theta = 1 - \frac{2}{\sqrt{\pi}} \int_0^\infty e^{-\eta'^2} d\eta' \tag{2.33}$$

The integral in Eq. (2.33) has a special name called the error function:

$$\text{erf}(\eta) = \frac{2}{\sqrt{\pi}} \int_0^\eta e^{-\eta'^2} d\eta' \tag{2.34}$$

The solution for θ, therefore, is

$$\theta = 1 - \text{erf}(\eta) = \text{erfc}(\eta) \tag{2.35}$$

where $\text{erfc}(\eta)$ is called the complementary error function. Written in terms of the original variables, Eq. (2.35) becomes

$$h(x, t) = h_0 \text{erf}\left(\frac{x}{2\sqrt{\kappa t}}\right) \tag{2.36}$$

Most solutions of the diffusion equation in infinite and semi-infinite domains can be written in terms of error functions.

Series solutions and similarity solutions are the two basic types of analytical approaches to solving the diffusion equation. In the following sections we will apply these equations to specific problems that arise in geomorphic applications.

2.2.4 Transient approach to steady state

In Section 2.2.1 we found the steady-state diffusive hillslope to be given by Eq. (2.13). Here we solve the time-dependent diffusion equation that describes the transient approach to steady state. We consider a diffusing hillslope of length L with constant uplift rate U:

$$\frac{\partial h}{\partial t} - \kappa \frac{\partial^2 h}{\partial x^2} = U \tag{2.37}$$

The boundary conditions for this problem are given by Eqs. (2.13) and (2.12). The initial condition is $h(x, 0) = 0$. The constant uplift term on the right side of Eq. (2.37) can be eliminated by introducing a new variable w that quantifies the difference from steady state. The variable w is defined by

$$w(x, t) = h(x, t) - \frac{U}{2\kappa}(L^2 - x^2) \tag{2.38}$$

The evolution of w is governed by the diffusion equation with the same boundary conditions on h. The initial condition for w is given by

$$w(x, 0) = -\frac{U}{2\kappa}(L^2 - x^2) \tag{2.39}$$

The solution for w is found by substituting Eq. (2.39) into Eq. (2.20) to obtain

$$w(x, t) = \frac{32}{\pi^3} \sum_{n=0}^\infty \frac{(-1)^n}{(2n+1)^3} \cos\left(\frac{(2n+1)\pi x}{2L}\right) e^{-\frac{\kappa n^2 \pi^2 t}{4L^2}} \tag{2.40}$$

Substituting Eq. (2.40) into Eq. (2.38) and rearranging for h gives

$$h(x, t) = \frac{UL^2}{2\kappa}\left[1 - \frac{x^2}{L^2} - \frac{32\kappa}{\pi^3 L^2} \sum_{n=0}^\infty \frac{(-1)^n}{(2n+1)^3}\right.$$
$$\left. \times \cos\left(\frac{(2n+1)\pi x}{2L}\right) e^{-\frac{\kappa n^2 \pi^2 t}{4L^2}}\right] \tag{2.41}$$

Examples of Eq. (2.41) are plotted in Figure 2.3a for values of κt ranging from $1000\,\text{m}^2$ to $2500\,\text{m}^2$. These results provide an estimate for the time scale necessary to achieve steady state. If we assume $\kappa = 1\,\text{m}^2/\text{kyr}$, then steady state is achieved after approximately 2.5 Myr of uplift for a hillslope 50 m in length. Interestingly, the time scale required to develop steady state does not depend on the uplift rate.

Figure 2.3b presents the same results as Figure 2.3a except that the hillslope length has been normalized to 1. In this approach the results depend only on a single nondimensional parameter $\kappa t/L^2$. This nondimensional framework is more flexible because we do not need to plot different solutions for hillslopes of different lengths. While Figure 2.3a is specific to hillslopes 50 m in length, Figure 2.3b allows us to estimate the time required to achieve steady state for any hillslope length. If $\kappa = 1\,\text{m}^2/\text{kyr}$ and $L = 100\,\text{m}$, for example, steady state is achieved (i.e. the relief has achieved 90% of the steady-state relief) after $\kappa t/L^2 = 1$ or $t = 10\,\text{Myr}$.

In this chapter we will usually present results in terms of actual length scales in order to develop intuition about the rates of landscape evolution. In practice, however, it is often best to solve the problem in a fully nondimensional way and then apply specific values of κ, L, and t to scale the results appropriately.

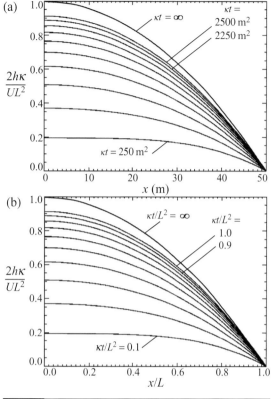

Fig 2.3 (a) Approach to steady state in a hillslope subject to uniform uplift or base level drop for a hillslope 50 m in length. (b) Same as (a) except that results are plotted in terms of dimensionless length x/L.

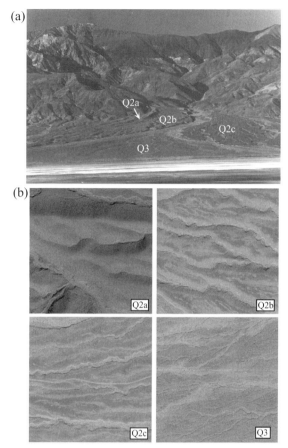

Fig 2.4 (a) Hanaupah Canyon alluvial fan, showing mid-Pleistocene (Q2a) to Holocene (Q3) terraces. (b) Aerial photographs of Q2a, Q2b, Q2c, and Q3 terraces, illustrating the correlation between age and degree of roundedness of the terrace slope adjacent to gullies.

2.2.5 Evolution of alluvial-fan terraces following fan-head entrenchment

In this section we consider the evolution of initially-planar hillslope terraces subject to rapid base-level drop. This example provides an opportunity to use both series and similarity solutions within the context of a real-world hillslope-evolution application.

First some background. In the Basin and Range province of the western United States, many alluvial fans have a suite of terraces that rise like a flight of stairs from the active channel. Figure 2.4 illustrates the classic alluvial fan terraces of Hanaupah Canyon in Death Valley, California. These terrace suites are generally considered to be caused by variations in the sediment-to-water ratio in channels draining the mountain front (Bull, 1979). When alluvial fans are in an aggrading mode, the ratio of sediment to water is high and sediment is deposited across the fan (or the available accommodation space in the entrenched zone of an older abandoned terrace) to form a low-relief deposit by repeated episodes of channel avulsion (Figure 2.5a). When the ratio of sediment to water decreases, such as when cooler and wetter climate conditions cause revegetation of hillslopes and anchoring of the available regolith, the sediment deposited during the previous period of high sediment flux is rapidly entrenched (Figure 2.5b). This entrenchment acts as a rapid base-level drop for the abandoned terrace and gullies quickly form at the distal end of the terrace and migrate headward (Figure 2.5c). At

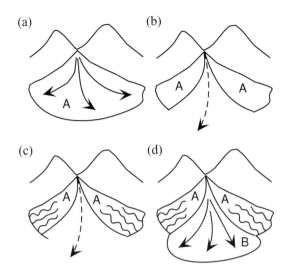

(a)   (b)   (c)   (d)

Fig 2.5 Schematic illustration of a cycle of aggradation and fan-head entrenchment. When alluvial fans are in an aggradational mode, the ratio of sediment to water is high and sediment is deposited across the available accommodation space to form a low-relief surface by repeated episodes of channel avulsion (a). When the ratio of sediment to water decreases, the sediment deposited during the previous high-sediment-flux period is rapidly entrenched leaving an abandoned terrace (b). This entrenchment acts as a rapid base-level drop for the abandoned terrace and gullies quickly form at the distal end of the terrace and migrate headward (c).

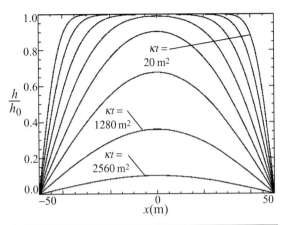

Fig 2.6 Plots of Eq. (2.42) for κt values of 50, 100, 200 ... 6400 m^2 and a terrace width of $2L = 100$ m. The early phase of the model is characterized by penetration of erosion into the scarp interior. After $\kappa t \approx 1000$ m^2 no planar remnant remains and the scarp profile can be approximated by a sine function with height decreasing linearly with time.

any point along a line parallel to the mountain front, a profile can be plotted that crosses one or more gullies. In young terraces, gully profiles are characterized by narrow valleys and broad, planar terrace treads. In older profiles, gully erosion penetrates farther into the terrace. Given a sufficiently old terrace, hillslope erosion will completely penetrate the terrace tread and no planar remnant will be preserved. The effects of erosional beveling can be seen in the aerial orthophotographs shown in Figure 2.4b. As the terrace age decreases from mid-Pleistocene (Q2a) to latest Pleistocene–early Holocene (Q3), more planar terraces are preserved and the degree of terrace rounding at gullies decreases markedly.

Equations (2.36) and (2.20) can be used to model the evolution of gully profiles assuming that gullies form rapidly following entrenchment and that the gully floor has a constant elevation

once the gully has formed. This approach provides a relative-age dating and correlation tool for fan terraces (Hsu and Pelletier, 2004). Series solutions (Eq. (2.20)) are the most comprehensive approach to this problem, but similarity solutions can be used to model young terraces (those with planar remnants) with no loss of accuracy.

First we consider series solutions. Substituting $f(x) = h_0$ into Eq. (2.20) gives

$$h(x, t) = \frac{4h_0}{\pi} \sum_{n=0}^{\infty} \frac{(-1)^n}{2n + 1}$$
$$\times \cos\left(\frac{(2n + 1)\pi x}{2L}\right) e^{-\frac{\kappa(2n+1)^2\pi^2 t}{4L^2}} \quad (2.42)$$

Equation (2.42) can be used to model terrace gully profiles through their entire evolution from backwearing scarp to downwearing gully remnant. That two-phase evolution is illustrated in Figure 2.6. The early phase of the model is characterized by penetration of the backwearing scarp into the terrace interior. After $\kappa t \approx 250$ m^2, no planar remnant remains and the scarp profile can be approximated as a sine function (or cosine function, depending on where $x = 0$ is defined) with a linearly-decreasing height over time.

The results of Figure 2.6 nicely illustrate the filtering aspects of diffusion. A series solution is composed of sine functions multiplied by an

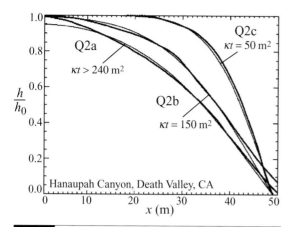

Fig 2.7 Gully profiles of the Q2a, Q2b, and Q2c fan terraces at Hanaupah Canyon, Death Valley, California. Best-fit morphologic ages to the diffusion profile given by Eq. (2.42) are 240 m², 150 m², and 50 m², respectively. Only a minimum age can be determined for Q2a because the height of the original terrace is unknown.

exponential damping term that is a function of wavelength. Diffusion acts to filter the small-wavelength components of the topography over time through the exponential term in the series solution. For large values of t, all of the small wavelength (i.e. large n) terms will be negligible and the only significant term will be the largest-wavelength sine function. This is a general result: all base-level-controlled hillslopes will approximate a sine function during their waning stages of evolution, regardless of their initial conditions.

Figure 2.7 illustrates normalized surveyed cross sections of the Q2a, Q2b, and Q2c terraces in Hanaupah Canyon along with their best-fit profiles to Eq. (2.42). The Q2c scarp is best represented by Eq. (2.42) with $\kappa t = 50$ m², while Q2b is best represented by $\kappa t = 150$ m². These κt values are sometimes referred to as morphologic ages. Only a minimum κt value can be determined for Q2a because no planar remnant remains and we do not know the height of the original terrace. Therefore, once the terrace profile is normalized, any κt value greater than ≈ 250 m² matches the observed profile equally. These results suggest that the Q2b terrace on the Hanaupah Canyon fan is approximately three times older than the Q2c terrace, assuming that κ values have been

relatively constant over these time periods. If we assume a κ value of 1 m²/kyr, the estimated age of the Q2c terrace is 50 kyr and the Q2b terrace is 150 kyr. These values are reasonably close to the estimated ages of these terraces based on available geochronology.

Normalization is appropriate in comparing diffusion-model predictions to natural landforms because the relative shape of the hillslope does not depend on the absolute height of the hillslope. In other words, the landform height can be scaled up or down and the relative hillslope shape predicted by the diffusion model will remain the same.

In the field, the length of scarp backwearing can be used as a tool to correlate terraces. The length of erosional penetration into a scarp can be estimated as $\lambda = \sqrt{\kappa t}$. This is a general result for all diffusion problems. This relationship implies that Q2c- and Q2b-aged terraces wear back by approximately 7 and 12 m, respectively, assuming $\kappa = 1$ m²/kyr.

The similarity solution can be used to estimate the age of the Q2c and Q2b terraces because planar remnants are preserved on these terraces. As such, the terrace width can be considered to be semi-infinite. To solve the problem with the similarity method, we consider a semi-infinite terrace of height h_0 subject to rapid base-level drop at $h = L$. The solution to this problem is

$$h(x, t) = h_0 \mathrm{erf}\left(\frac{L - x}{2\sqrt{\kappa t}}\right) \quad (2.43)$$

Equation (2.43) is identical to Eq. (2.42) for small values of κt.

2.2.6 Evolution of alpine moraines

Advancing alpine glaciers bulldoze rock and debris out along their margins. When the glacier retreats, a terminal moraine is formed. Terminal moraines are initially triangular in cross section. As such, the solution to the diffusion equation with a triangular initial condition provides a simple model of moraine evolution.

In the simplest case of a symmetrical moraine, only half of the moraine need be considered. The highest point of the moraine forms a divide located at $x = 0$ (Figure 2.8a). At the base of the moraine two different boundary conditions

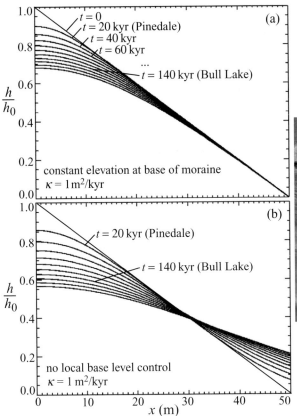

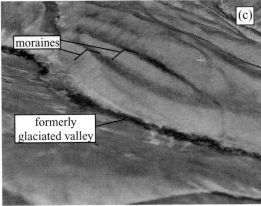

Fig 2.8 (a) Diffusion model of moraine evolution plotted for 20 kyr intervals up to $t = 200$ kyr, assuming a κ value of 1 m²/kyr, with a fixed-elevation boundary condition at the slope base. (b) Same as (a), but with a fixed-elevation boundary condition at $x = \infty$. (c) Virtual oblique aerial photograph of a series of terminal moraines in Owens Valley, Inyo County, California.

can be considered. First we consider the case of a fixed elevation at a distance L from the divide. As in previous examples, this boundary condition is most appropriate for cases in which a stable channel is present at the slope base. Substituting $f(x) = h_0(L - x)$ into Eq. (2.20) gives the solution

$$h(x, t) = \frac{8h_0}{\pi^2} \sum_{n=0}^{\infty} \frac{1}{(2n+1)^2}$$
$$\times \cos\left(\frac{(2n+1)\pi x}{2L}\right) e^{-\frac{\kappa(2n+1)^2\pi^2 t}{4L^2}} \quad (2.44)$$

Figure 2.8a shows plots of Eq. (2.45) at 20 kyr increments from $t = 20$ kyr to $t = 200$ kyr assuming a κ value of 1 m²/kyr. A Pinedale-age (approx.

20 ka) moraine, for example, is rounded at the top over a width of approximately 10 m. A Bull Lake moraine (approx. 140 ka), in contrast, has a rounded crest of approximately 30 m in width assuming a κ value of 1 m²/yr.

Second, we consider the case in which the moraine deposits sediment at its slope base (such as when the moraine is surrounded by an alluvial piedmont). In this case, sediment shed from the moraine will be deposited at the base of the moraine and on the surrounding piedmont. The boundary condition in this case is a fixed elevation of zero at $x = \infty$. For the case of an initially triangular hillslope, the solution is

$$h(x, t) = \frac{2h_0}{L}\left[(L - x)\text{erf}\left(\frac{L - x}{\sqrt{4\kappa t}}\right) + (L + x)\right.$$
$$\times \text{erf}\left(\frac{L + x}{\sqrt{4\kappa t}}\right) - 2x\text{erf}\left(\frac{x}{\sqrt{4\kappa t}}\right)$$
$$\left. + 2\sqrt{\frac{\kappa t}{\pi}}\left(e^{-\frac{(x-L)^2}{4\kappa t}} + e^{-\frac{(x+L)^2}{4\kappa t}} - 2e^{-\frac{x^2}{4\kappa t}}\right)\right]$$
$$(2.45)$$

Figure 2.8b illustrates profiles of Eq. (2.45) for the same ages as in Figure 2.8a. For the case with a depositional slope base the erosion of the moraine crest occurs at a slightly lower rate compared to the base with a fixed base level. The main effect of the deposition is to widen the moraine as sediment prograde out on the adjacent slope base.

2.2.7 Evolution of pluvial shoreline and fault scarps

The study of pluvial shoreline and fault scarps has had great historical importance in hillslope geomorphology. These landforms have been studied within the context of the diffusion model of hillslope evolution for over forty years. In that time, researchers have taken several different approaches to analyzing scarps. In this section, we present the analytic solutions to the diffusion equation appropriate for pluvial shoreline and fault scarps. In addition, however, we also compare and contrast past approaches in order to put the analytic solutions into the historical context of other approaches. By comparing different methods for scarp analysis and modeling, we are able to explore some of the subtleties involved in comparing model predictions with natural landforms. For example, how do we handle uncertainty in the landform initial condition at some point in the past? How do we handle measurement uncertainty and random variability in the landscape?

Pluvial shoreline scarps form by wave-cut action during a prolonged period of lake-level high stand. Following lake-level fall, scarps formed in unconsolidated alluvium evolve to an angle of repose by mass movements and then further evolve by diffusive hillslope processes.

Historical scarp-analysis methods come in two basic types: midpoint-slope methods and full-scarp methods. Mid-point slope methods can be further divided into two types. In the first type, the midpoint slope of a single scarp is inverted for diffusion age using the analytic solution to the diffusion equation (Andrews and Hanks, 1985). In the second approach, the midpoint gradient is plotted versus scarp offset for a collection of many scarps with different heights (Bucknam

and Anderson, 1979; Hanks and Andrews, 1989). In this approach, the far-field gradient (i.e. the gradient of the initially unfaulted surface) must first be subtracted from the midpoint gradient to take into account the effects of scarps cut into sloping surfaces. The plot of midpoint gradient minus the far-field gradient is compared to characteristic curves for the diffusion equation with different parameter values (e.g. Bucknam and Anderson, 1979; Hanks et al., 1984) to determine which parameters best match the observed data. Both methods are potentially unreliable because they reduce the information content of the scarp to the slope angle or gradient at a single point. The scarp midpoint is the least diagnostic point along the entire scarp profile, and it is the point that is also most sensitive to uncertainty in the initial scarp angle. As a scarp evolves, the greatest amount of erosion and deposition occurs where the magnitude of the hillslope curvature is greatest, near the top and bottom of the scarp. The scarp midpoint is a point of inflection with zero curvature, and hence this point changes the least of all points along the profile for young scarps. An alternative approach is to fit the entire scarp (or its derivative, the hillslope gradient) to analytical or numerical solutions of the diffusion equation (linear or nonlinear). Avouac and his colleagues (Avouac, 1993; Avouac and Peltzer, 1993) pioneered the use of this technique in fault-scarp studies in Asia. Mattson and Bruhn (2001) extended the technique to include uncertainty in the initial scarp angle.

Figure 2.9 presents an oblique aerial view of the shoreline in east-central Tule Valley as it cuts across several distinct alluvial fan terraces. This figure illustrates that scarp height is controlled primarily by the slope of the fan terrace, with the tallest scarps formed in steeply dipping alluvial fan terraces and debris cones close to the mountain front.

The geometry of a general scarp is defined by the scarp height $2a$, initial gradient α, and far-field gradient b (Figure 2.10a). The initial and far-field gradients can be combined for the purposes of scarp analysis into a single variable: the reduced initial gradient $\alpha - b$. The analytic solution to the diffusion equation for a general scarp

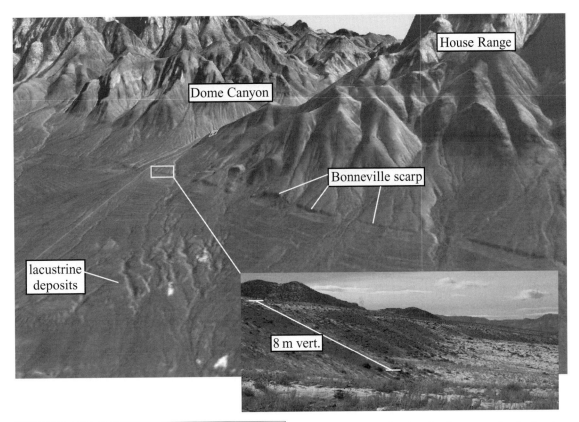

Fig 2.9 Virtual oblique aerial photograph (i.e. aerial photo draped over digital topography) of pluvial features on the east side of central Tule Valley. Scarp height is primarily controlled by the slope of the alluvial-fan and debris-cone deposits, i.e. wave-cut erosion of steeply dipping deposits tends to produce steep, tall scarps. Inset photo shows 8 m tall scarp at the entrance to Dome Canyon.

was given by Hanks and Andrews (1989) as

$$h(x, t) = (\alpha - b)\sqrt{\frac{\kappa t}{\pi}} \left(e^{-\frac{x+a/(\alpha-b))^2}{4\kappa t}} - e^{-\frac{x-a/(\alpha-b))^2}{4\kappa t}} \right)$$
$$+ \frac{(\alpha - b)}{2} \left(\left(x + \frac{a}{\alpha - b} \right) \text{erf} \left(\frac{x + a/(\alpha - b)}{\sqrt{4\kappa t}} \right) \right.$$
$$\left. - \left(x - \frac{a}{\alpha - b} \right) \text{erf} \left(\frac{x - a/(\alpha - b)}{\sqrt{4\kappa t}} \right) \right) + bx$$
$$(2.46)$$

Full-scarp analyses can be performed by comparing the measured elevation profile of a scarp to Eq. (2.46), or by taking the derivative of the elevation profile and working with the slope or gradient profile. Working with the gradient profile

is mathematically simpler because the derivative of Eq. (2.46) is a more compact expression:

$$\frac{\partial h(x, t)}{\partial x} - b = \frac{(\alpha - b)}{2} \left(\text{erf} \left(\frac{x + a/(\alpha - b)}{\sqrt{4\kappa t}} \right) \right.$$
$$\left. - \text{erf} \left(\frac{x - a/(\alpha - b)}{\sqrt{4\kappa t}} \right) \right) \quad (2.47)$$

The left-hand side of Eq. (2.47) is called the reduced gradient profile (i.e. the gradient profile reduced by b, in order to take the far-field gradient into account). Equations (2.46) and (2.47) are plotted in Figure 2.1 for a range of diffusion ages κt and reduced initial scarp gradients, $\alpha - b$. Figure 2.10b illustrates that the midpoint of the scarp is insensitive to age for young, tall scarps. Note that in Figure 2.9, the units of the x-axis are scaled to the scarp width x_0, which, in turn, is scaled to the scarp height through the relationship $x_0 = a/(\alpha - b)$. Figure 2.10c illustrates that, as the reduced initial gradient decreases, the age-diagnostic portion of the scarp becomes progressively concentrated away from the midpoint towards the top and bottom of the scarp.

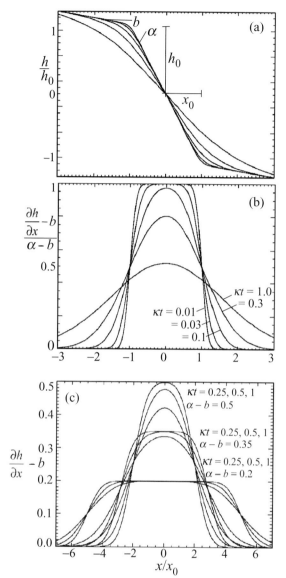

Fig 2.10 (a) Analytic solutions of the linear diffusion equation, starting from an initial gradient α, scarp height $2h_0$, and far-field slope b. The initial scarp width, $2x_0$, is related to height and reduced initial gradient by $x_0 = h_0/(\alpha\tilde{n}b)$. (b) Plots of reduced gradient corresponding to the profiles in (a). For small κt values (< 0.1), no change in the midpoint slope is observed, illustrating the insensitivity of the midpoint slope to diffusion age for young scarps. (c) Plots of reduced gradient versus distance along profile, for families of solutions corresponding to a range of morphologic ages ($\kappa t = 0.25$, 0.5, 1) and a range of reduced initial gradients ($\alpha - b = 0.2$, 0.35, 0.5).

These plots suggest that full-scarp methods are necessary to extract the independent controls of scarp age and initial gradient on scarp morphology.

Alternatively, a brute-force approach can be used which computes analytic or numerical solutions corresponding to a full range of model parameters and then plots the error between the model and measured profiles as a function of each parameter value (e.g. diffusion age and initial gradient). The diffusion age is then constrained by the model profile that has the lowest error relative to the measured data. This brute-force approach is most often used, and it has the advantage that the goodness-of-fit can be evaluated visually on a profile-by-profile basis. Avouac (1993) and Arrowsmith *et al.* (1998) pioneered this technique, assuming initial scarp angles between 30° and 35°. Mattson and Bruhn (2001) improved upon this approach by including a range of initial gradients in the analysis. Mattson and Bruhn's error plots are thus two-dimensional, including a range of both diffusion age and reduced initial angle in the analysis.

Figures 2.11a and 2.11b illustrate the full-scarp method on a short and a tall scarp, respectively, each selected from the set of Bonneville scarps. The reduced gradient is plotted as a function of distance along the profile. To the right of each plot is a grayscale map that illustrates the relative least-square error between the model and observed profiles for a range of diffusivity values and reduced initial gradients. In both maps, black colors represent the lowest mismatch and X marks the spot of the optimal fit. The location of this best-fit solution in the model parameter space simultaneously determines diffusivity and reduced initial gradient on a profile-by-profile basis. In the short-scarp example case, the best-fit solution corresponds to $\kappa = 0.95\,\mathrm{m^2/kyr}$ and $\alpha - b = 0.70$. The black band running across the grayscale map illustrates that, although the best-fit solution occurs for $\alpha - b = 0.70$, the solution is essentially independent of the reduced initial gradient. This makes physical sense because, for a short, narrow scarp, diffusive smoothing quickly reduces the maximum slope to a value far below the angle of repose. For a tall, broad scarp (Figure 2.11b), however, the scarp form is

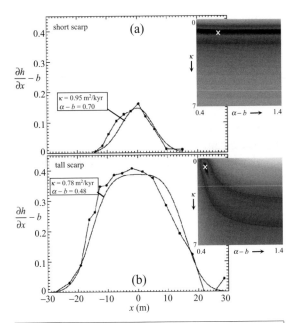

Fig 2.11 Full-scarp curve-fitting method illustrated for (a) short and (b) tall example Bonneville scarps. The reduced gradient is plotted as a function of distance from the scarp midpoint. Grayscale maps to the right of each plot illustrate the relative error between the observed and modeled scarp as a function of diffusivity κ and reduced initial gradient $\alpha - b$. In both maps, black colors represent the lowest mismatch and X marks the spot of the optimal fit. The location of this best-fit solution in the model parameter space simultaneously determines diffusivity and reduced initial gradient on a profile-by-profile basis. In the short-scarp example case in (a), the best-fit solution corresponds to $\kappa = 0.95 \, \text{m}^2/\text{kyr}$ and $\alpha - b = 0.70$. The black band running across the grayscale map illustrates that, although the best-fit solution occurs for $\alpha - b = 0.70$, the solution is essentially independent of the reduced initial gradient. This makes physical sense because, for a short, narrow scarp, diffusive smoothing quickly reduces the maximum slope to a value far below the angle of repose. For a tall, broad scarp (b), however, the solution is sensitive to the initial angle. The best-fit model parameters in this case have well-defined values for both diffusion age and reduced initial gradient ($\kappa = 0.78 \, \text{m}^2/\text{kyr}$ and $\alpha - b = 0.48$).

sensitive to the initial angle. The best-fit model parameters in this case have well-defined values for both diffusion age and reduced initial gradient ($\kappa = 0.78 \, \text{m}^2/\text{kyr}$ and $\alpha - b = 0.48$). Physically, fitting the full scarp works best because the scarp profile contains independent information about the diffusion age and the initial gradient. The diffusion age is determined primarily by the width

of the slope break at the top and bottom of the scarp. The reduced initial gradient is determined by the shape of the flat, central portion of the gradient profile shown in Figure 2.11b (also shown in the analytic solutions of Figure 2.1f).

2.2.8 Degradation of archeological ruins

In the summer of 2006 I performed field work in Tibet with Jennifer Boerner. Tibet is full of important archeological ruins, and I wondered whether the degradation of stone-wall ruins could be described by a diffusion model. Stone walls begin as consolidated, vertical structures. Over time scales of centuries to millennia following abandonment, however, mortar will degrade, stones will topple, and surface processes (creep, bioturbation, eolian deposition, etc.) will act to smooth the surface.

Figure 2.12 illustrates the application of the diffusion equation to stone ruins in western Tibet. Figure 2.12a illustrates the solution to the diffusion equation with an initially vertical structure. This equation is identical to the solution for the gradient profile of a scarp, Eq. (2.47):

$$h(x, t) = \frac{h_0}{2x_0} \left(\text{erf} \left(\frac{x + x_0}{\sqrt{4\kappa t}} \right) - \text{erf} \left(\frac{x - x_0}{\sqrt{4\kappa t}} \right) \right) \quad (2.48)$$

where h_0 is the initial height of the wall and x_0 is the half-width. Figure 2.12b presents an oblique view of one example of stone ruins in Tibet. These degraded walls are up to 0.5 m tall and have spread out to cover a width of several meters. Figure 2.12d presents a contour plot of a detailed topographic survey of a small section of one of these ruins. Figure 2.12c plots two topographic profiles extracted from Figure 2.12d. These plots are compared with Eq. (2.48) for κt values of $2.0 \, \text{m}^2$ and $1.3 \, \text{m}^2$. Radiocarbon ages indicate a site abandonment age of 1.7 ka at this location. Diffusivity values for these walls, therefore, are 1.2 and $0.8 \, \text{m}^2/\text{kyr}$, respectively. These values are broadly comparable to the representative value for the Basin and Range.

2.2.9 Radionuclide dispersion in soils

Thus far we have only considered applications of the diffusion equation to landform evolution. Many applications of the diffusion equation are applicable in other aspects of surface processes,

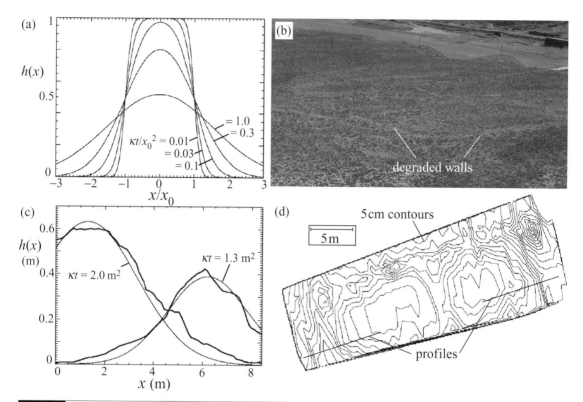

Fig 2.12 Diffusion modeling of the stone-wall degradation in western Tibet. (a) Solutions to the diffusion equation with an initially vertical structure. (b) Oblique view of degraded walls in Tibet. (c) Plots of topographic profiles perpendicular to the wall, with best-fit solutions corresponding to $\kappa t = 2.0\,\text{m}^2$ and $1.3\,\text{m}^2$. (d) Contour map of a microtopographic survey of degraded walls in western Tibet (near Kyunglung), with profile locations noted.

however. Here we consider the migration of radionuclides into the subsurface within a soil profile.

Beginning in the mid-1950s and continuing through the mid-1960s, above-ground nuclear tests introduced radioactive fallout ^{137}Cs into the soil (He and Walling, 1997). Fourteen ^{137}Cs profiles from the mapped area of Fortymile Wash fan were collected and analyzed (Table 2.1) in order to determine the rate of Cs migration into the soil over the past 50 years. The Fortymile Wash fan is of particular interest because it is the depozone for sediment eroded from the Nevada Test Site. Bulk samples were collected at 0–3 cm, 3–6 cm, and 6–9 cm. The fraction of total ^{137}Cs in the upper half of the profile typically varied between 80

and 95%, suggesting that very little ^{137}Cs has diffused below 6 cm in these profiles. The radionuclide profile shape may reflect the influence of both surface erosion/deposition and redistribution within the soil column. However, in the case of Fortymile Wash it is reasonable to assume that surface erosion and deposition were negligible during the past 50 yr. Except for the active channel, all of the older surfaces have not experienced significant flooding for several thousand years. Eolian deposition rates inferred for the Fortymile Wash fan by Reheis et al. (1995) also suggest that eolian erosion/deposition is less than or equal to several millimeters over the 50-yr time scale since the introduction of fallout ^{137}Cs. Due to the relative stability of the Fortymile Wash fan surfaces to both fluvial and eolian erosion/deposition, we expect the shape of the ^{137}Cs profile to predominantly reflect redistribution processes. The exception is the active channel. In this case the Cs profile may be strongly influenced by fluvial mixing in addition to infiltration and other mixing processes.

The equation describing the evolution of radionuclide concentration C by diffusive processes

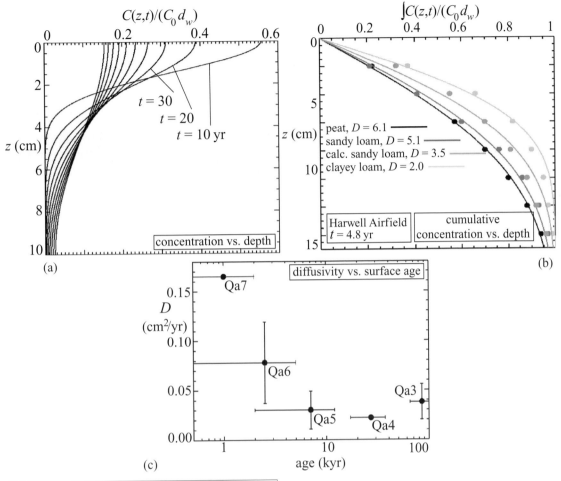

(a) Diffusion-model results for normalized concentration vs. depth from $t = 10$ to $100\,yr$ with $D = 0.1\,cm^2/yr$. (b) Cumulative ^{137}Cs concentration as a function of depth in four different soil types at Harwell, England (data from Gale (1964)), along with best-fit results for the diffusion model. (c) Plot of diffusivity values vs. surface age for the ^{137}Cs profiles on Fortymile Wash alluvial fan. Horizontal error bars represent the range of age estimates for each surface based on Whitney et al. (2004). Vertical bars represent the standard deviation of diffusivity values for all of the samples collected on each surface.

is given by the solution to the diffusion equation in a semi-infinite medium with no-flux boundary condition at the surface and a fallout mass $C_0 d_w$ input at $z = 0$ at $t = 0$ (Carslaw and Jaeger, 1959):

$$C(z, t) = C_0 d_w \frac{1}{\sqrt{\pi D t}} e^{-z^2/4Dt} \qquad (2.49)$$

where C_0 is the initial concentration within a thin surface layer of thickness d_w, z is depth in the soil profile, D is the diffusivity (called D in this section to emphasize that it is not a hillslope diffusivity), and t is time following deposition. Equation (2.49) also approximates the permeable soil layer as semi-infinite. This is an accurate approximation for man-made fallout profiles, even in a soil with a calcic horizon at depth, because radionuclides do not penetrate more than several cm into desert soils over decadal time scales. It should also be noted that radioactive decay need not be considered explicitly in this analysis because decay does not affect the relative radionuclide concentration at different depth intervals, only their absolute values. Figure 2.13a gives an example of the diffusion model (Eq. (2.49) using $D = 0.1\,cm^2/yr$ and $t = 10–100\,yr$).

Table 2.1 | ^{137}Cs data and inferred diffusivity values for geomorphic surfaces on the Fortymile Wash alluvial fan

Sample ID	total ^{137}Cs (pCi/g)	fraction at 3 cm	D (cm^2/yr)	Geomorphic Unit
Cs-071802-A	0.313	0.8274	0.047	Qa6
Cs-071802-B	0.271	0.5387	0.165	Qa7
Cs-071802-C	0.258	0.8100	0.052	Qa3
Cs-071802-E	0.208	0.7644	0.063	Qa3
Cs-071802-G	0.118	0.9593	0.021	Qa3
Cs-071802-I	0.388	0.9614	0.021	Qa4
Cs-071802-J	0.155	0.6387	0.108	Qa6
Cs-071802-K	0.340	0.9558	0.022	Qa5
Cs-071802-N	0.218	0.9082	0.031	Qa5
Cs-071802-P	0.245	0.9428	0.024	Qa4
Cs-071802-Q	0.205	0.9951	0.011	Qa5
Cs-071802-R	0.237	0.9578	0.021	Qa3
Cs-071802-S	0.285	0.8807	0.037	Qa3
Cs-071802-BB	0.321	0.7943	0.056	Qa5

Radionuclide concentrations measured in the field are bulk measurements rather than point measurements because the concentration within different depth intervals is measured. For the purposes of extracting model parameters from bulk measurements, it is most accurate to represent measured data cumulatively as the fraction of total concentration to a given depth. To compare the diffusion model predictions to this normalized cumulative curve, it is necessary to integrate Eq. (2.49) to give

$$\int_0^z \frac{C(\eta, t)}{C_0 d_w} d\eta = \text{erf}\left(\frac{z}{2\sqrt{Dt}}\right) \tag{2.50}$$

Figure 2.13b compares the diffusion model prediction to ^{137}Cs profiles measured during a 4.7-yr field experiment at Harwell Airfield in Oxfordshire, England (Gale *et al.*, 1964). Although these data are from a very different climatic regime than Fortymile Wash, the results illustrate the precision of the diffusion model in a way that profiles in desert soils cannot show because shallow penetration limits the spatial resolution of desert-soil profiles. To obtain the results in Figure 2.13b, the measured ^{137}Cs profiles were fit to Eq. (2.50). Diffusivity values are lowest for clayey and calcareous soils at Harwell, increasing in value in sandy soils and peat.

Table 2.1 summarizes the data and results from Fortymile Wash. To calculate D, the fraction of total activity at 3 cm was first computed by dividing the activity from 0–3 cm by the total activity from 0–6 cm. Equation (2.50) was then used to infer the value of the error function argument, equal to $z/\sqrt{4Dt}$, corresponding to the fraction of activity at 3 cm after 50 yr of diffusion following nuclear testing. A table of calculated error function values was used for this purpose. The value of the error function argument was then used to solve for D (column 3 in Table 2.1). Figure 2.13c illustrates the relationship between diffusivity values and geomorphic-surface age on the Fortymile Wash fan. The vertical bars are the standard deviation of D values obtained on different profiles of the same soil-geomorphic unit. The horizontal bars correspond to the age range for each unit based on the available age control (Whitney *et al.*, 2004). The plot exhibits an inverse relationship between diffusivity and age, with values increasing slightly on Qa3 (the oldest surface) relative to those on Qa4.

Radionuclides may be redistributed within the soil profile through entrainment and redeposition by infiltrating runoff and/or by bulk-mixing processes including freeze/thaw, wetting/drying, and bioturbation. The rate of transport by infiltration can be expected to correlate positively with hydrologic conductivity, and hence negatively with surface age based on the results of Young *et al.* (2004). As such, we can conclude that transport of radionuclides during infiltration is a significant component of transport.

For long time scales, the finite depth of soil profiles must be considered. In such cases, the concentration of radionuclides in the soil profile will be given by the series solution

$$C(z,t) = C_0 \left[\frac{d_w}{L} + \frac{2}{\pi} \sum_{n=1}^{\infty} \frac{1}{n} \sin\left(\frac{n\pi d_w}{L}\right) \right.$$
$$\left. \times \cos\left(\frac{n\pi z}{L}\right) e^{-n^2\pi^2 Dt/L^2} \right] \quad (2.51)$$

where L is the depth of the soil profile. In some cases, L may be the depth to an impermeable soil layer in the subsurface. In other cases, it may be more appropriate to take L to be the depth of the wetting front (i.e. the maximum distance to which radionuclides are transported by infiltrative transport).

2.2.10 Evolution of cinder cones

Volcanic cinder cones often have a high degree of radial symmetry and are composed largely of unconsolidated fallout tephra. As such, solutions to the diffusion equation in polar coordinates can be used to model their evolution. Cinder cones are particularly important landforms for this type of analysis because their initial conditions can often be precisely inferred by measuring the dip of subsurface bedding planes on opposite sides of the crater rim. The initial morphology of pluvial shoreline and fault scarps, in contrast, cannot be inferred with comparable precision because the initial mass-wasting phase of the scarp typically erases any memory of the initial shape. In addition, ^{40}Ar/^{39}Ar dating enables the eruption age of cinder cones to be precisely measured, and hundreds of cones in the western United States have been dated in this way. The framework for analyzing volcanic cones with analytic solutions to the diffusion equation was described by Culling (1963) and Hanks (2000). To date, however, no analytic solution has been compared to specific natural cones.

The diffusion equation with radial symmetry is given by

$$\frac{\partial h}{\partial t} = \frac{1}{r}\frac{\partial}{\partial r}\left(\kappa r \frac{\partial h}{\partial r}\right) \quad (2.52)$$

If κ is a constant, Eq. (2.52) becomes

$$\frac{\partial h}{\partial t} = \kappa\left(\frac{\partial^2 h}{\partial r^2} + \frac{1}{r}\frac{\partial h}{\partial r}\right) \quad (2.53)$$

The solution to Eq. (2.53) with an initial radially symmetric topography $f(r)$ and a constant-elevation boundary condition $h(a,t) = 0$ at $r = a$ is given by

$$h(r,t) = \frac{2}{a^2}\sum_{n=1}^{\infty} e^{-\kappa\alpha_n^2 t}\frac{J_0(\alpha_n r)}{J_1^2(\alpha_n a)}$$
$$\times \int_0^a rf(r)J_0(\alpha_n r)dr \quad (2.54)$$

where $\alpha_n, n = 1, 2, \ldots$ are the positive roots of $J_0(\alpha a) = 0$ and $J_0(r)$ and $J_1(r)$ are Bessel functions of the first and second kind (Culling, 1963). This solution utilizes two boundary conditions. First, at $r = 0$ Eq. (2.54) implicitly assumes

$$\left.\frac{\partial h}{\partial r}\right|_{r=0} = 0 \quad (2.55)$$

because Eq. (2.54) is a series comprised of even functions of radius (i.e. $J_0(r)$). The Bessel function $J_0(r)$ has a derivative of zero at $r = 0$, so a series comprised of a sum of these functions must also obey that condition. The second boundary condition is a constant elevation of zero at $r = a$. For simplicity we assumed the elevation of the surrounding alluvial flat equal to zero, but any constant value could be used. In hillslope geomorphology, a fixed-elevation boundary condition applies to a channel that is capable of transporting all of the sediment delivered by the hillslope, resulting in neither deposition nor erosion at the channel location. In the cinder cone case, a fixed elevation at $r = a$ could correspond to a channel that wraps around the base of the cone, but this is not a common occurrence. In most cases, volcanic cones are surrounded by alluvial flats that do not readily transport material from the base of the cone. Equation (2.54) can still be used for these cases, but a must be chosen to be much larger than the radius of the cone. This way, debris will be removed from the

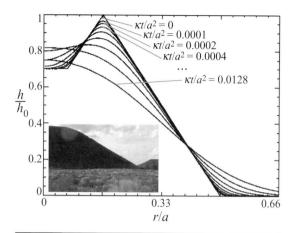

Fig 2.14 Plots of analytic solutions to the 3D diffusion equation for a volcanic cinder cone for a range of times following eruption.

cone and deposited on the surrounding flat. In these cases, the boundary condition at $r = a$ will simply serve to maintain the alluvial flat or piedmont at a constant elevation very far from the base of the cone.

For a volcanic cone of radius r_c, crater-rim radius r_r, colluvial fill radius of r_f, and maximum height of h_0 (Figure 2.14), $f(r)$ is given by

$$f(r) = \begin{cases} h_0\left(1 + \frac{r_f - r_r}{r_c - r_r}\right) & \text{if } r < r_f \\ h_0\left(1 + \frac{r - r_r}{r_c - r_r}\right) & \text{if } r_f \geq r < r_r \\ h_0\left(1 + \frac{r_r - r}{r_c - r_r}\right) & \text{if } r_r \geq r < r_c \\ 0 & \text{if } r \geq r_c \end{cases} \quad (2.56)$$

As a technical point, it should be noted that the colluvial fill radius r_f must be finite in order for Eq. (2.54) to apply because otherwise the boundary condition given by Eq. (2.55) is violated. Physically, r_f corresponds to the width of the colluvium that fills the crater shortly following eruption. Substituting Eq. (2.56) into Eq. (2.54) gives

The three integrals that appear in Eq. (2.57) can also be written as infinite series. In practice, however, it is more accurate to evaluate those integrals numerically because the series converge very slowly. Therefore, although closed-form analytic solutions for volcanic cones can be written down, calculating and plotting the solutions requires some numerical work.

First, we must solve for the roots of $J_0(\alpha a) = 0$. This is done numerically using a root-finding technique (Chapter 9 of Press *et al.* (1992) provides many techniques for doing this). The Bessel function has an infinite number of roots, so how many do we need to compute? For our purposes, the first one hundred roots are adequate, but more roots may be required for high-resolution profiles or very young cones. Second, we must evaluate the integrals in Eq. (2.57) numerically. To do this, the algorithms in Chapter 4 of Press *et al.* (1992) can be used. Appendix 1 provides a program for computing the integrals and performing the sums in Eq. (2.57).

Figure 2.14 illustrates radial profiles of Eq. (2.57) for the initial cone and at eight subsequent times from $\kappa t/a^2 = 0.0001$ to $\kappa t/a^2 = 0.0128$. Both axes are normalized. The y-axis is normalized to the initial cone height h_0. As in all diffusion problems, the solution can be scaled up or down in height with no change in the relative cone shape. The r-axis is scaled to the model domain length a.

In the cone's early-stage evolution, the greatest change occurs at the crater rim. This is not surprising since this is where the profile curvature is greatest. The crater rim is locally triangular and for early times the crater evolution is broadly similar to moraine evolution considered in Section 2.2.6. For intermediate times, the position of the crater rim migrates inward. This migration is associated with the additional advective term in Eq. (2.52) that is not present

$$h(r, t) = \frac{2h_0}{a^2} \sum_{n=1}^{\infty} e^{-\kappa \alpha_n^2 t} \frac{J_0(\alpha_n r)}{\alpha_n J_1^2(\alpha_n a)} \left[\left(\frac{r_f^2}{r_c - r_r} J_1(\alpha_n r_f) - \frac{2r_r}{r_c - r_r} J_1(\alpha_n r_r) + \left(1 + \frac{r_r}{r_c - r_r}\right) r_c J_1(\alpha_n r_c) \right) \right.$$
$$\left. + \frac{\alpha_n}{r_c - r_r} \left(2\int_0^{r_r} r^2 J_0(\alpha_n r_r) dr + \int_0^{r_f} r^2 J_0(\alpha_n r_f) dr + \int_0^{r_c} r^2 J_0(\alpha_n r_c) dr \right) \right]$$

$$(2.57)$$

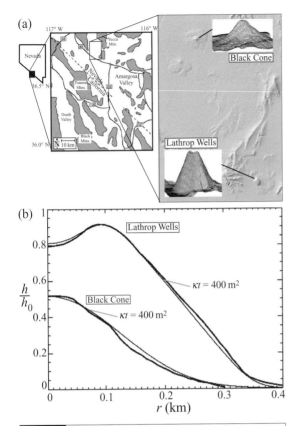

Fig 2.15 (b) Observed (thick line) and best-fit model profiles (thin line) for a relatively young (Lathrop Wells Cone) and a relatively old (Black Cone) cone in the Crater Flat volcanic field, Nevada (location map in (a)).

in 2D problems. Eventually, the crater is filled by diffusion with debris and the late-stage cone morphology is described by $J_0(ar)$ with decreasing amplitude over time. At this point, the exponential time-dependent term in Eq. (2.54) has filtered all of the high-frequency components in the topography and only the lowest-frequency term is significant in the series. The inset photo in Figure 2.14 shows the profile of an early Quaternary volcanic cone in the Cima volcanic field, California. The profile shape is similar to the late-stage profiles plotted in Figure 2.14.

Figure 2.15 compares the observed topography of two cones in the Crater Flat volcanic field, Amargosa Valley, Nevada, to best-fit model solutions. Lathrop Wells Cone has been radiometrically dated to be 77 ka (Heitzler *et al.*, 1999), while Black Cone has an age of 1.1 Ma (Valen-

tine *et al.*, 2005). The model solutions compare well to the observed profiles, and best-fit morphologic ages are approximately 400 and 4000 m^2, respectively. The ratios of the morphologic age to the radiometric age provide estimates for the time-averaged κ values for these cones: 5.2 and 3.6 m^2/kyr. These values are at the upper range of κ values inferred from pluvial shoreline and fault scarps in the southwestern US (Hanks, 2000).

This approach enables us to estimate how much height the cone has lost since its eruption. Figure 2.15 suggests that the crater rim of Lathrop Wells has been eroded by approximately 8% of its original height, or 9–10 m. Black Cone, in contrast, has been degraded by approximately 50% according to this approach.

A final word about Lathrop Wells should be noted. The dating of this cone has been the subject of considerable debate because it is the youngest volcano in the Yucca Mountain region, the site of the proposed high-level nuclear waste repository. Initial attempts to date the cone geomorphically concluded that the cone was no older than approximately 10 ka based on its complete lack of drainage channels. However, cinder cones are comprised of coarse (pebble-to-gravel-sized) tephra. As such, it is likely that little or no surface runoff can occur until sufficient time has passed for eolian dust to clog surface pore spaces. Therefore, runoff generation on these cones may be impossible for tens of thousands of years following eruption. In the absence of any radiometric dates for this cone, diffusion-morphologic dating would suggest that the cone is 20 ka to 80 ka assuming a conservative range of κ values from 0.5 to 2.0 m^2/kyr. This range includes the now-accepted radiometric age but does not include the original estimate of 10 ka. Of course, all morphologic-age estimates suffer from poor constraints on κ values. However, since κ values can be calibrated on a cone-by-cone basis as shown here, they offer great potential for better regional calibration of κ values and improved understanding of their underlying climatic and textural controls.

2.2.11 Delta progradation

The application of the diffusion equation to large-scale depositional landforms (i.e. deltas and alluvial fans) is problematic for two reasons. First, depositional landforms have complex internal processes that are not captured by the diffusion model. For example, deposition in deltas and alluvial fans takes place by a series of avulsion events. At short time scales, deposition is localized along one or more lobes that are active until the lobe becomes superelevated with respect to the surrounding terrain. A large flood can cause the lobe to shift rapidly to another portion of the fan or delta. If the diffusion equation applies at all to the large-scale evolution of these systems, it applies only at a time scale that averages the effects of many avulsion events. Second, variations in sediment texture can produce spatial variations in sediment mobility. The sediment texture in the proximal portion of an alluvial fan, for example, can be one to two orders of magnitude larger than sediment in the distal region (i.e. boulders compared to sand). Variations in channel morphology (and hence flow depth) and sediment texture both combine to create spatial variations in the effective diffusivity that are not well constrained.

Kenyon and Turcotte (1985) applied the 1D diffusion equation to model the progradation of a single delta lobe. Their model does not apply to the long-term 3D morphology of deltas. Here we review their approach and propose a model for the 3D evolution of deltas over geologic time scales. Both models are limited in that they neglect the flexural–isostatic subsidence of the basin in response to sediment loading. This limitation will be corrected in Chapter 5 where we consider the geometry of foreland basins within the context of a model that couples the diffusion equation to flexural subsidence.

The Kenyon and Turcotte model assumes that sediment is introduced to the basin at an elevation h_0 above the basin floor. As the basin fills, the position of the river mouth migrates towards the basin at a rate U_0. Generally speaking, solving diffusion problems with moving boundaries poses a considerable challenge. In this case, however, the moving boundary can be handled by introducing a new variable ξ given by

$$\xi = x - U_0 t \qquad (2.58)$$

The sediment source is fixed in this new coordinate system. Substitution of Eq. (2.58) into the

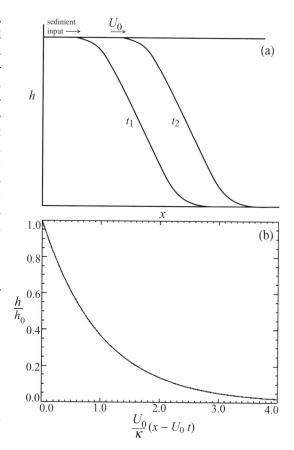

Fig 2.16 (a) Schematic diagram of the 2D model for a prograding delta lobe in the Kenyon and Turcotte (1985) model. (b) Solution for the profile shape of a prograding delta (Eq. (2.61)).

diffusion equation gives

$$-U_0 \frac{dh}{d\xi} = \kappa \frac{d^2 h}{d\xi^2} \qquad (2.59)$$

Equation (2.59) must be solved with boundary conditions $h(0) = h_0$ and $h(\infty) = 0$. The solution is

$$h(\xi) = h_0 e^{-\frac{U_0 \xi}{\kappa}} \qquad (2.60)$$

Substituting Eq. (2.58) into Eq. (2.60) gives

$$h(x, t) = h_0 e^{-\frac{U_0}{\kappa}(x - U_0 t)} \qquad (2.61)$$

The solution to this model is given in Figure 2.16. The solution is an exponential function with a width controlled by a balance between diffusion (which transport sediment away from the source) and the movement of the source. This solution does a good job of matching the observed stratigraphy of the southwest pass segment of the Mississippi River delta (Turcotte and Schubert, 2002). More generally, however, the exponential form of Eq. (2.61) is inconsistent with the generally sigmoidal nature of deltaic deposits.

Over geologic time scales delta lobes will avulse and a radial, 3D landform will be created. In order to model this 3D development, we consider the diffusion equation in polar coordinates (Eq. (2.52)). We also assume a fixed sediment source (in contrast to the moving sediment source of Kenyon and Turcotte) and a spatially variable diffusivity that is greatest near the river mouth and decreases with distance from the mouth as $1/r$:

$$\kappa(r) = \frac{\kappa}{r} \tag{2.62}$$

This spatially variable diffusivity represents the loss of transport capacity as rivers reach the ocean. The specific $1/r$ dependence we have assumed is not well constrained, but it provides a useful starting point for representing the decrease in sediment mobility as distance from the shoreline increases.

The 3D diffusion equation with radial symmetry is given by Eq. (2.52). Substituting Eq. (2.62) into Eq. (2.52) gives

$$\frac{\partial h}{\partial t} = \frac{\kappa}{r}\frac{\partial^2 h}{\partial r^2} \tag{2.63}$$

Equation (2.63) can be reduced to an ordinary differential equation by introducing the similarity variable

$$\eta = \frac{r^3}{9\kappa t} \tag{2.64}$$

In terms of η, Eq. (2.63) becomes

$$-\frac{d\theta}{d\eta} = \frac{d^2\theta}{d\eta^2} \tag{2.65}$$

where $\theta = h/h_0$. The solution to this equation is

$$\theta = e^{-\eta} \tag{2.66}$$

Transforming back to the original coordinates:

$$h(r, t) = h_0 e^{-\frac{r^3}{9\kappa t}} \tag{2.67}$$

Figure 2.17 presents plots of Eq. (2.67) for different values of κt. The model predicts the classic sigmoidal (S-shaped) geometry of deltaic

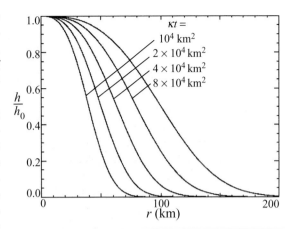

Fig 2.17 Profiles of the 3D radially symmetric model of delta formation (Equation (2.67)).

deposits, which is more consistent than the exponential form of the Kenyon and Turcotte model.

2.2.12 Dust deposition downwind of playas

The dust cycle in arid environments is characterized by a net eolian transfer of dust from playas to piedmonts (Pye, 1987). Understanding the processes and rates of this transfer is important for many basic and applied geologic problems. Dust transport in the atmosphere also provides a nice illustration of diffusion acting in concert with advection. In this section we describe the phenomenology of dust deposition in a well-studied field site in California, and then apply an analytic solution for the advection, diffusion, and gravitational settling of dust from complex, spatially-distributed sources, following the work of Pelletier and Cook (2005).

The study area is located in southern Amargosa Valley, California, where Franklin Lake Playa abuts the Eagle Mountain piedmont (Figure 2.18a). The water table in Franklin Lake Playa is less than 3 m below the surface (Czarnecki, 1997). Czarnecki identified several distinct geomorphic surfaces on Franklin Lake Playa that can be readily identified in LANDSAT imagery (Figure 2.18a). For the purposes of numerical modeling, the active portion of the playa was mapped based on Czarnecki's map of playa surfaces with significant dust-emitting potential (Figure 2.18a).

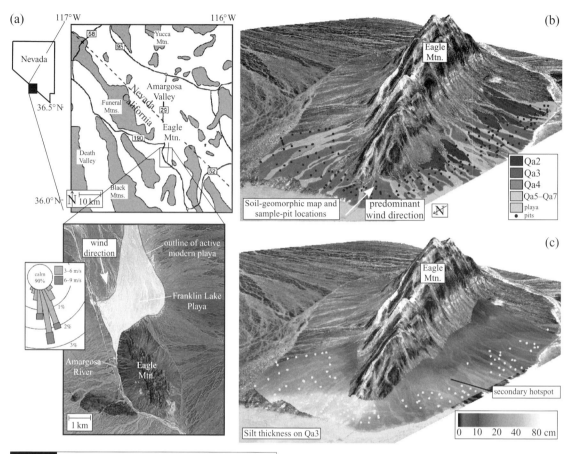

Fig 2.18 (a) Location map and LANDSAT image of Eagle Mountain piedmont and adjacent Franklin Lake Playa, southern Amargosa Valley, California. Predominant wind direction is SSE, as shown by the wind-rose diagram (adapted from January 2003–January 2005 data from Western Regional Climate Center, 2005). Calm winds are defined to be those less than 3 m/s. (b) Soil-geomorphic map and oblique aerial perspective of Eagle Mountain piedmont, looking southeast. Terrace map units are based on the regional classification by Whitney *et al.* (2004). Approximate ages: Qa2 – middle Pleistocene, Qa3 – middle to late Pleistocene, Qa4 – late Pleistocene, Qa5–Qa7 – latest Pleistocene to active. (c) Map of eolian silt thickness on Qa3 (middle to late Pleistocene) surface, showing maximum thicknesses of 80 cm close to the playa source, decreasing by approximately a factor of 2 for each 1 km downwind. Far from the playa, background values of approximately 20 cm were observed. For color version, see plate section. Modified from Pelletier and Cook (2005).

The Eagle Mountain piedmont acts as the depositional substrate for dust emitted from Franklin Lake Playa under northerly wind conditions (see wind-rose diagram in Figure 2.18a). Silt-rich eolian deposits occur directly underneath the desert pavements of Eagle Mountain piedmont, varying in thickness from 0 to 80 cm based on soil-pit measurements. The technical term for these deposits is cumulic eolian epipedons (McFadden *et al.*, 2005), but here we refer to them as eolian or silt layers for simplicity. These layers are predominantly composed of silt but also include some fine sand and soluble salts. The homogeneity of these deposits, combined with their rapid transition to gravelly alluvial-fan deposits below, makes the layer thickness a reasonable proxy for the total dust content of the soil.

Alluvial-fan terraces on Eagle Mountain piedmont have a range of ages. Eagle Mountain piedmont exhibits the classic sequence of Quaternary alluvial-fan terraces widely recognized in the southwestern United States (Bull, 1991). Mapping and correlation of the units was made based on the regional chronology of Whitney *et al.* (2004). In this example, we focus on silt thicknesses of the Qa3 terrace unit (middle to late Pleistocene) because of its extensive preservation and limited evidence of hillslope erosion. Silt thicknesses were measured at locations with undisturbed, planar terrace remnants to the greatest extent possible. A color map of silt layer thickness for this surface is shown in Plate 2.18b, draped over the US Geological Survey (USGS) 30 m digital elevation model and an orthophotograph of the area. Silt thickness is observed to decrease by roughly a factor of 2 for every 1 km of distance from the playa. Several kilometers downwind, a background thickness of 15–20 cm was observed on Qa3 surfaces. The strongly localized nature of

material deposited over time) and that Qa4 surfaces have received a higher average dust flux because they have existed primarily during the warm, dry Holocene. The Qa5 unit (latest Pleistocene) exhibited uniformly thin deposits underlying a weak pavement, independent of distance from the playa, suggesting that the trapping ability of young surfaces is limited by weak pavement development.

As discussed in Chapter 1, atmospheric transport of particulate matter can be modeled as a combination of turbulent diffusion, downwind advection, and gravitational settling. The 3D concentration field for a point source located at $(x', y', 0)$, obtained by solving Eqs. (1.17) and (1.18), is given by

$$c_p(x, y, z, x', y') = \frac{Q}{\sqrt{\frac{4\pi K(x-x')}{u}}} \exp\left(-\frac{u(y-y')^2}{4K(x-x')}\right) \left[\frac{\exp\left(-\frac{uz}{4K(x-x')}\right)}{\sqrt{\pi u K(x-x')}} - \frac{p}{uK}\exp\left(\frac{pz}{K} + \frac{p^2(x-x')}{Ku}\right)\right.$$

$$\left.\times \operatorname{erfc}\left(\sqrt{\frac{u}{4K(x-x')}}z + p\sqrt{\frac{x-x'}{Ku}}\right)\right] \qquad (2.68)$$

downwind deposition in this area suggests that Franklin Lake Playa is the source for nearly all of the eolian deposition on Eagle Mountain piedmont. Localized deposition also implies that dust deposition rates may vary regionally by an order of magnitude or more, down to spatial scales of 1 km or less.

Figure 2.19a illustrates plots of silt thickness as a function of downwind distance, for comparison with two-dimensional model results. This plot includes all measurements collected from terrace units late Pleistocene in age or older (Qa4–Qa2) within a 1 km wide swath of western Eagle Mountain piedmont. Data from the Qa4 and Qa2 units show a similarly rapid downwind decrease in silt thickness, illustrating that the Qa3 pattern is robust. Silt thicknesses were relatively similar on the three terrace units at comparable distances from the playa. This similarity was not expected, given the great differences in age between the three surfaces. This may be partly explained by the fact that Qa2 surfaces have undergone extensive hillslope erosion (and hence preserve only a portion of the total eolian

where Q is the source emission rate and erfc is the complementary error function. Equation (2.68) combines the two-dimensional solution of Smith (1962) with an additional term required to describe crosswind transport (Huang, 1999). Equation (2.68) assumes that the settling velocity q is small compared with the deposition velocity p. Stokes' Law (Allen, 1997) implies a settling velocity of less than 1 cm/s for silt particles (i.e. particles less than 0.05 mm) in air. Deposition velocities consistent with the spatial distribution of deposition on Eagle Mountain piedmont, however, are approximately 5 cm/s, or at least five times greater than q. For nonpoint sources, Eq. (2.68) can be integrated to give

$$c(x, y, z) = \int_{-\infty}^{\infty} c_p(x', y')dx'dy' \qquad (2.69)$$

Using Eqs. (2.68) and (2.69), the deposition rate on a flat surface is given by $pc(x, y, 0)$. Deposition on a complex downwind surface can be estimated as $pc[x, y, h(x, y)]$, where $h(x, y)$ is the elevation of the downwind topography. This approach is only an approximation of the effects of complex

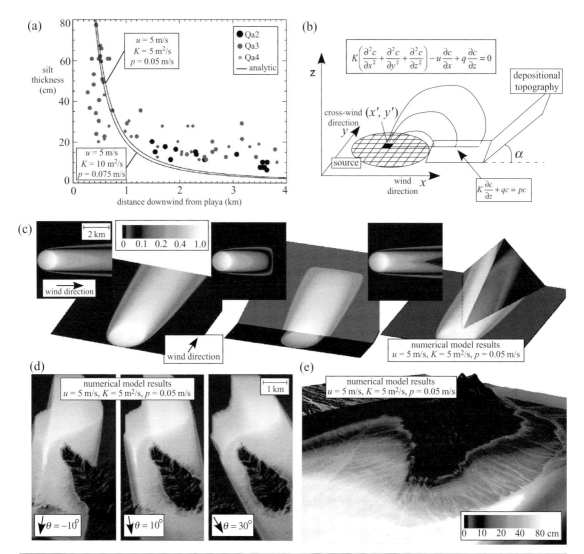

Fig 2.19 (a) Plot of eolian silt thickness versus downwind distance, with analytic solutions for the two-dimensional model for representative values of the model parameters. (b) Schematic diagram of model geometry. Depositional topography shown in this example is an inclined plane located downwind of source (but model can accept any downwind topography). (c) Color maps of three-dimensional model results, illustrating the role of variable downwind topography. In each case, model parameters are $u = 5$ m/s, $K = 5$ m^2/s, and $p = 0.05$ m/s. Downwind topography is, from left to right, a flat plane, an inclined plane, and a triangular ridge. Width and depth of model domain are both 6 km. (d) Color maps of three-dimensional model results for deposition downwind of Franklin Lake Playa, illustrating the role of variable wind direction. From left to right, wind direction is $\theta = -10°$ (0° is due south), 10°, and 30°. (e) Map of three-dimensional model results obtained by integrating model results over a range of wind conditions, weighted by the wind-rose data in Figure 2.18a. For color version, see plate section. Modified from Pelletier and Cook (2005).

topography because mass loss from the plume is still assumed to occur along a horizontal plane in Eq. (2.68). However, by using the elevated plume concentration – i.e., $c[x, y, h(x, y)]$ – to estimate the deposition term pc, the model approximates

the spatially variable deposition that occurs as the plume intersects with complex topography.

Two-dimensional (2D) model solutions identify a fundamental length scale for downwind deposition. Smith (2003) showed that Eq. (2.68) can be approximated by a power-law function along the plume centerline:

$$c(x, 0, 0) = \frac{Q}{2\sqrt{\pi}\, pL_p} \left(\frac{x}{L_p}\right)^{3/2} \quad (2.70)$$

where L_p is a characteristic length scale given by $L_p = uK/p^2$. Figure 2.19a, for example, presents two plots of Eq. (2.70) with different sets of model parameters. These parameter sets were chosen to have different values of K and p but similar values of L_p. The similarity of the resulting plots indicates that the pattern of 2D dust deposition is not sensitive to particular parameters; it is only sensitive to the scaling parameter L_p. Therefore, raising the value of K and p simultaneously, as in Figure 2.19a, leads to solutions that are equally consistent with observations downwind of Franklin Lake Playa. This nonuniqueness makes model calibration (in 2D or 3D) difficult. The model is best calibrated using wind-speed data to constrain u and K to the greatest extent possible, because these values can be independently determined using readily-available data. The deposition velocity, in contrast, cannot be readily determined. By constraining u and K, however, model results can then be run for a range of values of p to determine the value most consistent with observations.

Model results are shown in Figure 2.19c to illustrate the role of complex downwind topography in controlling deposition patterns. In each case, a circular source with uniform Q was considered with model parameters $u = 5$ m/s, $K = 5$ m^2/s, and $p = 0.05$ m/s. From left to right, the downwind topography was considered to be flat, an inclined plane, and a triangular ridge. Relative to flat topography, the inclined plane leads to a short plume, and the triangular ridge diverts and bisects the plume.

Long-term dust transport does not correspond to a single wind direction as assumed in the simplified cases of Figure 2.19c. The effects of multiple wind directions can be incorporated into the model by integrating a series of model runs with a range of wind directions, each weighted by the frequency of that wind direction in the wind-rose diagram. Figure 2.19d illustrates model results for the Eagle Mountain region with wind directions from $\theta = -10°$ to $+30°$ ($\theta = 0°$ is due South). Figure 2.19e illustrates the integrated model corresponding to the wind-rose diagram of Figure 2.18a. The thickness values mapped in Figure 2.19e correspond to a uniform Q value of 1.0 m/kyr (assuming a Qa3 surface age of 50 ka), obtained by scaling the model results for different values of Q to match the maximum thickness values between the model and observed data. The modeled and observed deposition patterns match each other in most respects. However, the secondary hot spot located on the western piedmont of the study area is not reproduced by the model. This feature could result from more westerly paleowinds transporting more dust eastward from the Amargosa River relative to modern conditions. Alternatively, it could result from enhanced hillslope erosion of Qa3 surfaces in this area.

2.3 | Numerical techniques and applications

2.3.1 Forward-Time-Centered-Space method

Analytic solutions are elegant but they are often only loosely applicable to real problems. Linear fault scarps and volcanic cones are special cases where the landform symmetry allows for an analytic approach because a 2D landform is reduced to 1D. Numerical solutions are generally needed in order to model the evolution of landforms that lack symmetry or are controlled by complex base-level boundary conditions.

In order to solve partial differential equations on a computer, we must discretize the equations in both space and time. Discretization in space means that we represent the spatial domain as a discrete set of grid points (e.g. from position x_1 to x_n for 1D problems). Discretization in time means that the solution will be solved forward in discrete intervals of time. First, we must determine how to best represent first, second, and higher derivatives in space within this discrete framework. The starting point is the Taylor expansion:

$$h(x + \Delta x, t) = h(x, t) + \Delta x \left.\frac{\partial h}{\partial x}\right|_{x,t} + \frac{(\Delta x)^2}{2} \left.\frac{\partial^2 h}{\partial x^2}\right|_{x,t}$$
$$+ O\left((\Delta x)^2\right) \quad (2.71)$$

The simplest formula for the first derivative is obtained by neglecting second-order $((\Delta x)^2)$ terms and rearranging Eq. (2.71) to obtain the forward-difference approximation:

$$\frac{h(x + \Delta x, t) - h(x, t)}{\Delta x} = \left.\frac{\partial h}{\partial x}\right|_{x,t} + O(\Delta x) \quad (2.72)$$

This is called forward difference because we look forward to $x + \Delta x$ to compute the difference. Equation (2.72) is not unique, however, because we can also calculate the first derivative using a backward difference:

$$\frac{h(x, t) - h(x - \Delta x, t)}{\Delta x} = \left.\frac{\partial h}{\partial x}\right|_{x,t} + O(\Delta x) \quad (2.73)$$

We can also use a centered-difference approximation:

$$\frac{h(x + \Delta x, t) - h(x - \Delta x, t)}{2\Delta x} = \left.\frac{\partial h}{\partial x}\right|_{x,t} + O((\Delta x)^2)$$
$$(2.74)$$

The second derivative can be calculated by using a forward difference to calculate the slope in the positive direction and subtracting the backward difference in the negative direction:

$$\frac{1}{\Delta x}\left[\frac{h(x + \Delta x, t) - h(x, t)}{\Delta x} - \frac{h(x, t) - h(x - \Delta x, t)}{\Delta x}\right]$$
$$= \frac{h(x + \Delta x, t) - 2h(x, t) + h(x - \Delta x, t)}{(\Delta x)^2}$$
$$= \left.\frac{\partial^2 h}{\partial x^2}\right|_{x,t} + O((\Delta x)^2) \quad (2.75)$$

The discretization of the diffusion equation, using Eq. (2.75) and introducing abbreviated discrete notation $h(i\Delta x, n\Delta t) = h_i^n$, is given by

$$\frac{h_i^{n+1} - h_i^n}{\Delta t} = D\left(\frac{h_{i+1}^n - 2h_i^n + h_{i-1}^n}{(\Delta x)^2}\right)$$
$$h_i^{n+1} = h_i^n + \frac{D\,\Delta t}{(\Delta x)^2}(h_{i+1}^n - 2h_i^n + h_{i-1}^n) \quad (2.76)$$

This Forward-Time-Centered-Space (FTCS) method is a useful means for solving the 1D diffusion equation for small grids. This scheme is numerically stable provided that the time step is less than or equal to $(\Delta x)^2/2D$. Equation (2.76) is also known as an explicit scheme because the value of each grid point is an explicit function of the grid-point values at the previous time step. The FTCS method is limited by the fact that it is stable only for small timesteps.

The FTCS method starts with an initial value h_i^0 everywhere on the grid. In addition to the initial condition, boundary conditions on the value of h or its first derivative must also be specified at both ends of the grid. These boundary conditions may be constant or may vary as a function of time. Equation (2.76) is then applied sequentially from one end of the grid to the other to specify the grid of values at the next time step.

The FTCS method is generally not useful for large grids or higher dimensions because very small step sizes must be taken in order to maintain stability. To see this, consider the sequence of grid points in Figure 2.20. Initially, the values of h are zero everywhere on the grid. At time zero, the left-side boundary condition on h is raised to a finite value. The right-side boundary condition is $h_L^n = 0$. During the first time step, all of the grid points remain zero except for the second grid point from the left. The second derivative of h, $(h_0^0 - 2h_1^0 + h_2^0)/(\Delta x)^2$ is positive, which results in a finite value for h_1^1. Note that if the time step is large enough, the value of h_1^1 can actually rise higher than that of h_0^0. If this happens, the model will become unstable and will continue to diverge from the correct solution in subsequent timesteps. In order to maintain stability, the timestep Δt must be less than $(\Delta x)^2/2D$. The time required for diffusion to propagate a distance L is $\approx L^2/D$. Therefore, the number of time steps required to model diffusion is $\approx L^2/(\Delta x)^2$. To model the diffusion equation in 1D on a grid with 1000 grid points, therefore, requires a minimum of about one million time steps. Once we move to 2D and 3D problems with tens or hundreds of thousands of grid points, one million time steps becomes computationally unfeasible on even the fastest computers. Another related disadvantage of the FTCS method is that information travels very slowly from one end of the grid to the other. Figure 2.20 illustrates that it would take 1000 time steps for the grid points on the right side of the grid to feel any effect of the boundary condition on the left side of the grid. This slow propagation of information

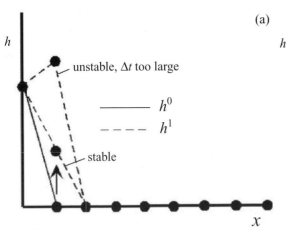

(a)

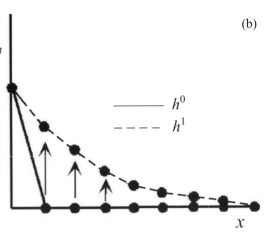

(b)

explicit method:

$$h_i^{n+1} = h_i^n + \frac{D\Delta t}{(\Delta x)^2}(h_{i+1}^n - 2h_i^n + h_{i-1}^n)$$

implicit method:

$$h_i^{n+1} = h_i^n + \frac{D\Delta t}{(\Delta x)^2}(h_{i+1}^{n+1} - 2h_i^{n+1} + h_{i-1}^{n+1})$$

Fig 2.20 Illustration of the difference between (a) explicit numerical schemes and (b) implicit schemes. In explicit schemes, grid values at the previous time steps are used to solve for values at the next time step. In the example, the left-side boundary, h_0^0, is abruptly raised at $t = 0$. In the explicit scheme, only the first grid point to the right, h_1^1 will feel the effects of this change at the boundary in the following time step. If the value of the time step is chosen to be too large, the explicit scheme can overcorrect and make h_1^1 larger than h_0^1. In the implicit scheme, the effect of raising the left boundary is simultaneously felt throughout the grid no matter how large it is, resulting in a more stable solution.

is one reason why the time step must remain so low.

So, if the FTCS is so poor, why use it at all? The answer is that it is easy to program and, in some ways, more versatile than other schemes we will discuss. In the next two sections we utilize the FTCS method to solve for the evolution of hillslopes with spatially variable diffusivity and nonlinear sediment transport.

2.3.2 Evolution of hillslopes with spatially-varying diffusivity

On hillslopes where slope wash is the predominant transport process, the classic diffusion equation cannot be used. In such cases it is appropriate to model hillslope sediment flux as increasing linearly with distance from the divide (i.e.

Eq. (2.45)). Substituting Eq. (2.45) into Eq. (2.2) gives

$$\frac{\partial h}{\partial t} = \kappa \left(x \frac{\partial^2 h}{\partial x^2} + \frac{\partial h}{\partial x} \right) \tag{2.77}$$

Here we consider the slope-wash-dominated version of instantaneous base-level drop of an initially flat terrace considered in Section 2.2.5.

Figure 2.21a compares hillslope profiles obtained for the nonuniform diffusion equation (Eq. (2.77)) with those of the classic diffusion equation (Eq. (2.20)). For early times, the two models show no difference. For later times, however, hillslope erosion occurs more rapidly in the nonuniform diffusion case, but otherwise there is little difference in the profile shape.

These results are broadly consistent with those of Carson and Kirkby (1972), who compared hillslope profiles for uniform and nonuniform cases. Figure 2.21b is based on the classic figure from their book, illustrating the signature hillslope forms for different hillslope processes. In this figure, Carson and Kirkby argued that slopes governed by the uniform diffusion equation ($m = 0$ and $n = 1$ in their terminology) were characteristically convex downward. In contrast, the slope-wash-dominated case (i.e. $m = 1$ and $n = 0$) was characterized by a straight profile with no curvature.

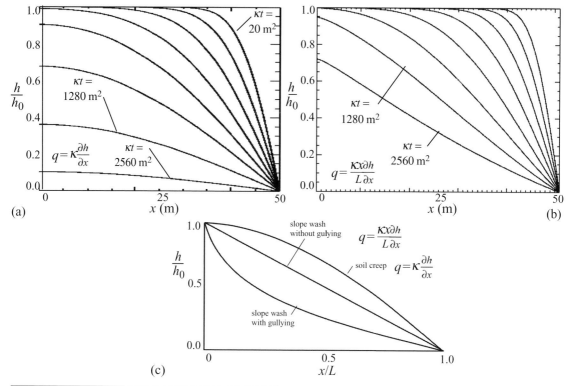

(a)

(b)

(c)

Fig 2.21 (a) Comparison of hillslope evolution of an initially horizontal terrace with instantaneous base-level drop for homogeneous diffusion (i.e. flux is given by Eq. (2.2)) and inhomogeneous diffusion (i.e. flux is also proportional to distance from the divide, or Eq. (2.45)). The two evolution equations yield almost identical hillslope forms, but the rate of erosion is higher in the inhomogeneous case if the same κ value is used in both cases. (b) Characteristic slope profiles based on the characteristic forms of Carson and Kirkby (1972). These authors argued that uniform and nonuniform diffusion result in distinctly different hillslope forms, in contrast to the model results in (a).

2.3.3 Evolution of hillslopes with landsliding

The literature on scarp analysis has included a vigorous debate on linear versus nonlinear diffusion models. For example, Andrews and Bucknam (1987) first proposed that the flux of material is proportional to the gradient for small values, but increases towards infinity as the gradient reaches a critical value S_c:

$$q = -\kappa \frac{\partial h/\partial x}{1 - \left(\frac{|\partial h/\partial x|}{S_c}\right)^n} \tag{2.78}$$

where n is a constant. Substituting Eq. (2.78) into Eq. (2.2) gives

$$\frac{\partial h}{\partial t} = \kappa \frac{\partial h/\partial x}{1 - \left(\frac{|\partial h/\partial x|}{S_c}\right)^n} \left[1 + \frac{n|\partial h/\partial x|^n}{S_c^n \left(1 - \left(\frac{\partial h/\partial x}{S_c}\right)^n\right)} \right] \tag{2.79}$$

Andrews and Hanks (1985) proposed using Eq. (2.79) with $n = 1$, while Andrews and Bucknam (1987), Roering et al. (1999), and Mattson and Bruhn (2001) assumed $n = 2$. Appendix 1 provides a program to solve for hillslope evolution governed by Eq. (2.79).

Figure 2.22 presents the nonlinear hillslope response (Eq. (2.79)) to instantaneous base-level drop with an initially planar terrace. The initial phase of the model is an instantaneous relaxation of the hillslope to the critical angle S_c at time $t = 0$. During the first 1 kyr of the model (i.e. $\kappa t = 1\,\mathrm{m}^2$ assuming $\kappa = 1\,\mathrm{m}^2/\mathrm{s}$), little hillslope rounding takes place and the hillslope wears back to a slightly lower angle. As time proceeds, downwearing gives way to diffusive rounding of the hillslope crest. The rate of hillslope evolution decreases rapidly with the overall relief

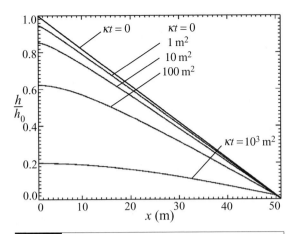

Fig 2.22 Plots of nonlinear hillslope profiles (Eq. (2.79)) at times $\kappa t = 1\,\mathrm{m}^2$ to $10^3\,\mathrm{m}^2$ following instantaneous base-level drop of an initially planar terrace.

because of the strongly nonlinear relationship between flux and hillslope gradient. This slowing is evident in the very large range of time scales (varying over three orders of magnitude) between the first and the last profile shown. Clearly this

evolution is very different from that of the linear diffusion equation. Appendix 1 provides a program to solve for hillslope evolution governed by Eq. (2.79). This program may be used as a template to solve hillslope evolution problems with a broad range of flux equations, initial conditions, and boundary conditions.

2.3.4 2D Evolution of alluvial-fan terraces
Figure 2.23 illustrates solutions to the 2D diffusion equation using the FTCS method. The model domain is 100 m on each side and the hillslope is assumed to diffuse with $\kappa = 1\,\mathrm{m}^2/\mathrm{kyr}$. Figure 2.23a illustrates the channel mask grid that identifies which grid points are to be kept at a fixed elevation. At $t = 0$ a planar alluvial terrace is abruptly incised according to the channel mask grid. This example is designed to be a 2D analog of the abrupt terrace entrenchment considered in Section 2.2.5. Figures 2.23b–2.23d illustrate the terrace morphology after (b) 50 kyr, (c) 120 kyr, and (d) 500 kyr following entrenchment. The motivation for this example is the observation that

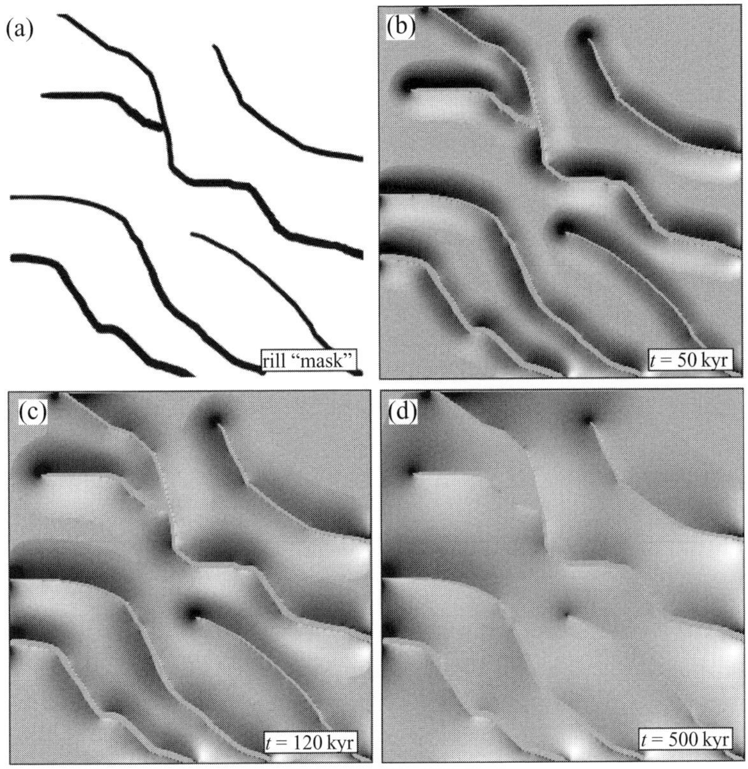

(a)

rill "mask"

(b)

$t = 50\,\mathrm{kyr}$

(c)

$t = 120\,\mathrm{kyr}$

(d)

$t = 500\,\mathrm{kyr}$

Fig 2.23 Solution to the diffusion equation with $\kappa = 1\,\mathrm{m}^2/\mathrm{kyr}$ in the neighborhood of a series of gullies (shown in (a)) kept at constant base level and a model domain of $0.01\,\mathrm{km}^2$. This model represents the evolution of an alluvial-fan terrace abruptly entrenched at time $t = 0$. After (b) 50 kyr, diffusional rounding of the terrace near gullies has penetrated $\approx \sqrt{\kappa t}$ or 7 m into the terrace tread and planar terrace treads are still widely preserved. After (c) 120 kyr approximately 11 m of rounding has taken place. Finally, after (d) 500 kyr, erosional processes have removed all planar terrace remnants and a rolling ridge-and-ravine topography remains.

latest Pleistocene alluvial fan terraces in the western US often have extensive preservation of planar terraces. As terraces increase in age to several hundred thousand years or longer, they eventually lose all planarity and become ridge-and-ravine topography. Appendix 1 provides a program for modeling the evolution of terraces using the FTCS method to obtain the results in Figure 2.23.

2.3.5 Implicit method

Fortunately, there are ways to solve diffusion problems much more efficiently than with the FTCS method. For example, we can rewrite the right side of Eq. (2.76) in terms of the new time step $n + 1$ rather than the previous time step n:

$$h_i^{n+1} = h_i^n + \frac{\kappa \Delta t}{(\Delta x)^2} \left(h_{i+1}^{n+1} - 2h_i^{n+1} + h_{i-1}^{n+1} \right) \quad (2.80)$$

At first glance Eq. (2.80) may seem nonsensical: grid values at time $n + 1$ appear on both sides of the equation. So, how can we solve for h_i^{n+1} on the left side of Eq. (2.80) if it has to be input on the right side first? The answer is that Eq. (2.80) can be rearranged to form a matrix equation:

$$\begin{pmatrix} 1 & 0 & 0 & 0 \\ -\alpha & (1 + 2\alpha) & -\alpha & 0 \\ 0 & -\alpha & (1 + 2\alpha) & -\alpha & \cdots \\ 0 & 0 & \alpha & (1 + 2\alpha) \\ & & \cdots & \end{pmatrix}$$

$$\times \begin{pmatrix} h_0^{n+1} \\ h_1^{n+1} \\ h_2^{n+1} \\ h_3^{n+1} \\ \cdots \end{pmatrix} = \begin{pmatrix} h_0^n \\ h_1^n \\ h_2^n \\ h_3^n \\ \cdots \end{pmatrix} \quad (2.81)$$

where $\alpha = \kappa \Delta t / (\Delta x)^2$. Equation (2.81) is a matrix equation of the form

$$\mathbf{A} \mathbf{h}^{n+1} = \mathbf{h}^n \quad (2.82)$$

which has the solution

$$\mathbf{h}^{n+1} = \mathbf{A}^{-1} \mathbf{h}^n \quad (2.83)$$

The order of the matrix in Eq. (2.83) is the same as the number of grid points in the model. Solving the diffusion equation in 1D using Eq. (2.83), therefore, requires inverting a 1000×1000 matrix. That may sound ominous, but, in fact, solving Eq. (2.83) is much easier than solving a

general matrix equation, because the matrix to be inverted in Eq. (2.83) is a tridiagonal matrix. Tridiagonal matrices have zero values everywhere except for on the main diagonal and one row above and below it. Such matrices can be inverted very quickly using a short-cut version of Gaussian elimination (Press *et al.*, 1992). This approach to solving diffusion problems is often called an implicit method, because the unknown grid values at time $n + 1$ appear on both sides of Eq. (2.80) rather than just on one side as in explicit methods.

The computational advantage of the implicit method is hard to overstate. In the explicit FTCS scheme, the new values for each grid point are calculated based on the old (out-dated) information from the previous time step. This is the reason why it takes 1000 time steps for the right side of the grid to know anything about the left side. In the implicit method, however, all of the new values everywhere on the grid are calculated simultaneously. Figure 2.20b, for example, illustrates how an instantaneous jump in the boundary condition at one end of the grid is felt immediately at the next time step throughout the grid. In the implicit scheme, grid points instantly receive information from every other grid point no matter how large the grid is. This difference makes the implicit method far more stable than explicit methods. In fact, the implicit method is stable for *any* size time step. The accuracy of the solution, however, still depends on the time step. In practice, it is useful to run two versions of the same implicit simulation with time steps that differ by a factor of 2. The difference between the two solutions provides an estimate of the accuracy of the results for the larger time step.

The implicit method can be made more accurate by averaging the second derivative at time $n + 1$ and time n:

$$h_i^{n+1} = h_i^n + \frac{\kappa \Delta t}{2(\Delta x)^2} \left[\left(h_{i+1}^{n+1} - 2h_i^{n+1} + h_{i-1}^{n+1} \right) \right.$$
$$\left. + \left(h_{i+1}^n - 2h_i^n + h_{i-1}^n \right) \right] \quad (2.84)$$

This is known as the Crank–Nicholson method and it is a work horse for solving diffusion problems. The implicit method was first used in a landform evolution context by Fagherazzi *et al.*

(2002), but its use is still surprisingly limited given the power of the method.

2.3.6 Alternating-Direction-Implicit method

The implicit method is particularly powerful in 2D and 3D because the advantage of the implicit scheme grows with the number of grid points and grid sizes grow geometrically with the number of dimensions. In 2D, the implicit method solves the equation

$$
h_{i,j}^{n+1} = h_{i,j}^{n} + \frac{\kappa \Delta t}{(\Delta x)^2} \left(h_{i+1,j}^{n+1} + h_{i,j+1}^{n+1} \right.
$$
$$
\left. -4h_{i,j}^{n+1} + h_{i-1,j}^{n+1} + h_{i,j-1}^{n+1} \right) \quad (2.85)
$$

Unfortunately, Eq. (2.85) does not yield a tridiagonal matrix because grid points are dependent on points to their right and left but also up and down. The up and down linkages in Eq. (2.85) result in a much more complex matrix compared to the simple tridiagonal matrix.

The power of the implicit method can still be brought to bear on this problem, however, by solving the problem repeatedly as a series of 1D implicit problems. Each row of the grid is calculated as a 1D implicit problem with the grid values in the up and down directions ($h_{i,j+1}^{n+1}$ and $h_{i,j-1}^{n+1}$) treated as constants. Then, the directions are reversed: each column is calculated as a 1D implicit problem with the grid values in the right and left directions treated as constants using the output from the implicit calculations done previously. This process is repeated several times, solving for rows and then columns, until a prescribed level of accuracy has been achieved as indicated by very small improvements to the calculated values. One additional advantage of this Alternating-Direction-Implicit (ADI) scheme is that it can handle complex boundary conditions almost effortlessly. If, for example, we wished to solve the diffusion equation in the neighborhood of a meandering river, we could input a mask of grid values that are kept constant, assuming that the river acts as a constant base level for the surrounding hillslope. Then, the ADI technique can be used and the fixed elevations of the river can simply be flagged so that the value at $n+1$ is kept equal to the value at n.

As a concrete example of the ADI technique, we consider a model for hillslope evolution in Hanaupah Canyon. In this model, we input the USGS 30 m/pixel DEM of Hanaupah Canyon as our starting condition. We then solve the 2D diffusion equation with $\kappa = 10\,\text{m}^2/\text{kyr}$ on all pixels where the contributing area is less than a threshold value of $0.1\,\text{km}^{-2}$. The results of this model are shown in Figure 2.24 for a simulation of 10 Myr duration using time steps of 1 Myr. The ADI technique is particularly useful for solving 2D and 3D problems with complicated boundary conditions, as in this case, where each channel pixel is a fixed-elevation boundary. It should be noted that this example is a bit contrived because the hillslopes in Hanaupah Canyon are, as noted previously, dominated by mass movements and therefore are not well represented by the diffusion equation. Nevertheless, there are many other areas of low relief where this approach would be a good model for hillslope evolution. Code for implementing this model is presented in Appendix 1.

Exercises

2.1 Many glaciers around the world have responded to global warming by increasing their rates of basal sliding. Assuming that basal sliding occurs as a result of warmer atmospheric temperatures conducted through the ice, estimate the lag time between atmospheric warming and glacier response for a glacier 100 m thick.

2.2 Consider a long alluvial ridge of height 100 m, width 1 km, and a sinusoidal cross-sectional topographic profile (i.e. $h(x) = h_0 \sin(\pi x/\lambda)$). Assuming a diffusivity of $\kappa = 1\,\text{m}^2/\text{kyr}$, what will be the height of the ridge 1 Myr in the future?

2.3 As time passes, fault and pluvial shoreline scarps become more subdued features of the landscape. Given diffusivity values prevalent in the western US of $\kappa \approx 1\,\text{m}^2/\text{kyr}$, fault scarps 10 kyr in age are readily apparent. Older scarps, however, eventually become too diffused to identify. Assuming an initial angle of $30°$, how old is the oldest identifiable scarp in the western US? Assume that scarps can be identified if their slope values are greater than 2% (relative to the far-field slope).

2.4 You have been contracted to conduct a preliminary seismic hazard assessment for an area that has fault scarps of unknown age with a range of

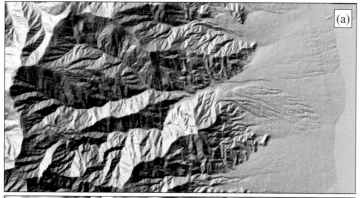

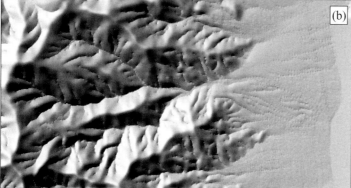

Fig 2.24 Example application of the ADI technique to modeling hillslope evolution in Hanaupah Canyon. (a) Shaded relief of US Geological Survey DEM of Hanaupah Canyon. (b) Solution with 10 Myr of hillslope evolution assuming $\kappa = 10\,m^2/kyr$ and channels with fixed elevations. Channels were defined to be all pixels with a contributing area greater than 0.1 km^2.

heights. You wish to determine which scarps are likely Holocene in age. Develop an expression for the minimum midpoint angle required for a scarp to be Holocene in age, based on the height of the scarp, $2a$, and the regional diffusivity κ.

2.5 The fallout of man-made radionuclides from atmospheric nuclear testing in the 1950s and 1960s provided a pulse of tracer particles that scientists have used to quantify transport rates in many different settings. Although ocean currents are very complex, the vertical distribution of tritium in the ocean is approximately diffusive. Plot the relative concentration of tritium in the ocean as a function of depth assuming that a pulse of tritium was deposited in the ocean in 1960 and $\kappa = 100\,m^2/yr$.

2.6 In this chapter we used Fourier series methods in 1D only, but they can also be used in 2D for simple geometries. Use the Fourier series method to solve for the shape of a rectangular ridge of initial height H, length L, and width W as a function of time. Plot the solutions in contour or shaded relief for several different time values assuming $\kappa = 1\,m^2/kyr$.

2.7 Using a spreadsheet program, implement the FTCS scheme for the diffusion equation for

any initial condition. Begin by constructing two columns for x and $h(x, 0)$. Choose at least ten grid points and use any values you wish for x and $h(x, 0)$ (i.e. you may choose to construct a hypothetical fault scarp). Construct a third column in your spreadsheet for $h(x, \Delta t)$, where the value of Δt is chosen to be small enough to ensure stability. Assume that the value of h is constant at the two ends of the domain. By copying the $h(x, \Delta t)$ column nine times and modifying the links accordingly, plot the solutions for $h(x, 2\Delta t)$, $h(x, 3\Delta t) \ldots h(x, 10\Delta t)$.

2.8 Older moraines often have a lag of coarse material on their crests. Develop a model for this coarsening based on the diffusion equation with a κ value that depends on mean grain diameter. For example, assume that the moraine parent material is 30.

2.9 Identify a radiometrically dated cinder cone in the western US using the published literature. Download USGS Seamless 10 m/pixel DEM (http://seamless.usgs.gov) for an area that includes the cone and extract radial topographic profiles of the cone. Determine a local κ value for the cone using the Fourier–Bessel solutions to determine a best-fit profile using different values of κ.

2.10 Desertification is lowering water levels in many lakes worldwide (e.g. Dead Sea). As lakes act as the base level for channels draining into them, lake-level drops will likely trigger channel incision on the margins of these lakes. Model the evolution (analytically or numerically) of a channel longitudinal profile with diffusivity $\kappa = 10\,\mathrm{m}^2/\mathrm{yr}$ subject to continuous base-level drop of 1 m/yr. How far from the shoreline will the effects of base level be felt 10 and 100 yr into the future?

2.11 Using available DEM data (e.g. http://seamless.usgs.gov) extract a topographic profile across the Sierra Nevada Mountains. Compute the temperature profile (analytically or numerically) at a shallow depth h in the crust beneath the profile. Assume a geothermal gradient of 20 °C/km and a mean annual surface temperature of 0 °C.

Chapter 3

Flow routing

3.1 Introduction

Digital Elevation Models (DEMs) play an important role in modern research in Earth surface processes. First, DEMs provide a baseline dataset for quantifying landscape morphology. Second, they enable us to model the pathways of mass and energy transport through the landscape by hillslope and fluvial processes. Given the importance of DEMs in a broad range of geoscientific research, the ability to digitally process DEMs should be a part of every geoscientists' toolkit.

Flow-routing algorithms lie at the heart of DEM analysis. Flow-routing algorithms are used to model the pathways of mass and energy through the landscape. There is no unique flow-routing method because different constituents move through the landscape in different ways. Water, for example, moves through the landscape somewhat differently than sediment. Also, flow-routing models are necessarily simplified models of transport. Indeed, the only ideal model for water flow in the landscape is the full solution to the Navier–Stokes equations of fluid dynamics. Modeling the Navier–Stokes equations in a complex topographic environment, however, is beyond the scope of any computer. As such, all flow-routing methods involve some approximation to the real physics of mass and energy transport. The key, then, is to develop a suite of flow-routing algorithms that partition mass and energy down-slope in ways that mimic the complex processes of actual fluid flow. Modelers can then choose which algorithm represents the best compromise between realism, computational speed, and ease of use.

Flow-routing algorithms have application to a wide range of geomorphic problems. Most landscape-evolution models, for example, assume that erosion or sediment transport is a function of the discharge or drainage area routed through each pixel. As such, most landscape evolution models require a flow-routing algorithm in order to quantify erosion and sediment transport. Flow-routing algorithms are also important for modeling contaminant transport in fluvial systems. In such cases, flow-routing algorithms are used to transport contaminants from source regions and mix them with uncontaminated sediments downstream. In this chapter we describe the common flow-routing algorithms used in DEM analysis and present three detailed case studies of their application.

3.2 Algorithms

3.2.1 Single-direction algorithms

The simplest flow-routing algorithms are single-direction algorithms. These algorithms assume that all incoming flow is routed in the direction of steepest descent. By "flow," we usually mean discharge or some proxy for discharge such as contributing area. These models are general, however, and may apply to any constituent that moves through the hillslope and fluvial system.

The D8 algorithm (O'Callahan and Mark, 1984) works by searching the eight neighbors (including diagonals) for the steepest down-slope gradient. All flow from the central pixel is then delivered to that pixel. To use this algorithm to compute the contributing area at each point within a grid, each pixel is first initialized to have a contributing area equal to $(\Delta x)^2$, where Δx is the pixel width. Then, moving through the grid in a systematic manner (e.g. upper-left to lower-right) the path of steepest descent is computed from each pixel to the DEM boundary, incrementing the value of the contributing area by $(\Delta x)^2$ in each pixel along each path.

One problem with the D8 algorithm is that flow pathways predicted by this algorithm have segments that are constrained to be multiples of 45°. Since flow pathways in nature can take on any orientation, at a minimum a good flow-routing algorithm should not depend on the rectangular nature of the underlying DEM. In practice, the D8 algorithm tends to produce overly straight flow pathways since the local hillslope or channel aspect (i.e. orientation) must deviate by at least 45° in order for the D8 algorithm to resolve a change in flow direction.

The $\rho 8$ algorithm (Fairfield and Laymarie, 1991) attempts to overcome the problem of overly straight flow pathways in the D8 algorithm. The $\rho 8$ algorithm randomly assigns flow from the center grid cell to one of its down-slope neighbors (including diagonals) with a probability proportional to gradient. Although this algorithm produces fewer straight flow pathways, it still restricts pathway segments to be multiples of 45°. Also, the fact that this algorithm does not produce a unique answer is a problem for many applications.

In addition to the restricted orientations of the D8 and $\rho 8$ algorithms, a larger problem exists. Both of these models implicitly assume convergent flow everywhere on the landscape. In nature, divergent flow is common on hillslopes and in distributary channel environments (e.g. alluvial fans) where flow spreads out with distance downstream rather than becoming more concentrated downstream. In the early days of DEM analysis, many DEMs were so coarse (e.g. 90 m/pixel) that they did not resolve hillslopes or channel beds. Single-direction flow-routing algorithms worked well on many of these early DEMs

because fluvial topography is characterized by tributary flow in many areas at scales larger than 90 m. Today, however, high-resolution DEMs constructed by LIDAR and low-elevation photogrammetry commonly have resolutions of 1 m/pixel. Accurate flow routing in these high-resolution DEMs requires multiple-direction flow-routing algorithms that partition flow from a central point into multiple down-slope pixels.

3.2.2 Multiple-direction algorithms

The Multiple-Flow-Direction (MFD) algorithm of Freeman (1991) was the first algorithm to partition flow into multiple down-slope pixels. In this algorithm, flow from a central pixel is partitioned into all down-slope pixels, weighted by slope. To implement this technique, the pixels must be processed in a specific order. In this and most other multiple-flow-direction algorithms, pixels must be processed from the highest elevations to the lowest. Since mass and energy move downhill, processing pixels from highest to lowest elevation ensures that all of the flow coming into a pixel from upstream has been computed before the flow is partitioned down-slope. To implement this algorithm, an index list must first be created that ranks the DEM pixels from highest to lowest. This ranked list can be computed very efficiently using the `quicksort` algorithm (Press *et al.*, 1992).

Mathematically, the MFD algorithm can be written as

$$f_i = \frac{\max(0, S_i^p)}{\sum_{j=1}^{8}(\max(0, S_j^p))} \tag{3.1}$$

where f_i is the fraction of incoming flow directed to each of the neighboring pixels, S is the slope or gradient between the central point and each of its neighbors (indexed by i and j) and p is a free parameter. To calibrate this method, Freeman (1991) used Eq. (3.1) to predict flow on a conical surface for various values of p. For $p = 1.0$, Freeman found that flow was preferentially directed towards diagonal pixels. Using a slightly higher value of $p = 1.1$, this effect was eliminated. As such, $p = 1.1$ should be used for best results. Codes for implementing the MFD algorithm and for hydrologically correcting DEMs (i.e. removing spurious pits and flats) prior to flow routing are available in Appendix 2.

The difference between single and multiple-direction flow-routing algorithms is illustrated in Figure 3.1. Figure 3.1b and 3.1c illustrate the contributing area for Hanaupah Canyon, Death Valley, California using the steepest-descent (D8) and MFD methods, respectively. These algorithms result in very different spatial distributions of flow on the alluvial fan and on hillslopes, where divergent flow is important. Both flow-routing methods are approximations to the true distribution of flow, but the MFD method is more realistic because it enables the flow distribution to depend continuously on the topographic shape while the steepest-descent method only depends on the drainage network topology. For example, the steepest-descent method does not distinguish between gentle, broad valleys and steep, narrow valleys as long as they have the same drainage-network topology. The MFD method, in contrast, is sensitive to both topology and hypsometry. Not surprisingly, the behavior of landform evolution models in which erosion and sediment transport are a function of drainage area depends on the flow-routing algorithm used in the model (e.g. Pelletier, 2004d).

In addition to MFD, two additional multiple-direction algorithms are widely used. Both of these algorithms are similar to the steepest descent algorithm in that they assume that flow within a pixel has a well-defined direction. However, rather than forcing the flow direction to be a multiple of $45°$ as in the steepest descent algorithm, both the DEMON and D∞ algorithms compute a flow direction that is allowed to acquire any value between 0 and $360°$. The D∞ algorithm (Tarboton, 1997) calculates the direction of steepest descent based on the slopes of eight triangular facets determined by the central pixel and pairs of adjacent neighboring pixels. Flow is then partitioned between two of the eight directions nearest to the computed aspect, weighted by their angular difference from the aspect. For example, if the flow direction for a certain pixel is determined to be $10°$ north of east, then 35/45ths of the flow is delivered to the pixel to the east and 10/45ths of the flow is delivered to the pixel to the northeast. In the DEMON algorithm

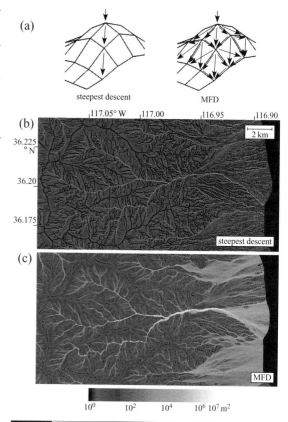

Fig 3.1 Comparison of steepest-descent and MFD flow-routing algorithms. (a) Steepest descent: a unit of precipitation that falls on a grid point (shown at top) is successively routed to the lowest of the eight nearest neighbors (including diagonals) until the outlet is reached. Precipitation can be dropped and routed in the landscape in any order. MFD: all incoming flow to a grid point is distributed between the down-slope pixels, weighted by bed slope. To implement this algorithm efficiently, grid points should be ranked from highest to lowest, and routing should be done in that order to ensure that all incoming flow from up-slope is accumulated before downstream routing is calculated. (b) Map of contributing area calculated with steepest descent for Hanaupah Canyon, Death Valley, California, using USGS 30 m DEMs. Grayscale is logarithmic and follows the legend at figure bottom. (c) Map of contributing area computed with bifurcation routing, for the same area and grayscale as (b). Multiple-flow-direction routing results in substantially different and more realistic flow distribution, particularly for hillslopes and areas of distributary flow. For color version, see plate section. Modified from Pelletier (2004d).

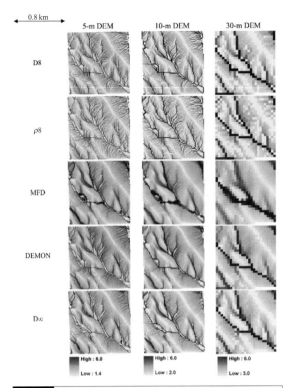

Fig 3.2 Maps of log(A) estimated using the five algorithms discussed in this chapter, using a high-resolution DEM of the north field, an agricultural field in northeastern Colorado. From Erskine *et al.* (2006). Reproduced with permission of the American Geophysical Union.

(Costa-Cabral and Burges, 1994), grid elevation values are taken as pixel corners and flow directions are based on the aspect of a plane surface fit to each pixel. As in the D∞ algorithm, flow is then partitioned into the two neighbors closest to the flow direction.

Several studies (e.g. Erskine *et al.*, 2006) have conducted comparison tests of different flow-routing algorithms used for computing contributing area. In general, these studies endorse the use of multiple-flow-direction algorithms over single-flow-direction algorithms for all applications. Figure 3.2 illustrates that multiple-flow-direction algorithms generate more realistic flow pathways than single-direction methods.

There is no widespread agreement regarding which of the multiple-flow-direction algorithms is best. In introducing his D∞ algorithm, Tarboton (1997) argued that the MFD algorithm

creates too much "dispersion" (i.e. flow spreads out more quickly with downstream distance than is realistic). There is no clear basis for concluding that MFD overpredicts the spread of flow, however. In creating his MFD algorithm, Freeman (1991) began with the hypothesis that flow on a simple, symmetric landform such as a cone must obey the symmetry of the cone. By assuming that there is one principal direction of flow for all pixels, the DEMON and D∞ algorithms violate that symmetrical flow hypothesis. On divides, for example, where flow tends to be directed roughly equally in two opposite directions, the DEMON and D∞ algorithms unrealistically force the flow to occur in one principal direction. The same bias occurs in channels that bifurcate in two directions separated by more than 45°.

Figure 3.3 compares the D∞ and MFD algorithms for a 1 m/pixel DEM of a crater wall on Mars. This landform is dominated by a steep (20–35°) northwest-oriented slope dissected by longitudinal channels 1 m deep. Both of these algorithms yield similar spatial distributions for A, but the MFD algorithm does create a somewhat "smoother" distribution, consistent with the fact that flow down such a slope will generally be partitioned between three down-slope pixels (i.e. north, northwest, and west) rather than only two as in the D∞ method. So, which map is correct? In order to answer that question, we should remember the goal of flow-routing algorithms: to mimic the flow of water (or some other constituent of mass or energy) through the landscape. Viewed through this lens, we can conclude that *both* algorithms are limited by the fact that they do not include any information on flow depth. In relatively planar terrain such as a steep crater slope, the spread of flow across and down the slope will depend on the ratio of the flow depth to the channel depth. If flow depths are comparable to or larger than channel depths, flow will be broadly distributed across the slope. Conversely, if flow depths are small relative to channel depths, flow will be relatively concentrated in channels, as predicted by both Figure 3.3b and 3.3c. In Section 3.4, we will discuss an algorithm based on the MFD model that successively "fills" the topography up to a specified flow depth in order to quantify the effects of

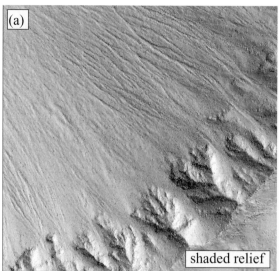

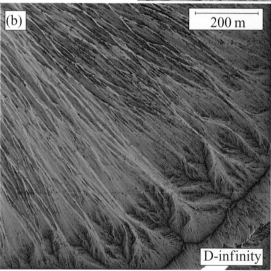

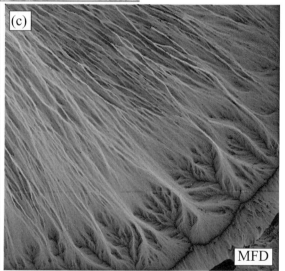

Fig 3.3 Grayscale maps of $\log(A)$ estimated using (b) the D∞ and (c) MFD algorithms, using a high-resolution DEM of a crater wall on Mars. Shaded relief image shown in (a), with illumination from the northeast.

a finite flow depth on the spatial distribution of flow.

3.3 | "Cleaning up" US Geological Survey DEMs

The most widely used DEMs in the United States are the US Geological Survey's 10 m/pixel and 30 m/pixel DEMs. Despite their widespread use, these DEMs suffer from well-known artifacts in low-relief terrain. Figure 3.4, for example, illustrates a shaded-relief image of Hanaupah Canyon fan in Death Valley. On the alluvial fan, where the along-dip topography is locally planar, the topography appears to have "bands" or "steps" in this image. This banding is an artifact of the way that US Geological Survey DEMs are constructed.

USGS DEMs are made by interpolating 7.5 min USGS contour maps. In high-relief areas, contours are closely spaced and DEM values are quite accurate at the scale of 10 or 30 m. On alluvial fans and other low-relief areas where slopes are a few percent or less, contours can be spaced by 100 m or more. Interpolating contours in areas with such low-density information is problematic. DEMs have two kinds of problems

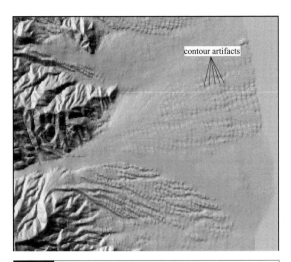

contour artifacts

Fig 3.4 Shaded relief image of a portion of the USGS 30 m/pixel DEM of Hanaupah Canyon fan, Death Valley, California, showing contour artifacts common in low-relief areas.

in low-relief areas. First, microtopographic variations will not be resolved in the original contour map, and hence they will also not be resolved in any DEM constructed from the contour map. Second, USGS DEMs are constructed by interpolating between contours along N–S and E–W grid lines. This interpolation method is simple to implement but results in most of the banding present in USGS DEMs. A more natural interpolation method would interpolate between contours along flow lines rather than along artificial N–S and E–W grid lines. In this section, we explore the use of flow-line interpolation methods for "cleaning up" some of the contour artifacts present in USGS DEMs.

Our study site is the Marshall Gulch watershed near Summerhaven, Arizona. Figure 3.5a shows a shaded-relief image of the 10-m resolution USGS DEM of the Marshall Gulch watershed. Contour artifacts are clearly present in this DEM. In order to develop and test a new method for contour interpolation, we first need to backtrack from the USGS DEM to the original contour data that was used to create it. We can do this by first creating a contour map from the DEM with the same contour spacing as the original topographic map. The USGS contour map for this area has a 20 ft contour interval. Using this interval, we created a digital contour map that is nearly identical

to the published contour map of the area. Then, we encoded each of the contour pixels in the digital contour map with elevation values from the DEM. The result is the elevation-coded raster dataset of contours shown in Figure 3.6a.

The idea of the new interpolation method is that variations in slope occur most gradually along flow lines. As such, the best estimate of the elevation values between contours can be obtained with a weighted interpolation along flow lines. In order to perform this interpolation, the distance between each pixel and its upstream and downstream contours must be computed. Figure 3.6b presents a grayscale map of the minimum distance to the downstream contour. In general, fluvial topography evolves in order to minimize downstream transport distances. As such, flow lines will generally correspond to pathways with the shortest distance down-slope. To produce this minimum-downstream-distance map, a local search algorithm was used that calculates the distance between each pixel and all possible flow pathways down-slope from that pixel to the first contour line. The distance along each pathway was computed but only the minimum value was retained. In the up-slope direction it is more appropriate to use the *maximum* distance to the upstream contour. This distance was calculated using a similar search algorithm and is illustrated in Figure 3.6c.

Figure 3.5b illustrates the DEM resulting from a weighted linear interpolation of elevation values along flow lines. The algorithm minimizes some but not all of the contour artifacts present in the USGS DEM and generally provides for crisper topographic detail. Figures 3.5a and 3.5c can be compared to the "actual" topography in Figure 3.5c. This shaded-relief map was produced from a 1 m/pixel LIDAR DEM of Marshall Gulch. Most of the small-scale detail present in this image is missing in both of the contour-derived DEMs. This underscores the fundamental limitations of any DEM derived from relatively coarse (20 ft) contour data. Clearly, LIDAR data is superior where it is available. One reason why some contour artifacts remain in the DEM is that a linear interpolation along flow lines, while superior to a linear interpolation along N–S and E–W grid lines, nevertheless produces a slope break at contour lines that may not be present in the

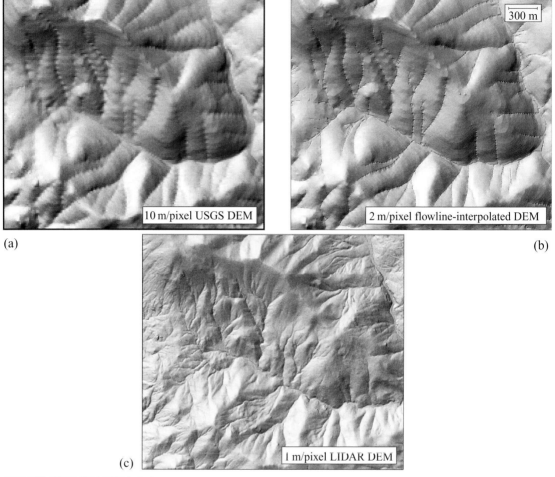

(a)

(b)

(c)

Fig 3.5 Shaded-relief maps of Marshall Gulch DEMs.
(a) USGS 10 m/pixel DEM, (b) flow-line-interpolated
2 m/pixel DEM using method described in this section,
(c) 1 m/pixel LIDAR DEM.

actual topography. As such, the DEM of Figure 3.5b could be further improved by using a spline interpolation involving two or more contours in the upstream or downstream directions.

3.4 Application of flow-routing algorithms to estimate flood hazards

A principal drawback of all of the flow routing methods we have considered is that they do not incorporate flow depth. Clearly, the flow paths of water and water-borne constituents depend on how much water is flowing across the landscape. When the flow depth in a channel exceeds bankfull capacity, for example, flow will be diverted onto the floodplain. Since the MFD and other multiple-direction flow-routing methods route flow according to bed slope and do not incorporate flow depth, they implicitly predict flow pathways for cases of negligible flow depth. In many geomorphic applications it is necessary to constrain water and sediment pathways for a range of event sizes, not just for those corresponding to low flow conditions. Furthermore, in many cases modeling the detailed behavior of individual flood events would be unrealistic. For such applications it would be useful to have a flow-routing method that retains the simplicity of the MFD method yet is able to resolve the different flow pathways that occur when flow

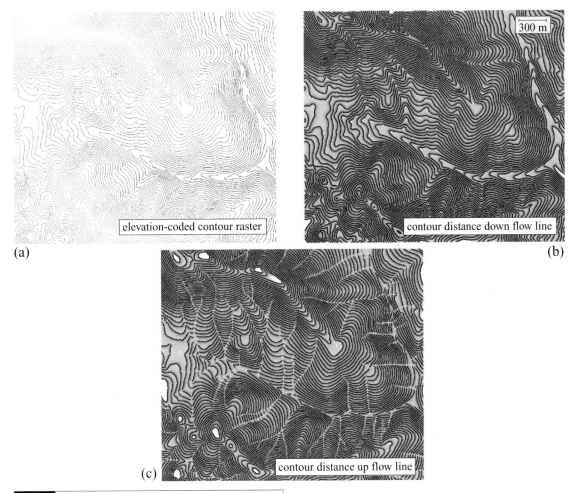

(a)

elevation-coded contour raster

300 m

contour distance down flow line

(b)

(c)

contour distance up flow line

Fig 3.6 Intermediate products used to create flow-line-interpolated DEM. (a) Elevation-encoded contour raster dataset, (b) grayscale map of minimum distance to downstream contour, (c) grayscale map of maximum distance to upstream contour. Datasets in (b) and (c) provide weighting factors for interpolating between contours.

depths exceed the local cross-sectional area of the channel.

The MFD model can be implemented in an iterative fashion to account for the effects of flow depths on flow pathways. The basic idea is to first determine flow pathways in low-flow conditions using the basic MFD routing method. Then, the flow is rerouted on a DEM that includes the original bed topography plus a small increment of flow determined by a prescribed maximum flow depth together with the spatial distribution of flow computed by the first MFD iteration. Then,

the MFD model is run again on the new partially filled DEM. This process is repeated using small increments of flow until the prescribed maximum flow depth is reached. This algorithm effectively mimics the flood behavior of a slowly rising hydrograph. Done properly, this method can be shown to converge to a unique solution that satisfies the flow continuity equation as the number of iterations is increased.

Figure 3.7 illustrates the results of successive MFD flood routing on a high-resolution (1 m/pixel) photogrammetric DEM of Fortymile Wash in Amargosa Valley, Nevada. The initial condition for each of these runs is an upstream discharge prescribed at the location of the main channel at the northern boundary of the DEM. The discharge value is first translated into a flow depth using Manning's equation. Then, the MFD algorithm is run successively with larger

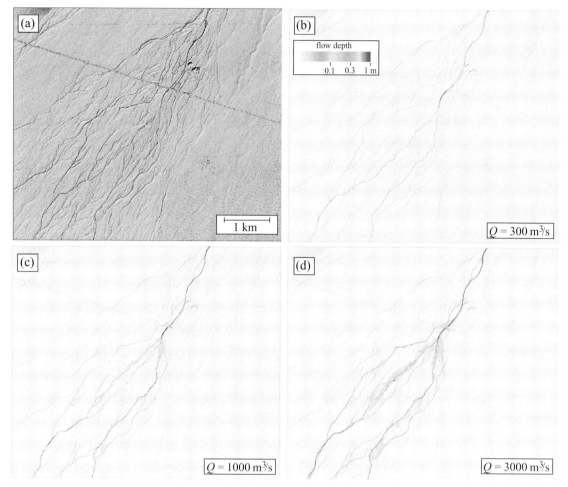

Fig 3.7 Illustration of a successive flow-routing algorithm applied to Fortymile Wash alluvial fan in Amargosa Valley, Nevada. (a) Shaded-relief image of a high resolution (1 m/pixel) photogrammetric DEM of Fortymile Wash, (b)–(d) flow depths predicted by the model for different values of the prescribed upstream discharge: (b) 300 m³/s, (c) 1000 m³/s, and (d) 3000 m³/s. For color version, see plate section.

increments of flow until the input flow depth corresponding to the upstream discharge is achieved. This algorithm is useful at identifying distributary channels and adjacent overbank areas that are activated when certain threshold flow depths are exceeded. This technique makes a number of simplifying assumptions about the steadiness of the flow, and hence cannot be used for detailed flood-hazard assessment. Nevertheless, it is a simple, fast method for identifying flow pathways and their thresholds of

inundation, and can be used on a reconnoitering basis to determine relative flood risk. Successive flow-routing analysis can be performed for a single master channel network by specifying flow at an upstream boundary, or a uniform runoff depth can be specified for the entire DEM in order to model distributed flow routing. Appendix 2 provides code for implementing the distributed flow routing case.

3.5 | Contaminant transport in channel bed sediments

Predicting the transport and fate of contaminants in fluvial systems is a critical aspect of applied geomorphology. Worldwide, many river systems have channel-bed sediments with elevated heavy-metal concentrations as a result of centuries of ore-mining activity (James, 1989; Miller, 1997). Similarly, radionuclide contamination of channel-bed sediments is a potential human health hazard downstream of areas of high-level nuclear processing and testing (Reneau *et al.*, 2004).

In this section, we explore a model for the prediction of contaminant distributions in fluvial channel sediments. The motivation for developing the model was the need to quantify the dispersal of radionuclide-contaminated sediments in the event of a volcanic eruption at the proposed nuclear-waste repository at Yucca Mountain. Underground storage of nuclear waste primarily poses a hydrologic risk associated with contaminated groundwater. However, volcanism in the Yucca Mountain region could pose a significant geomorphic hazard if an eruption were to occur. Probabilistic volcanic hazard analyses constrain the annual risk of an eruption through the proposed repository to be a very low 1.5×10^{-8} (CRWMS M&O, 1996). In the event of an eruption, individuals living on the Fortymile Wash alluvial fan 18 km south of the repository (i.e. the closest residents to the repository; Figure 3.8) could be affected by radionuclide-contaminated tephra deposited as fallout or redistributed from upstream. Under most wind direction scenarios (i.e. southerly winds), primary fallout tephra is concentrated to the north of the proposed repository location and entirely within the Nevada Test Site (BSC, 2004a,b). In such cases, fluvial redistribution of contaminated tephra from the primary fallout location to the Fortymile Wash alluvial fan could be significant and hence must be evaluated.

In collaboration with scientists at Los Alamos National Laboratory, my graduate students Stephen DeLong, Mike Cline, and I developed a model that quantifies the concentration of tephra and radionuclides in channels of the Fortymile Wash alluvial fan as a result of hillslope and fluvial redistribution processes in the event of a volcanic eruption through the repository. The model uses the MFD flow-routing algorithm to route contaminated and uncontaminated sediments through the fluvial system

in order to map the fraction of contaminated sediments in all the channels of the Fortymile Wash drainage basin. The contaminant fraction computed for channels in the area of the critical population provides a basis for estimating the potential dose rate to that population in the event of an eruption.

The model divides the Fortymile Wash drainage basin into two domains: (1) the tributary drainage basin of Fortymile Wash and (2) its distributary alluvial fan where the nearest population resides (Figure 3.8). The Fortymile Wash drainage basin includes the proposed repository location and the area of most of the primary fallout in the event of a volcanic eruption through the repository. The drainage basin and the fan are divided at the fan apex. The model assumes that primary fallout is mobilized and transported toward the alluvial fan if it falls on slopes steeper than a threshold gradient or on channels greater than a threshold stream power. To do this calculation, the model performs a spatially distributed analysis of the slopes and stream powers for the entire Fortymile Wash drainage basin using an input 30-m resolution US Geological Survey Digital Elevation Model (DEM).

Before the mobilized tephra and radionuclides are deposited on the alluvial fan, they are transported through the alluvial channel system of Fortymile Wash where mixing with uncontaminated channel sediments results in the dilution of contaminated tephra. During extreme flood events, contaminated tephra deposited in channels or mobilized into channels from adjacent hillslopes will be entrained and potentially mixed with uncontaminated channel-bed sediments within the scour zone. The cumulative effect of many floods is to mix the tephra with any uncontaminated channel-bed sediments located beneath the tephra but within the scour zone both locally and with distance downstream. Quantifying this dilution process requires a spatially distributed model of long-term flood scour depths throughout the drainage basin. The MFD flow routing model, together with empirical relationships between discharge and scour depth, provide the means for mapping the scour depths necessary for the model.

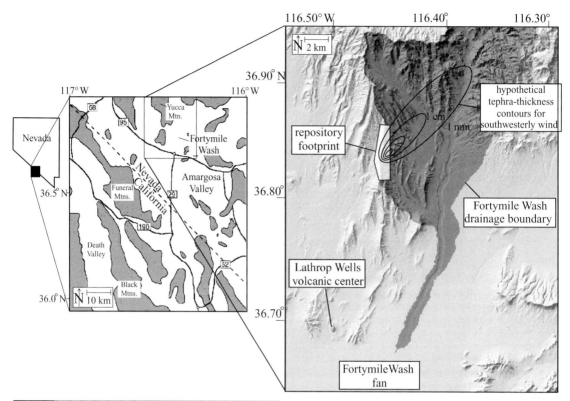

Fig 3.8 Location map of the study areas in Amargosa Valley, Nevada. Shaded relief image includes Lathrop Wells volcanic center, the footprint of the proposed nuclear-waste repository at Yucca Mountain, and the southern portion of the Fortymile Wash drainage basin (i.e. the site of primary fallout for tephra with potential for redistribution to the Fortymile Wash fan where the nearest population resides. From Pelletier *et al.* (2008). Reproduced with permission of Elsevier Limited.

Previous work on the dilution and mixing of fluvial-system contaminants predict one-dimensional (1D) concentration of contaminants in channel systems. These models are 1D in the sense that vertical and cross-channel concentration variations are not considered. The classic dilution-mixing model of Hawkes (1976) and Marcus (1987) assumes that the primary process of downstream dilution is the mixing of contaminants with uncontaminated sediments delivered from upstream. Mathematically, the model states that the concentration at a point downstream from a contaminant source is an average of the contaminant concentration at source and

non-source areas weighted by their relative upstream areas:

$$C = \frac{A_s}{A_s + A_{ns}} C_s + \frac{A_{ns}}{A_s + A_{ns}} C_{ns} \quad (3.2)$$

where C is the contaminant concentration at a point downstream, A_s and A_{ns} are the source and non-source areas upstream from the point, and C_s and C_{ns} are the source and non-source concentrations. The value of C_{ns} is usually zero, but in cases where contaminants have a finite natural background level, or in cases where non-fluvial (e.g. eolian) processes have transported contaminants from source to non-source areas in the basin, C_{ns} may be greater than zero. Equation (3.2) reduces to an exponential form if the source area is located in the basin headwaters and the upstream area is written as a function of x, the along-channel distance:

$$C = C_{ns} + C_s \exp(-x/x_s) \quad (3.3)$$

where x_s is a length scale controlled by the relative areas of source and non-source regions and the along-channel distance through the source area. Equation (3.3) is an approximation of Eq. (3.2); it describes the general downstream-dilution trend but does not capture the abrupt concentration decrease that occurs as large tributaries enter the main channel. Equation (3.2) has been tested in many contaminated rivers (e.g. Hawkes, 1976; Marcus, 1986) and its accuracy is often remarkable considering the simplicity of the model relative to the complexity of the processes involved.

In addition to mixing of the contaminant with uncontaminated sediments from upstream, dilution can also occur by mixing of contaminants with local channel-bed sediments downstream. In this process, a flood wave moves though the channel causing net erosion during the growing phase of the flood, turbulent mixing of the contaminant with uncontaminated channel-bed sediments during the flood, and redeposition of the sediment–contaminant mixture as the flood wave passes. The depth of contaminant mixing by this process is the scour depth, i.e. the thickness of sediments that undergo active transport during extreme flood events. The scour depth, in turn, is proportional to the square root of the unit discharge (discharge per unit channel width) (Leopold et al., 1966). A numerical model that maps unit discharge and uses Leopold et al.'s empirical relationship to predict scour depth, therefore, provides a basis for estimating the dilution effect caused by contaminant mixing with channel-bed sediments downstream. In our model, the mobile portion of the bed (including both contaminants and uncontaminated channel-bed sediments) is assumed to be lifted from the bed, mixed by turbulence, and deposited back onto the bed downstream. The dilution effect caused by this process can be quantified as a function of source and nonsource upstream areas as in the classic model.

The classic dilution-mixing and scour-dilution-mixing models are compared schematically in Figure 3.9. The basic operation of the classic model is illustrated by the diagram in the lower left corner of Figure 3.9a. The lengths of arrows coming into the central point represent the sediment yield from the two upstream tributaries (one contaminated and one uncontaminated). The model predicts the contaminant concentration downstream of the

confluence to be an average of the upstream concentrations, weighted by the upstream sediment yields or basin areas. The model makes no prediction regarding the vertical distribution of contaminants.

The basic operation of the scour-dilution-mixing model is illustrated by the diagram in the lower left corner of Figure 3.9b. The scour-dilution-mixing model is conceptually similar to the classic model because the contaminant concentration downstream is an average of upstream concentrations. However, the vertical component of mixing is explicitly included, and the local contributions of contaminant and uncontaminated channel-bed sediments are incorporated into the mixture. The basic assumption of the model is that the concentration at each point is a mixture of contaminant and channel-bed sediments upstream from the point. As such, the contaminant concentration at a channel point is a ratio of the total contaminant volume delivered from upstream to the total mobile-bed volume. One significant advantage of the scour-dilution-mixing model is that it predicts the vertical distribution of contaminants as well as the surface concentration. In addition, by using spatially distributed routing of runoff and contaminants, the model can predict cross-channel distributions and can be applied to distributary environments. In this sense, the model predicts a 3D distribution rather than the 1D distribution of the classic model.

The scour-dilution-mixing model can be readily implemented using a raster-based framework. In this framework, the contributing-area grid is first initialized with the value of the pixel area, $(\Delta x)^2$. Second, the MFD flow-routing algorithm of Freeman (1991) is used to calculate the contributing area routed through each pixel in the DEM. Third, a critical stream-power threshold is used to distinguish between hillslopes and channels in the DEM, where the stream power is defined as the product of the local slope S and the square root of the drainage area A (Montgomery and Dietrich, 1988). The critical stream power is related to the drainage density X through the relationship $1/X$. The value of the drainage density can be directly measured in the field or by using digital map products. Pixels with stream-power

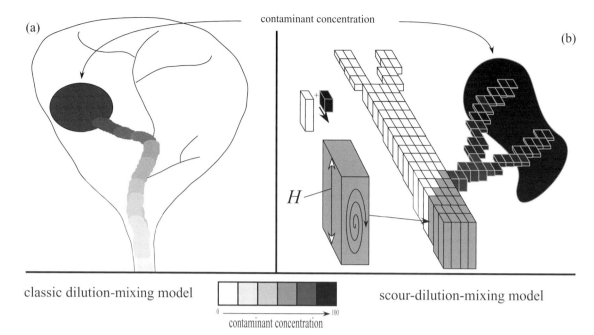

contaminant concentration

(a)

(b)

H

classic dilution-mixing model

scour-dilution-mixing model

0 ⟶ 100
contaminant concentration

Fig 3.9 Schematic diagram of (a) classic dilution-mixing model, and (b) scour-dilution-mixing model. (a) The basic operation of the classic model is illustrated in the lower left. Here, lines with arrows represent upstream contributing area, a proxy for sediment yield. Concentrations downstream of a confluence are equal to the average of the upstream concentrations weighted by contributing area. In a drainage basin, concentrations decrease rapidly with distance downstream from a localized contaminant source, with abrupt decreases at major confluences (illustrated by grayscale pattern). (b) In the scour-dilution-mixing model, the vertical and cross-channel distributions of contaminant are also considered. At each pixel (i, j), the total volumes of upstream contaminant and uncontaminated channel-bed sediments are mixed with the local volumes of contaminant and uncontaminated sediments. The contaminant distribution at that pixel is defined by that concentration value and the local scour depth. The resulting mixture is transported downstream where the same operation is repeated.

values less than $1/X$ are made zero in the grid, leaving finite values in the remaining pixels. Fourth, the contributing-area grid is converted to a unit-discharge grid using empirical data (i.e. a flood-envelope curve), which, in turn, is converted to a scour/mixing-depth grid using the empirical square-root relationship documented by Leopold *et al.* (1966). Following this series of operations, the resulting grid represents the

estimated scour depth in each channel pixel. As a simple example, assume that the flood-envelope curve shows the peak unit discharge for extreme floods to be linearly proportional to contributing area. Then, if the scour depth is observed to be 1 m (e.g. using a scour chain emplaced prior to a recent extreme flood) at a channel pixel with a contributing area of 100 km^2, then the scour depth at any channel pixel in the basin will be equal to the square root of the normalized contributing area at that pixel (i.e. the contributing area divided by 100 km^2) multiplied by 1 m.

The model routes the contaminant (expressed as an equivalent thickness for each pixel) downstream using the same flow-routing algorithm used for routing contributing area. Finally, as the contaminant is routed downstream, the model calculates the concentration at each pixel as the ratio of total contaminant volume upstream divided by the total volume of scour depth for all active channels upstream. Mathematically, the model states that the concentration at a pixel (i,j) is given by

$$C_{i,j} = \frac{\min\left[\left(\sum_{up} h_{l,m}\right) + h_{i,j} \cdot \left(\sum_{up} H_{l,m}\right) + H_{i,j}\right]}{\left(\sum_{up} H_{l,m}\right) + H_{i,j}}$$

(3.4)

where $h_{l,m}$ is the effective contaminant thickness at pixel (l,m), $H_{l,m}$ is the scour depth, and the sums are computed over all pixels (l,m) upstream from pixel (i,j).

The fundamental tool for the scour-dilution-mixing model is the MFD algorithm. In this algorithm, contributing area, contaminants, and scour depths are routed from each pixel to all of its neighboring pixels weighted by topographic slope. This algorithm is implemented by initializing the grid with the quantity to be routed. In the case of contributing area, the grid is initialized with the area of that pixel, Δx^2 (e.g. 900 m^2 for a 30-m resolution DEM). In the case of contaminant routing, each pixel is equal to the effective thickness of contaminant in channels of the source region. The MFD or bifurcation-routing algorithm then ranks all grid points in the basin from highest to lowest in elevation. Routing is then calculated successively at grid points in rank order from highest to lowest, thereby ensuring that all incoming areas of contaminant have been accounted for before downstream routing is performed.

To test the model, we consider the concentration of basaltic sediment in channels draining the Lathrop Wells tephra sheet. This tephra sheet provides a concentrated source of ash and lapilli that acts as a contaminant source for downstream channels. The tephra found in channels downstream from the tephra sheet is comparable in texture to the non-basaltic sand of downstream channel beds. As such, the model assumption that contaminant material and uncontaminated channel sediments are transported at similar rates is likely to be a good approximation in this case. In addition, the proximity of this area to Yucca Mountain (i.e. 18 km south of the proposed repository location) makes it an ideal analog (e.g. similar climate, hydrology, and geomorphology) for tephra redistribution following a hypothetical volcanic eruption at Yucca Mountain. The physical volcanology of the Lathrop Wells volcanic center was described in detail by Valentine et al. (2005). At 77 000-years old (Heizler et al., 1999), the Lathrop Wells cone is the youngest basaltic volcano in the Yucca Mountain region. It is the southern-most surface expression of the Plio–Pleistocene-age Crater Flat Volcanic

Zone (CFVZ) (Crowe and Perry, 1990). Its relative youth and location with respect to Yucca Mountain has made it a focus of intense study as a possible analog for an eruption through the proposed nuclear waste repository.

There are two principal drainages that transport tephra from Lathrop Wells (Figure 3.10). The western drainage system transports material from the exposed tephra sheet on the northwest side of Lathrop Wells cone west and south into Amargosa Valley. The eastern drainage system heads near the northern margin of the tephra sheet and transports material around the eastern side of the Lathrop Wells cone and adjacent lava flows. Twenty five samples of channel-bed sediments were collected along these two channel systems in order to evaluate the significance of the dilution process and validate the scour-dilution-mixing model. In small channels on the tephra sheet, the bottom of the scour zone was clearly visible and ranged from 12 to 29 cm in depth in channel-bed-sediment sample pits (Figure 3.11). In larger channels downstream from the tephra sheet, the bottom of the scour zone was not visible within our sample pits (which were limited to 30–40 cm in depth). Samples were taken by uniformly sampling pit-wall material to a depth of 30 cm, or to visible scour depth if less than 30 cm. Samples were sieved and material larger than silt size was separated into basaltic and non-basaltic fractions with a magnet and visually checked for complete separation using a stereomicroscope. The basaltic material was strongly magnetic and total separation error was estimated to be less than 5% by mass. Non-basaltic material was composed of Miocene welded tuff and eolian sand. The fraction of basaltic sediment by mass was calculated for each sample. This value was converted to a fraction by volume (for comparison to the volumetric dilution-mixing model) using an average measured basalt density of 1.5 g/cm^3 (a relatively low value for rock because of its high vesicularity) and a tuff density of 2.7 g/cm^3.

Mixing occurs along both the eastern and western drainages according to the trend evident in the plots of basalt concentration as a function of distance from the channel head in Figure 3.10b. These data indicate that the

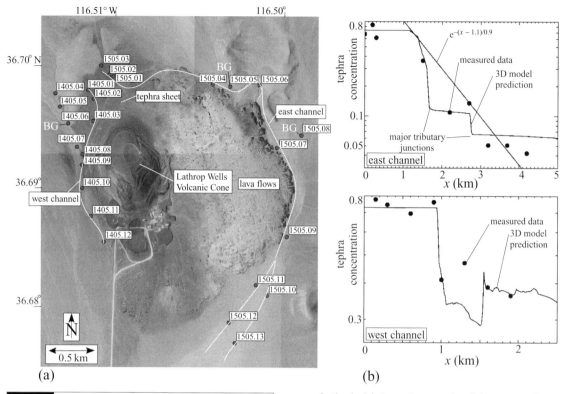

(a)

(b)

Fig 3.10 (a) Sample location map for tephra concentration in the east and west channels draining the Lathrop Wells tephra sheet. (b) Comparison plots of measured and modeled predictions for tephra concentration as a function of distance along-channel (note logarithmic scale on vertical axis, needed to visualize data over 1–2 orders of magnitude). Also graphed are best-fit exponential curves. Modified from Pelletier *et al.* (2008).

concentration of basaltic tephra is reduced by more than 50.

The Lathrop Wells tephra sheet was mapped by Valentine *et al.* (2005), providing a preliminary source map for input to the scour-dilution-mixing model for this application. The source map is complicated in this case by the fact that tephra from the 77 ka eruption dispersed tephra in all directions. Samples collected in the study reveal three separate groups of values. On the tephra sheet itself, contaminant concentrations vary between 61% and 83%, with an average value of 73%. Therefore, 73% was used as the value for the source concentration in the model. Off of the tephra sheet, background concentrations

are relatively high to the north of the cone where most of the tephra was deposited (11.1% and 6.8% in samples SD091405.06 and SD091505.05, respectively) and lower to the south (3.0% in sample SD091505.08). These background samples were collected in large drainages that provide representative average values over large areas. In order to resolve the two background values in the model, a ternary mask (Figure 3.12a) was used with a source concentration of 73%, a distal source value of 9% (north of the primary tephra sheet) and background value of 3% (everywhere else).

Additional inputs to the model are the threshold stream power and a 10-m resolution US Geological Survey DEM (Figure 3.12a). The value for the stream-power threshold was chosen to be 0.015 km by forward modeling. In this forward-modeling procedure, a map of incised channels was made by comparing the stream power at each point in the grid to a threshold value. This threshold was varied in value to produce different incised-channel maps. Maps were then overlain and visually compared with US Geological Survey orthophotographs to identify which value

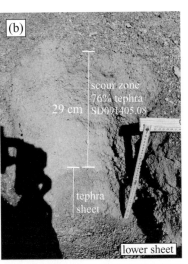

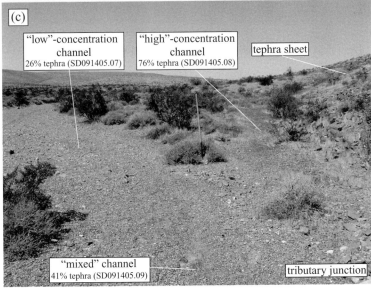

Fig 3.11 Field photos illustrating the three-dimensional pattern of contaminant dilution near Lathrop Wells tephra sheet. (a) and (b) Channel pits on the tephra sheets expose a fluvially mixed scour zone ranging from 12 to 29 cm in thickness. Two types of deposits occur beneath the scour zone: the tephra sheet itself (exposed on the upper and lower sheet) and debris-flow deposits comprised predominantly by Miocene volcanic tuff and eolian silt and sand transported from the upper tephra sheet. (c) The effects of dilution visible as high-concentration channels draining the tephra sheet (channel at right, 76% tephra) join with low-concentration channels (at left, 26% tephra). Tephra concentration in these channels correlates with the darkness of the sediments, with the dark-colored channel at right joining with the (larger) light-colored channel at left to form a light-colored channel downstream. For color version, see plate section. From Pelletier *et al.* (2008). Reproduced with permission of Elsevier Limited.

of the critical stream power produced the most accurate map.

In the first step of the model run, all pixels of the contributing area grid were initialized to 10^{-4} km² (i.e. the pixel area for a 10-m-resolution DEM). The MFD algorithm was then used to calculate the contributing-area grid. In the second step, the contributing-area grid was converted to a unit discharge map using the regional flood-envelope curve of Squires and Young (1984). These authors found the maximum discharge to be proportional to the contributing area to the 0.57 power using available stream gage data and paleoflood estimates from a range of drainage basins in the Yucca Mountain region. Maximum discharge is used because the largest floods control the long-term scour/mixing depth. The unit-discharge map was then used to create a relative map of scour/mixing depth in channels (Figure 3.12c) using the square-root dependence of Leopold *et al.* (1966) and a stream-power threshold of 0.015 km. Only a relative grid of scour/mixing depth was needed in this case because the scour depths in the source area were used to calculate the effective volume of contaminant routed downstream, and the scour depths downstream of the source region were used to dilute that contaminant. Since the contaminant

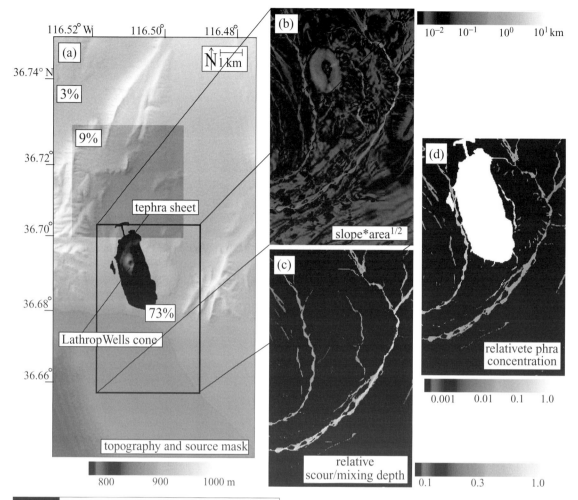

Fig 3.12 Model prediction for tephra concentration and scour/mixing depth downstream from the tephra sheet of the Lathrop Wells volcanic center. (a) Model inputs include a 10-m resolution US Geological Survey Digital Elevation Model (DEM) of the region and a grid of source and background concentrations. Input concentration values are: 73% (Lathrop Wells tephra sheet), 9% (distal source region), and 3% (background). (b) Map of stream power in the vicinity of the source region. Scale is logarithmic, ranging from 10^{-2} to 10^{1} km. (c) Map of scour/mixing depth (scale is quadratic, ranging from 10 cm to 1 m). (d) Map of tephra concentration (scale is quadratic, ranging from 0.001 to 1). For color version, see plate section. From Pelletier *et al.* (2008). Reproduced with permission of Elsevier Limited.

concentration is a ratio of these quantities, the concentration was unaffected by scaling the scour depth up or down by a constant factor. In all cases where the source is specified as a concen-

tration (as opposed to an effective thickness), the absolute value of the scour depth is not needed to predict concentration values downstream; only relative values are needed. The model predicted tephra concentrations (expressed as a fraction) in all channels draining the Lathrop Wells tephra sheet. This map has a quadratic gray scale varying from < 0.01 (black) to 1.0 (white).

The numerical model prediction was compared to the measured tephra concentrations by extracting tephra concentrations along channel profiles of the western and eastern channels in Figure 3.10b. Along both channels, concentrations are near their maximum values for approximately the first 1.0 km downstream from the channel head. Along the eastern channel (Figure 3.10b, top graph), abrupt concentration declines are associated with major tributary junctions at 1.6 and 2.7 km from the channel head. The first major step down is associated with an increase in discharge and scour depth associated with a relatively uncontaminated distributary channel entering the channel from the northeast just upstream of the U-shaped channel bend. Another abrupt drop in concentration occurs approximately 2.7 km from the channel head (just downstream of sample SD091515.07) where another large tributary enters. As the channel curves around the southeast corner of the Lathrop Wells lava field, it bifurcates into two distributary channels. Each pair of samples located at similar distances from the channel head were averaged to comprise the final two points on the curve. The overall downstream-dilution pattern can be roughly approximated by an exponential function with a characteristic length of 0.9 km. For every 0.9 km downstream, therefore, the tephra concentration decreases by approximately a factor of $1/e$. Along the western channel, concentrations follow a similar pattern in the bottom graph of Figure 3.10b. Note that this profile extends only about half as far from the channel head as the eastern channel, due to the presence of the mine road which disturbs the western channel about 2.5 km from the channel head.

The ability of the model to represent observed trends of downstream dilution at Lathrop Wells provides a basis for using the model to predict dilution patterns following a hypothetical eruption at Yucca Mountain. To do this, a spatially distributed model was used to calculate the volume of tephra fallout mobilized from steep slopes and active channels (Figure 3.13). The model accepts input from the ASHPLUME numerical model (e.g. Figure 3.13a, assuming a southwesterly wind). The ASHPLUME model is a numerical advection-dispersion-settling model that considers variable grain sizes and wind-speed profiles from the ground to 10 km altitude. The model then calculates the DEM slopes and contributing areas (Figures 3.13b and 3.13c). Fallout tephra is assumed to be mobilized if it lands on pixels with hillslope gradients greater than a threshold value equal to S or with a stream power greater than a threshold value equal to $1/X$, where X is the drainage density. Field observations at the San Francisco volcanic field suggest that hillslopes above a threshold gradient of approximately 0.3 m/m (17°) are susceptible to mass movements and/or rilling within 1 kyr after an eruption. The threshold stream power for the Fortymile Wash drainage basin was estimated to be 0.05 km (corresponding to a drainage density of $X = 20\,km^{-1}$), based on forward modeling and comparison with USGS orthophotographs. For simplicity, contaminated tephra mobilized from steep slopes and channels was assumed to be transported into the channel system instantly following the eruption, although in nature this process will take place over decades to centuries.

As an example, input to the tephra redistribution model was constructed based on the fallout from a hypothetical eruption with a northeast-directed plume computed using the ASHPLUME numerical model (Figure 3.13a) (Jarzemba, 1997; Jarzemba et al., 1997; BSC 2004a,b). The hypothetical fallout distribution used in Figure 3.13a corresponds to an eruption power of $9.55 \times 10^9\,W$ (a moderate-to-large magnitude eruption for the Crater Flat volcanic field), a duration of approximately 3 days, a wind speed of approximately 5 m/s, and an eruption velocity of 2 m/s. The contributing-area grid is used to calculate the scour-depth grid in several steps for this application. First, contributing area is related to peak discharge on a pixel-by-pixel basis using the regional flood envelope curve of Squires and Young (1984), as in the Lathrop Wells example. Second, discharge is related to scour depth on a pixel-by-pixel basis using unit discharge with the square-root dependence of Leopold et al. (1966). Together, these relationships imply that scour depth is proportional to contributing area to the 0.29 power. To determine the proportionality constant needed to relate contributing area to a specific value of scour depth, the model uses measurements of the US Geological Survey during the 1995 flood at the Narrows section of Fortymile Wash (Beck and Hess, 1996) located close to the fan apex. The contributing area at this location is the ratio of the basin area, 670 km², to the width, 44 m. The maximum scour depth at this location was measured to be at least 1.14 m, and had an average value of 0.72 m. The model runs of this paper used 1 m as a representative average scour

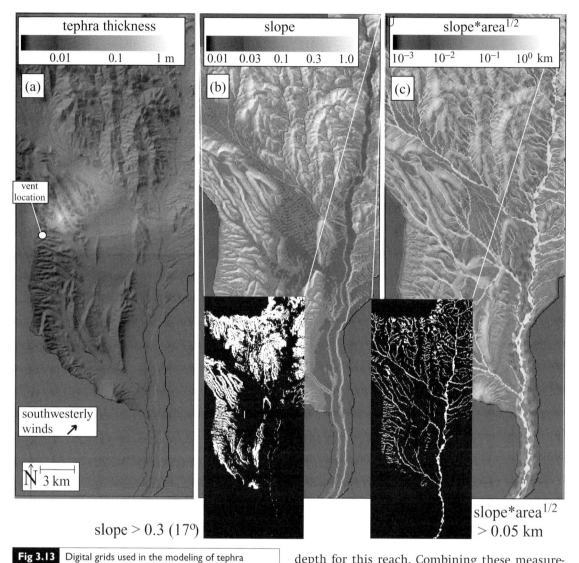

tephra thickness
0.01   0.1   1 m

slope
0.01 0.03 0.1 0.3 1.0

slope*area$^{1/2}$
10^{-3} 10^{-2} 10^{-1} 10^{0} km

(a)

vent location

southwesterly winds

N 3 km

(b)

(c)

slope > 0.3 (17°)

slope*area$^{1/2}$
> 0.05 km

Fig 3.13 Digital grids used in the modeling of tephra redistribution following a volcanic eruption at Yucca Mountain. (a) Shaded relief image of DEM and Fortymile Wash drainage basin (darker area). Tephra from only this drainage basin is redistributed to the RMEI location. (b) Grayscale image of DEM slopes within the Fortymile Wash drainage basin. (c) Black-and-white grid of areas in the drainage basin with slopes greater than 17°. (d) Black-and-white grid of active channels in the drainage basin (defined as pixels with contributing areas greater than 0.05 km^2). All tephra deposited within the black areas of (c) and (d) are assumed to be mobilized by mass movement, intense rilling, or channel flow. For color version, see plate section. Modified from Pelletier et al. (2008). Reproduced with permission of Elsevier Limited.

depth for this reach. Combining these measurements yields

$$H_{i,j} = H_{apex} \left(\frac{A_{i,j}}{670} \frac{44}{\Delta x} \right)^{0.29} \tag{3.5}$$

where $H_{i,j}$ is the scour depth at any point in the basin upstream from the apex, $H_{apex} = 1$ m is the representative scour depth at the fan apex based on measurements by the US Geological Survey, and $A_{i,j}$ is the contributing area at point (i,j).

The scour-dilution-mixing model predicts relatively high tephra concentrations along the northeast-directed plume track (Figure 3.14b). As tephra from these channels moves downstream, tephra in channels that are integrated into the

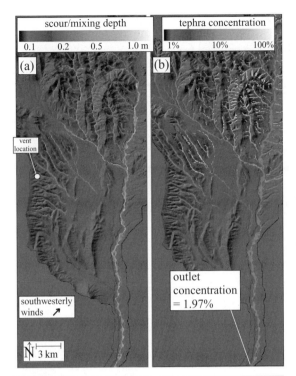

Fig 3.14 Digital grids output by the modeling of tephra redistribution following a potential volcanic eruption at Yucca Mountain. (a) Map of scour-depth grid, calculated from contributing area grid. (b) Grayscale map of tephra concentration in channels, calculated using the scour-dilution-mixing model. For color version, see plate section. Modified from Pelletier *et al.* (2008). Reproduced with permission of Elsevier Limited.

main Fortymile Wash becomes progressively diluted with each successive inflow of relatively uncontaminated tributary sediments. Note that some of the tephras draining from channels in the vicinity of Yucca Mountain are locally deposited in alluvial flats and do not reach the main Fortymile Wash. At the basin outlet the effects of dilution mixing and localized deposition of tephra limit the tephra concentration to 1.97% in this example. The effects of variable wind direction on contaminant concentrations can be determined (Figure 3.15) by running the scour-dilution-mixing model on a sequence of rotated plumes of the same size and shape as the example in Figure 3.14. Outlet tephra concentrations for these rotated plumes vary from a minimum of less than 0.1% to a maximum of approximately nearly 2% for the southwesterly wind case illustrated in Figure 3.14. The primary control on tephra concentration at the outlet is the topography in the primary fallout region (which controls the volume of tephra mobilized on steep slopes and in active channels). The highest outlet tephra concentrations occur for southerly and westerly winds (Figure 3.15), because these wind conditions transport tephra north and east from the repository location to areas with a relatively high density of tephra-transporting steep slopes and active channels within the Fortymile Wash drainage basin.

Exercises

3.1 Download a DEM of any area dominated by fluvial processes. Implement the `fillinpitsandflats` routine (Appendix 2) to filter out pits and flats in the DEM. Implement the MFD flow-routing method to calculate the contributing area A for each pixel and plot the results as a grayscale or color map.

3.2 After completing the above exercise, develop a program module that distinguishes hillslope and channel pixels in the DEM based on a threshold value of contributing area (i.e. assume channel pixels are those with $A > A_c$). Create maps of the channel network corresponding to several different values of A_c and compare the map to observed channel-head positions in the DEM.

3.3 After completing the above exercise, use the FTCS or ADI technique to solve the 2D diffusion equation for the evolution of the hillslope pixels in the DEM. Assume that the channel network has a fixed elevation through time. This exercise will require using a mask grid to identify the channel pixels and to ensure that only hillslope pixels are varied through time.

3.4 Modify the `fillinpitsandflats` routine to map the area of all closed depressions in a DEM by including a counter that is incremented during each loop through the routine. Download a USGS DEM of an area that includes depressions (e.g. karst, glacial lakes, etc.) and use the modified routine to map the depressions.

3.5 Download a USGS Seamless DEM of an alluvial channel or alluvial fan. Use the successive flooding algorithm with MFD routing to model the inundated area associated with some hypothetical flood depth. This will require that you identify a pixel or set of pixels and prescribe an input flow depth.

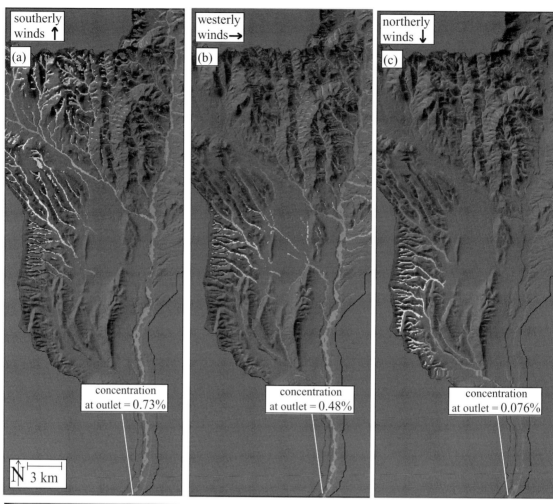

southerly winds ↑

(a)

westerly winds→

(b)

northerly winds ↓

(c)

concentration at outlet = 0.73%

concentration at outlet = 0.48%

concentration at outlet = 0.076%

N 3 km

tephra concentration

1%   10%   100%

Fig 3.15 Maps of tephra concentration in channel sediments following a hypothetical volcanic eruption at Yucca Mountain with (a) southerly, (b) westerly, and (c) northerly winds. Tephra concentration at the basin outlet varies from a maximum of 0.73% (for southerly winds) to a minimum value of 0.076% (for northerly winds) for this eruption scenario. Southerly winds result in higher concentrations because the high relief of the topography north of the repository is capable of mobilizing more tephra than other wind-direction scenarios. For color version, see plate section. Modified from Pelletier *et al.* (2008). Reproduced with permission of Elsevier Limited.

3.6 Develop a numerical model based on the MFD routing algorithm that estimates the time to peak discharge for a flash flood. Begin by initializing the flow grid with a uniform depth of runoff (e.g. 1 cm) for a DEM of your choice. As runoff is routed from each pixel to its neighbors (proceeding from high elevations to low elevations in rank order), estimate the transit time between successive grid points using a flow velocity calculated by Manning's equation or assuming critical flow. Assume no infiltration. At each pixel, multiple up-slope neighbors will contribute runoff, and each neighbor will have a different transit time. Adopt the transit time associated with the greatest incoming discharge. Integrate the transit times for each step to estimate the time to peak discharge at the basin outlet.

Chapter 4

The advection/wave equation

4.1 | Introduction

In this chapter we will focus on the advection equation, which in its simplest form is given by

$$\frac{\partial h}{\partial t} = c \frac{\partial h}{\partial x} \tag{4.1}$$

In general, advection equations involve a lateral translation of some quantity (e.g. temperature, topography, chemical concentration, etc.). The coefficient c in Eq. (4.1) has units of length over time and represents the speed at which the quantity h is advected. In the context of landform evolution, the advection equation is used to model retreating landforms, including cliffs, banks, and bedrock channel knickpoints. The advection equation is closely related to the classical wave equation which is given by

$$\frac{\partial^2 h}{\partial t^2} = c^2 \frac{\partial^2 h}{\partial x^2} \tag{4.2}$$

Equation (4.1) looks like the square root of the classical wave equation. In the classical wave equation, disturbances propagate as waves in both directions (i.e. $+x$ and $-x$). In the basic advection equation, disturbances propagate in one direction only. In 2D, the simple advection equation has the form

$$\frac{\partial h}{\partial t} = \vec{c} \cdot \vec{\nabla} h \tag{4.3}$$

In this case, the advection coefficient is a vector with components in the x and y directions.

Figure 1.7 contrasts the evolution of a hillslope or channel governed by the diffusion and advection models. In the diffusion model, which is applicable to certain transport-limited hillslopes and alluvial channels, the rate of erosion or deposition is proportional to the *change* in gradient, or curvature. In the diffusion model, material removed from the upper portion of the slope break is deposited on the lower portion, leading to a smoother topography over time. In the advection model, in contrast, the rate of erosion is directly proportional to elevation gradient, and deposition is not allowed. In the advection model, slope breaks undergo a lateral translation upstream with no change in shape. The basic stream-power model of bedrock channel evolution with $m = 1/2$ and $n = 1$:

$$\frac{\partial h}{\partial t} = K_w Q^{\frac{1}{2}} \frac{\partial h}{\partial x} \tag{4.4}$$

is simply an advection equation with spatially variable advection coefficient. Conceptually, the stream-power model says that the action of bedrock channel incision can be quantified by "advecting" the topography upstream with a local rate proportional to the square root of discharge. It should be noted that $\partial h/\partial t$ in Eq. (4.4) is always negative (i.e. erosion). Therefore, if x is the distance from the divide, $\partial h/\partial x$ is always negative and no negative sign is required in front of K_w.

Advection equations have many applications in Earth surface processes. The transport of heat, sediment, and chemical species in any geological fluid (e.g. the atmosphere, ocean, lakes, and rivers) includes some element of advective transport. Many physical processes combine effects of diffusion and advection. Heat is advected in the ocean, for example, as a result of well-established currents (e.g. the Gulf Stream). Superimposed on that net transport is a spreading of heat due to the cumulative effects of many small eddies superimposed on the current. This spreading is diffusive in nature. Not surprisingly, the advection equation features prominently in nearly all climate models, where the transport of heat and chemical species around the globe must be modeled with high precision. In the last few decades, some of the best new techniques for solving the advection equation (e.g. semi-Lagrangian methods) have originated in the climate and weather modeling communities.

4.2 | Analytic methods

The behavior of the simple advection equation (Eq. (4.1)) is to propagate disturbances in one direction with velocity c. As such, the analytic solution can be written as

$$h(x, t) = f(x + ct) \tag{4.5}$$

where $h(x, 0) = f(x)$ is the initial condition for h. In more complex advection equations where the advection coefficient is not constant but is a prescribed function of space and time, the method of characteristics can be used to determine analytic solutions in many cases. The idea of the method of characteristics is to take advantage of the lateral translation inherent in advection problems. Rather than using the local shape of the function to evolve the system forward in time (as in explicit approaches to the diffusion equation), the method of characteristics seeks to track the *trajectories* of individual segments of the solution laterally and through time. In many cases, these trajectories are described by simple functions that can be quantified analytically.

In general, the advection coefficient c in the advection equation is a function of space and time that coevolves with h. In the stream-power model, for example, the advection coefficient $K_w A^m$ varies through time as individual grid points compete for drainage. Thermally-driven convection is another example. In that case, the fluid flow advects the temperature field, and the

temperature field, in turn, controls the fluid flow through the effects of temperature on density and hence buoyancy. In such cases, numerical approaches are generally required.

The methods we will describe in this chapter can be used to solve a large set of equations known as "flux-conservative" equations. Flux-conservative equations have the form

$$\frac{\partial h}{\partial t} = -\frac{\partial F}{\partial x} \tag{4.6}$$

where F is the flux. Equation (4.6) reduces to the simple advection equation, for example, if $F = -ch$ and c is a constant.

4.2.1 Method of characteristics

The method of characteristics can be used to solve any linear advection equation and even some nonlinear advection equations. Here we present the method as it is applied to the general linear advection equation of the form

$$a(x, t)\frac{\partial h}{\partial x} + b(x, t)\frac{\partial h}{\partial t} + c(x, t)h = 0 \tag{4.7}$$

subject to the initial condition $h(x, 0) = f(x)$. The goal of the method of characteristics is to change the coordinates from (x, t) to a new coordinate system (x_0, s) in which the advection equation (a PDE) is transformed into an ODE along certain curves in the $x - t$ plane. If we choose a new coordinate s such that

$$\frac{dx}{ds} = a(x, t) \tag{4.8}$$

and

$$\frac{dt}{ds} = b(x, t) \tag{4.9}$$

then the advection equation becomes, in the new coordinate system,

$$\frac{dh}{ds} = \frac{dx}{ds}\frac{\partial h}{\partial x} + \frac{dt}{ds}\frac{\partial h}{\partial t} = a(x, t)\frac{\partial h}{\partial x} + b(x, t)\frac{\partial h}{\partial t} \tag{4.10}$$

Therefore, along the "characteristics curves" given by Eqs. (4.8) and (4.9), the advection equation becomes an ODE of the form

$$\frac{dh}{ds} + c(x, t)h = 0 \tag{4.11}$$

The strategy for applying this series of equations for specific linear advection equations is as

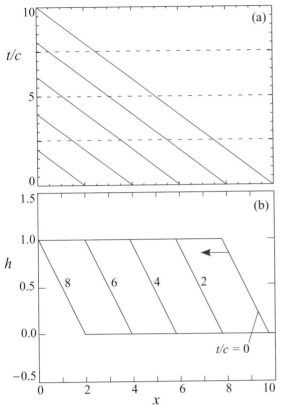

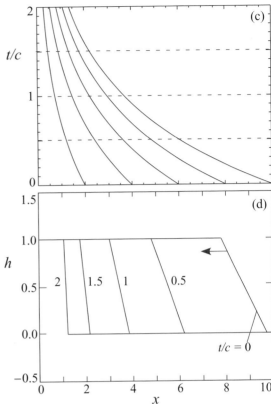

Fig 4.1 Application of the method of characteristics to solve Eqs. (4.1) and (4.12). (a) and (b) Plots of (a) characteristic curves and (b) model evolution for the simple advection equation (Eq. (4.1)) for a hypothetical channel knickpoint. (c) and (d) Plots of (c) characteristics curves and (d) model evolution for the advection equation with coefficient cx. In this case, the velocity of the knickpoint decreases and the gradient steepens as the knickpoint is propagated into the drainage headwaters.

follows. First, we solve the two characteristic equations, Eqs. (4.8) and (4.9), using integration. The constants of integration are constrained by setting $x(0) = x_0$ and $t(0) = 0$. Second, we solve the ODE (Eq. (4.11)) with the initial condition $h(0) = f(x_0)$, where x_0 are the initial points of the characteristic curves along the $t = 0$ axis in the $x - t$ plane. Third, we solve for s and x_0 in terms of x and t and substitute these values into $h(x_0, s)$ to get the solution of the original PDE.

First we consider the simple advection equation with constant coefficient c. In this case, $b(x, t) = 1$ and $c(x, t) = 0$. Therefore, the

characteristic equation (Eq. (4.9)) becomes $dt/ds = 1$, and integration gives the solution $t = s$. Next, the characteristic equation (Eq. (4.8)) becomes $x'(t) = c$ with the initial condition $x(0) = x_0$. The solution, by integration, gives the characteristic curves $x = x_0 + ct$. Next, the solution to the ODE (Eq. (4.11)), $h'(t) = 0$ with initial condition $h(0) = f(x_0)$, is $h(t) = f(x_0)$. The solution of the PDE is, transforming back to the original coordinates, $h(x, t) = f(x_0) = f(x + ct)$. Figure 4.1a plots the solution along with the characteristic curves. By writing the characteristic curves as $t = -x/c - x_0/c$, we can see that the characteristic curves are parallel lines with slope $-1/c$ in the $x - t$ plane.

Next we consider a case in which the advection coefficient varies with distance x. As a concrete example, consider the evolution of a bedrock channel evolving according to the stream-power model with $K_w Q^{1/2} = cx$, where x is distance from the divide:

$$\frac{\partial h}{\partial t} = cx \frac{\partial h}{\partial x} \tag{4.12}$$

This equation assumes that the square root of discharge increases linearly with distance from the divide, which is appropriate for a humid drainage system (i.e. $Q \propto A$) where drainage area A increases as the square of distance from the divide (i.e. an approximately semi-circular basin). Equation (4.12) corresponds to the general, first-order PDE with $b(x, t) = 1$, $c(x, t) = 0$, and $a(x, t) = cx$. As in the simple advection equation, the characteristic equation (Eq. (4.9)) yields $t = s$. The second characteristic equation to solve is $x'(t) = cx$ with initial condition $x(0) = x_0$. Solving it gives the characteristic curves $x = x_0 \exp(-cx)$. As before, $h(t) = f(x_0)$. Finally, the characteristic curves are given by $x = x_0 \exp(cx)$, so $x_0 = x \exp(-cx)$. Transforming back to the original variables gives $h(x, t) = f(x_0) = f(x \exp(-cx))$. Figure 4.1b illustrates solutions of this equation with the characteristic curves also shown. In this model, the knickpoint propagates more slowly and becomes steeper as it moves upstream into areas with lower stream power.

In the first set of papers to present the method of characteristics in the concept of landscape evolution, Luke (1972, 1974) showed how the method could be used to solve the nonlinear advection equation

$$\frac{\partial h}{\partial t} = c \left(\frac{\partial h}{\partial x} \right)^2 \tag{4.13}$$

for a knickpoint initial condition. Luke (1972, 1974) showed that in nonlinear equations like Eq. (4.13), the slope profile develops "shocks" due to the fact that certain slope segments are advected faster than others. In these nonlinear cases, the method of characteristics can be extended to incorporate these effects.

4.3 | Numerical methods

Methods for solving the advection equation can be classified as either Eulerian or Lagrangian. Eulerian methods use a fixed grid while Lagrangian methods use an adaptive grid that follows the characteristic curves of the system. In this book we will consider Eulerian methods exclusively. Although Lagrangian methods are useful for

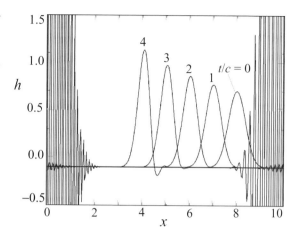

Fig 4.2 Instability of the FTCS method applied to the simple advection equation for a Gaussian (bell curve) initial condition (results obtained from modified version of Matlab code distributed by Marc Spiegelman). The solution becomes wildly unstable after the model has advected the Gaussian a distance of only a few times greater than its width.

some specific applications, Lagrangian methods are hard to use when the characteristic curves are complex. To understand how a Lagrangian method works, imagine the mixing of dye in a stirred fluid. As the dye becomes stretched and mixed into the fluid, the Lagrangian method produces a tortuous grid that follows the fluid. As the system evolves, determining neighboring grid points (to model viscous interactions, for example) requires complex bookkeeping. Eulerian methods are the most versatile, provide the same output format regardless of the specific model runs, and are the method of choice for most applications.

4.3.1 Failure of the FTCS method
In Chapter 2, we introduced the concept of finite differencing with the FTCS method. Applied to the basic advection equation, the FTCS method becomes

$$\frac{h_i^{n+1} - h_i^n}{\Delta t} = c \left(\frac{h_{i+1}^n - h_{i-1}^n}{2\Delta x} \right) \tag{4.14}$$

Unfortunately, the FTCS method is inherently unstable when it is applied to advection equations. Figure 4.2 illustrates the evolution of a Gaussian initial condition being advected to the right. In diffusion problems, the smoothing nature of the equations results in a stable method (for sufficiently small time steps). The advection equation lacks any inherent smoothness, therefore small numerical errors introduced by the FTCS method continually grow over time until they swamp the actual solution. Figure 4.2 also illustrates what numerical instabilities often look like: zig-zag functions that oscillate wildly from one grid point to the next.

4.3.2 Lax method

The instability of the FTCS method can be cured by a simple change: replace the term h_i^n in the time derivative by its average value:

$$h_i^n \rightarrow \frac{1}{2}(h_{i+1}^n + h_{i-1}^n) \qquad (4.15)$$

With this change, the FTCS discretization becomes

$$h_i^{n+1} = \frac{1}{2}(h_{i+1}^n + h_{i-1}^n) + \frac{c\Delta t}{2\Delta x}(h_{i+1}^n - h_{i-1}^n) \qquad (4.16)$$

This scheme turns out to be stable for sufficiently small time steps (i.e. $\Delta t \leq \Delta x/c$). The stability of the Lax method and other discretization schemes can be analyzed using the von Neumann stability analysis (Press *et al.*, 1992). The stability criterion $\Delta t \leq \Delta x/c$ (also called the Courant condition) is broadly applicable to many advection equations. In cases where c varies in space and time, the Lax method will be stable everywhere if the value of Δt is chosen to ensure stability where c has its largest value. Conceptually, the Courant condition ensures that changes in the value of h at one point on the grid travel along the grid more slowly than the advection velocity c. If the time step is too large, changes propagate along the grid faster than the advection velocity, and this unphysical condition will cause the solution to become unstable.

It turns out that there is a price to be paid for the stability introduced by the Lax method. The Lax method introduces a small but nonzero diffusive component to the solution. In a sense, the Lax method is not solving the advection equation at all, but rather an advection–diffusion equation. To see this, let's rewrite Eq. (4.16) so that

it has the same form as Eq. (4.14):

$$\frac{h_i^{n+1} - h_i^n}{\Delta t} = c\left(\frac{h_{i+1}^n - h_{i-1}^n}{2\Delta x}\right)$$
$$+ \frac{1}{2}\left(\frac{h_{i+1}^n - 2h_i^n + h_{i-1}^n}{\Delta t}\right) \qquad (4.17)$$

Equation (4.17) is the FTCS representation of the equation

$$\frac{\partial h}{\partial t} = c\frac{\partial h}{\partial x} + \frac{(\Delta x)^2}{2\Delta t}\frac{\partial^2 h}{\partial x^2} \qquad (4.18)$$

Equation (4.18) illustrates that the Lax method introduces a small diffusion term into the solution.

The problem of numerical diffusion turns out to be less than fatal, however. Smolarkiewicz (1983) noted that since we can quantify the magnitude of the numerical diffusion, we can make a correction to reverse its effect. We describe this "Smolarkiewicz correction" procedure below in the context of upwind-differencing. In addition, the numerical diffusion effect can be minimized by taking a half-step in time. This procedure is called the two-step Lax–Wendroff method.

4.3.3 Two-step Lax–Wendroff method

Thus far we have focused on the effects of different methods of spatial discretization. In all of our methods, the updating of h_i between time step n and $n+1$ happens in a single "jump." In this section we explore an implementation of the Lax method that uses two steps to perform the updating (i.e. the temporal discretization). In this intermediate step, one defines intermediate values $h_{i+1/2}$ at the half time steps $t_{n+1/2}$ and half grid points $x_{i+1/2}$. These intermediate values are calculated using the Lax method:

$$h_{i+1/2}^{n+1/2} = \frac{1}{2}(h_{i+1}^n + h_i^n) - \frac{\Delta t}{2\Delta x}(F_{i+1}^n - F_i^n) \quad (4.19)$$

Using Eq. (4.19), the user then calculates the fluxes at the half grid points and half time steps, $F_{i+1/2}^{n+1/2}$. Then, the values of h_i^{n+1} are calculated by the space-centered expression

$$h_i^{n+1} = h_i^n - \frac{\Delta t}{\Delta x}(F_{i+1/2}^{n+1/2} - F_{i-1/2}^{n+1/2}) \qquad (4.20)$$

The two-step Lax–Wendroff method eliminates most of the numerical diffusion of the Lax method.

4.3.4 Upwind-differencing method

Another simple but powerful numerical method for the advection equation is called upwind differencing. Upwind differencing involves calculating the slope along the direction of transport:

$$\frac{h_i^{n+1} - h_i^n}{\Delta t} = c_i^n \begin{cases} \dfrac{h_{i+1}^n - h_i^n}{\Delta x} & \text{if } c_i^n > 0 \\[2mm] \dfrac{h_i^n - h_{i-1}^n}{\Delta x} & \text{if } c_i^n < 0 \end{cases} \quad (4.21)$$

In the FTCS technique, the "centered-space" gradient is calculated by taking the difference between the value of the grid point to the left (i.e. at $i-1$) and the value to the right (i.e. at $i+1$) of the grid point being updated. This approach is prone to instability because it does not make use of the value at the grid point i itself. As such, the difference $h_{i-1}^n - h_{i+1}^n$ can be small even if the value of h_i^n is wildly different from the values on either side of it. In effect, the FTCS method creates two largely decoupled grids (one with even i, the other with odd i) that drift apart from each other over time. In the upwind method, that problem is corrected by calculating gradients using only one adjacent point (i.e. $h_{i-1}^n - h_i^n$ or $h_i^n - h_{i+1}^n$). Which direction to choose? If the flux of material is moving from left to right, then physically it makes sense that the value of h_i^{n+1} should depend on h_{i-1}^n, not on h_{i+1}^n. If the flux of material is in the opposite direction, h_i^{n+1} should depend on h_{i+1}^n. The upwind-differencing method implements this approach.

Upwind differencing also suffers from the numerical diffusion of the Lax method. In order to reverse the effects of numerical diffusion, we follow the procedure of Smolarkiewicz (1983). Given a small numerical diffusion characterized by the equation

$$\frac{\partial h}{\partial t} = \frac{\partial}{\partial x}\left(D_{num}\frac{\partial h}{\partial x}\right) \quad (4.22)$$

where D_{num} is a numerical diffusivity, the effects of numerical diffusion can be written as

$$\frac{\partial h}{\partial t} = \frac{\partial(V_d h)}{\partial x} \quad (4.23)$$

where

$$V_d = \begin{cases} -\dfrac{D_{num}}{h}\dfrac{\partial h}{\partial x} & \text{if } h > 0 \\[2mm] 0 & \text{if } h = 0 \end{cases} \quad (4.24)$$

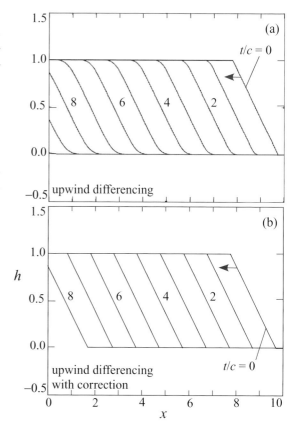

Fig 4.3 Solution to Eq. (4.1) for an initial condition of a hypothetical knickpoint (a) with upwind-differencing and (b) including the Smolarkiewicz correction. Results obtained using modified version of Matlab code distributed by Marc Spiegelman. Without the correction, the knickpoint in (a) gradually acquires a rounded top and bottom. With the correction, the initial knickpoint shape is preserved almost exactly as it is advected upstream.

This formulation shows that the effects of numerical diffusion can be written in terms of a diffusion velocity. The Smolarkiewicz correction involves taking a small advective step with anti-diffusion velocity $-V_d$ with the upwind differencing method following the original upwind differencing step (Eq. (4.21)) with velocity c_i^n. For the upwind-differencing method, $D_{num} = (|c_i^n|\Delta x - \Delta t(c_i^n)^2)/2$.

Figure 4.3 illustrates the solution to the basic advection equation with upwind differencing with and without Smolarkiewicz correction. The advantage of upwind differencing with Smolarkiewicz correction is that it represents a good compromise between accuracy and ease of implementation that makes it the method of choice for many advection problems. The primary limitation of the method is the relatively small time step (given by the Courant condition), which effectively limits the grid resolution that can be implemented in a reasonable amount of computing time.

In the context of landform evolution, solving the advection equation to model bedrock channel evolution is significantly simpler than many of the examples we have given here. In fluvial landscapes, slopes are "advected" upslope, so upwind differencing is simply a matter of calculating the local channel gradient in the downslope direction. The numerical diffusion of the upwind-differencing method still applies, however, so a Smolarkiewicz correction is needed for precise results. When solving advection problems more generally, such as the transport of heat in a complex flow field for long time periods, great care must be taken in order to determine the flux directions correctly in order to obtain stable, precise results.

4.4 | Modeling the fluvial-geomorphic response of the southern Sierra Nevada to uplift

In this section we consider a detailed case study of the application of bedrock incision models to modeling the propagation of knickpoints and escarpments at mountain-belt scales. This application illustrates the use of the upwind-differencing method combined with the MFD flow-routing method discussed in Chapter 3. This application also provides the motivation for an in-depth comparison of the behavior of the stream-power and sediment-flux-driven bedrock incision models. Codes for implementing the 2D bedrock drainage network model described in this section are given in Appendix 3.

The uplands of the Sierra Nevada are a slowly eroding plateau perched 1.5 km above narrow river canyons. If we assume that a low-relief landscape near sea level was uplifted recently and

that river knickpoints have not yet reached the upper plateau surface, the large-scale morphology of the Sierra Nevada suggests that 1.5 km of late Cenozoic surface uplift occurred (Clark et al., 2005). This conclusion has been challenged, however, by stable-isotope paleoaltimetry of Eocene gravels (Mulch et al., 2006) and the antiquity of the Sierra Nevadan rain shadow (Poage and Chamberlain, 2002), which both support the conclusion of a high (>2.2 km) Sierra Nevada in early Eocene time. A high early Eocene Sierra Nevada is not necessarily inconsistent with late Cenozoic surface uplift if the range was significantly denuded in the intervening period. Given the low erosion rates observed in the high Sierra today, however, these results are difficult to reconcile.

A large number of studies clearly indicate that significant late Cenozoic stream incision and rock uplift occurred throughout much of the Sierra Nevada. Unruh (1991) documented 1.4° of post-late-Miocene westward tilting of the Sierra Nevada based on the stratigraphy of the San Joaquin Valley. Stock et al. (2004, 2005) utilized cosmogenic isotope studies of cave sediments to document a pulse of relatively high channel incision rates (≈ 0.3 mm/yr) between 3 and 1.5 Ma in the South Fork Kings River and nearby drainages of the southern Sierra Nevada. What is currently unknown is whether significant surface uplift accompanied this late Cenozoic rock uplift. The relationship between rock uplift and surface uplift remains uncertain for two principal reasons. First, surface uplift triggers knickpoint retreat along mainsteam rivers, but the time lag between incision at the range front and incision tens of kilometers upstream is not well constrained. Thermochronologic data of Clark et al. (2005), for example, provide a 32-Ma maximum age for the onset of stream incision at their sample localities, but range-wide surface uplift could have occurred earlier. Second, rock uplift, stream-incision, and local surface uplift can all occur in the absence of range-wide surface uplift. As stream incision removes topographic loads from the crust, isostatic rebound raises slowly eroding upland plateau remnants to elevations much higher than the original, regionally extensive surface. As such, no simple relationship exists between the timing of local surface uplift and range-wide surface uplift.

Cosmogenic nuclide measurements provide important constraints on rates of landscape evolution in the Sierra Nevada. Small *et al.* (1997) and Stock *et al.* (2005) measured erosion rates of approximately 0.01 mm/yr on bedrock-exposed hillslopes of the Boreal Plateau. Riebe *et al.* (2000, 2001) calculated basin-averaged erosion rates of ≈ 0.02 mm/yr (largely independent of hillslope gradient) on the Boreal Plateau, increasing to ≈ 0.2 mm/yr in steep watersheds draining directly into mainstem river canyons. Stock *et al.* (2004, 2005) also measured mainstem incision rates along the South Fork Kings and nearby rivers as they incised into the Chagoopa Plateau. These authors found incision rates to be <0.07 mm/yr between 5 and 3 Ma, accelerating to 0.3 mm/yr between 3 and 1.5 Ma, and decreasing to ≈ 0.02 mm/yr thereafter. These studies provide direct constraints for parameters of hillslope and channel erosion in the numerical model.

It is not currently known whether water or sediment acts as the primary erosional agent in bedrock channels. The answer likely depends on the lithology of the eroding rock. Field studies, for example, indicate that sediment abrasion is likely to be most important in massive bedrock lithologies like granite, while water can play a large role in plucking blocks out of jointed sedimentary rocks (Whipple *et al.*, 2000). The classic method for quantifying bedrock channel erosion, the stream-power model, assumes that water is the primary erosional agent in bedrock channels.

Sklar and Dietrich (2001, 2004) proposed an alternative "saltation–abrasion" model in which bedrock channel erosion is controlled by the sediment flux delivered from upstream. A simplified version of their model is obtained by replacing drainage area with sediment flux in the stream-power model to obtain a sediment-flux-driven erosion model:

$$\frac{\partial h}{\partial t} = U - K_s Q_s^{\frac{1}{2}} \left| \frac{\partial h}{\partial x} \right| \qquad (4.25)$$

and introducing a new coefficient of erodibility K_s. This version of the sediment-flux-driven model incorporates the "tools" effect of saltating bedload but not the "cover" that occurs when channels are sufficiently overwhelmed with sediment to cause a reduction in erosion rates. Eq. (4.25) also does not explicitly include the effects of grain size on abrasional efficiency. Therefore, Eq. (4.25) is only an approximation to the considerable complexity of the saltation-abrasion process.

Figure 4.4 compares the behavior of two numerical landform evolution models incorporating stream-power erosion (Figures 4.4a–4.4f) and sediment-flux-driven erosion (Figures 4.4g–4.4l) for a vertically uplifted, low-relief plateau 200 km in width. Uplift occurs at a constant rate $U = 1$ m/kyr for the first 1 Myr of each simulation. The values of $K_w = 3 \times 10^{-4}$ kyr^{-1} and $K_s = 3 \times 10^{-4}$ (m kyr)$^{-1/2}$ used in these runs would predict identical steady state conditions if uplift was assumed to continue indefinitely. In the model, hillslope erosion rates are prescribed to be 0.01 mm/yr and flexural-isostatic rebound is included. Isostatic rebound was estimated by assuming regional compensation over a prescribed flexural wavelength and averaging the erosion rate over that wavelength. The flexural wavelength was assumed to be 200 km (the width of the model domain). The average erosion rate E_{av} was computed for each time step and isostatic rock uplift U_i rate was calculated using

$$U_i = \frac{\rho_c}{\rho_m} E_{av} \qquad (4.26)$$

where ρ_c and ρ_m are the densities of the crust and mantle, respectively.

First, consider the behavior of the stream-power model illustrated in Figures 4.4g–4.4l. Uplift initiates a wave of bedrock incision in which channel knickpoints propagate rapidly through the drainage basin. In this model, knickpoints reach the drainage headwaters after 25 Myr and the maximum elevation at that time is nearly 3 km (Figure 4.4l). Following 25 Myr, the range slowly erodes to its base level. In the sediment-flux-driven model (Figures 4.4a–4.4f), knickpoint migration occurs at far slower rates due to the lack of cutting tools supplied from the slowly-eroding upland plateau. In this model, knickpoints require 60 Myr to reach the drainage headwaters and the maximum elevation at that time is nearly 4 km. Incision of lowland channels

sediment-flux-driven model

stream-power model

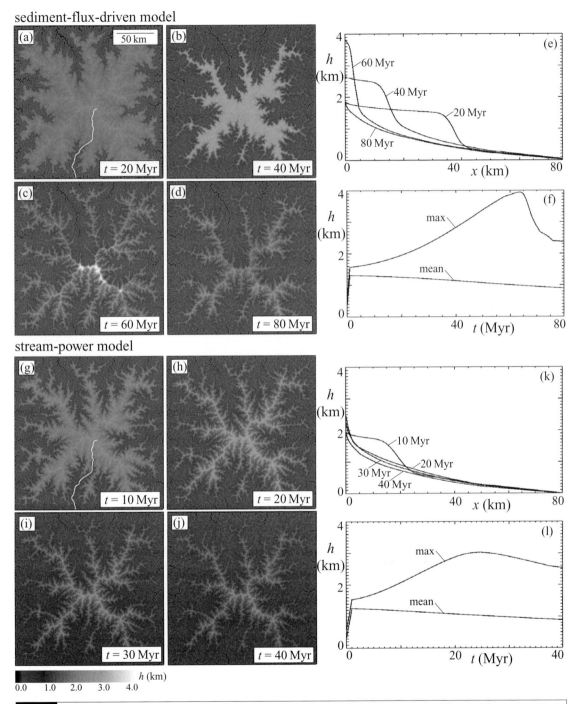

Fig 4.4 Model results comparing the (a)–(f) sediment-flux-driven and (g)–(l) stream-power models following 1-km uniform block uplift of an idealized mountain range. In the stream-power model, knickpoints rapidly propagate into the upland surface, limiting the peak elevation to ≈ 3 km at 25 Myr following uplift. In the sediment-flux-driven model, knickpoints propagate slowly due to the limited cutting tools supplied by the slowly eroding upland plateau, which raises plateau remnants to nearly 4 km 60 Myr following uplift. Modified from Pelletier *et al.* (2007c). Reproduced with permission of Elsevier Limited.

drives regional rock uplift in both models. In the sediment-flux-driven model, however, the low rates of channel incision on the upland plateau cause surface uplift of the relict surface remnants to higher elevations than in the stream-power model. The magnitude of this flexural-isostatic rebound effect is equal to the inverse of the relative density contrast between the crust and the mantle, $\rho_c/(\rho_m - \rho_c)$, which is between 4 and 5 for typical continental crust (Turcotte and Schubert, 2002). Typical flexural wavelengths in mountain belts are 100–200 km. Therefore, if the extents of individual surface remnants are equal to or smaller than this wavelength, canyon cutting nearby will cause substantial local surface uplift of these remnants up to four times higher than that of the original, regionally extensive plateau, even as the mean elevation of the range is decreasing. In the limit where the upland erosion rate goes to zero, a vertical escarpment would form in the sediment-flux-driven model that would retreat laterally at a rate equal to the rate of bedrock weathering (i.e. typically < m/kyr). At such rates, knickpoint retreat of 100 km requires more than 100 Myr. Appendix 3 provides codes for implementing the sediment-flux-driven model with flexural-isostatic rebound.

Given the predominantly granitic lithology of the Sierra Nevada, the results of the sediment-flux-driven model are most relevant to this application. Here we assume an initially low-relief, low-elevation landscape similar to the Boreal Plateau of the modern high sierra. However, no explicit assumption is made about the timing of initial uplift of the range. The initial topography input to the model respects the modern drainage architecture of the southern Sierra Nevada but replaces the modern stepped topography with a uniformly low-relief, low-elevation landscape (Figure 4.5a). To construct the initial topography, downstream flow directions were first computed using a 500-m resolution DEM of the southern Sierra Nevada. Second, a constant base-level elevation of 200 m a.s.l. was identified based on the modern range-front elevation. Starting from pixels in the DEM with this base-level elevation, DEM pixels with higher elevations were then "reset," enforcing a uniform hillslope gradient of 5%. The initial model topography produced in this way

has a mean elevation of approximately 0.5 km, a maximum elevation of 1 km, and local relief similar to that of the modern Boreal Plateau. All pixels draining to areas north, east, and south were "masked" from the model domain so that only the evolution of west-draining rivers was considered. The drainage divide at the crest was treated as a free boundary.

Hillslopes and channels in the model are distinguished on the basis of a stream-power threshold (Tucker and Bras, 1998; Montgomery and Dietrich, 1999). If the product of slope and the square root of contributing area is greater than a prescribed threshold given by $1/X$, where X is the drainage density, the pixel is defined to be a channel, otherwise it is a hillslope. The value of X was chosen to be 0.005 km^{-1} on the basis of observed drainage densities on the upland Boreal Plateau observed on a 30-m resolution DEM. Hillslope erosion in the model occurs at a prescribed rate E_h independent of hillslope gradient. Diffusion or nonlinear-diffusion models are commonly used to model hillslope evolution in numerical models, but cosmogenic erosion rates measured on the Boreal Plateau (Riebe et al., 2000) support the slope-independent erosion model used here. The value of E_h is constrained by cosmogenic erosion rates on bare bedrock surfaces to be approximately 0.01 mm/yr (Small et al., 1997; Stock et al., 2004, 2005). This value applies only to the upland low-relief surface, but in practice hillslope pixels in the model only occur on the upland landscape because of the large gradients and/or drainage areas that develop in the incised river canyons below the upland landscape. Hillslopes play an important role in the sediment-flux-driven model by supplying sediment to propagate knickpoints, but the model results are not sensitive to the particular value of X as long as it is within a reasonable range of 0.002 km^{-1} to 0.01 km^{-1}.

Bedrock channel erosion occurs in all channel pixels during each time step of the model according to Eqs. (4.4) or (4.25) implemented with upwind differencing. Active uplift in the model was assumed to occur as uniform, vertical, block uplift of $U = 1$ m/kyr during each uplift pulse. Slope–area relationships in steady-state bedrock rivers indicate that the ratio $m/n = 0.5$ is typical.

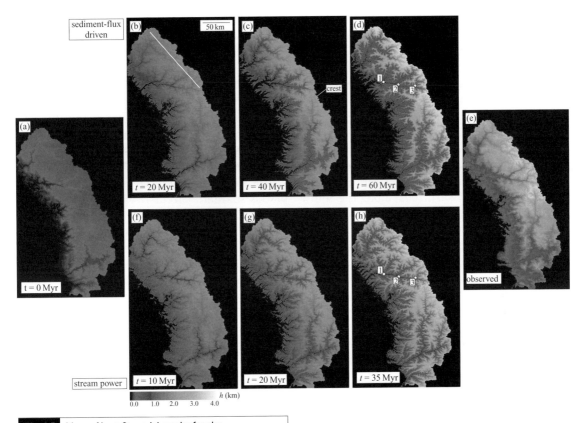

Fig 4.5 Maps of best-fit model results for the sediment-flux-driven model (b)–(d) and stream-power model (f)–(h) starting from the low-relief, low-elevation surface illustrated in (a). The actual modern topography is shown in (e). The best-fit uplift history for the sediment-flux-driven model occurs for a 1-km pulse of uplift starting at 60 Ma ($t = 0$) and a 0.5 pulse starting at 10 Ma ($t = 50$ yr out of a total 60 Myr). The best-fit uplift history for the stream-power model occurs for a 1-km pulse of uplift at 35 Ma and a 0.5 km pulse at 7 Ma. Also shown are transect locations and stream location identifiers corresponding to the erosion-rate data in Figure 4.6. For color version, see plate section. Modified from Pelletier *et al.* (2007c). Reproduced with permission of Elsevier Limited.

This m/n ratio implicitly includes the effect of increasing channel width with increasing drainage area. In this model experiment, n values of 1 and 2 were used, scaling m accordingly to maintain a constant $m/n = 0.5$ ratio. Sediment flux is computed in the sediment-flux-driven model using erosion rates computed during the previous time step. Sediment flux and drainage area are both routed downstream on hillslopes and in

channels using a bifurcation-routing algorithm that partitions incoming fluxes of sediment or drainage area between each of the downstream neighboring pixels, weighted by slope. Time steps in the model are variable (i.e. small time steps are needed when stream gradients are high following uplift in order to ensure numerical stability) but range from about 1 to 10 kyr.

Geomorphic and stratigraphic observations suggest that isostatic rock uplift takes place by westward tilting along a hinge line located west of the range front. Sediment loading of the San Joaquin Valley and the presence of a major high-angle fault along the eastern escarpment suggest that isostatic rebound is a maximum at the range crest and decreases uniformly towards the hinge line (Chase and Wallace, 1988). Therefore, we assume here that isostatic uplift has a mean value equal to Eq. (4.26) but is spatially distributed using a linear function of distance from the range crest:

$$U_i = \frac{L}{<x_c>} \left(1 - \frac{x_c}{L}\right) \frac{\rho_c}{\rho_m - \rho_c} E_{av} \qquad (4.27)$$

where x_c is the distance from the range crest along an east–west transect, L is the distance between the crest and the hinge line, and $< x_c >$ is the average distance from the range crest. For the results of this section I assumed $L = 100\,\text{km}$ (Unruh, 1991) and $\rho_c(\rho_m - \rho_c) = 5$.

Calibration of K_w and K_s was performed for each uplift history by forward modeling the response to a late Cenozoic pulse of uplift and adjusting the values of K_w and K_s to match the maximum incision rate of 0.3 mm/yr measured by Stock *et al.* (2004, 2005) along the South Fork Kings River. For each uplift history considered, the values of K_s and K_w were varied by trial and error until the maximum value of late Cenozoic incision matched the measured value of 0.3 mm/yr on the South Fork Kings River to within 5%. Late Cenozoic incision rates in the model increase monotonically with K_s and K_w, so trial values of K_s and K_w were raised (or lowered) if predicted incision rates were too low (or high). All uplift histories considered in the model experiments assumed two pulses of range-wide surface uplift with the late Cenozoic phase constrained to occur on or after 10 Ma and on or before 3 Ma to be consistent with firm constraints imposed by previous studies. The magnitude of uplift during each pulse was tentatively constrained using the offset of the two major river knickpoints of the North Fork Kern River, which suggests an initial pulse of $\approx 1\,\text{km}$ surface uplift corresponding to the upstream knickpoint, followed by a smaller ($\approx 0.5\,\text{km}$) pulse corresponding to the downstream knickpoint. Forward modeling indicates that knickpoint relief during the model remains nearly equal to the magnitude of the range-wide surface uplift that triggered knickpoint creation except for late stages of the model when the uplift of isolated plateau remnants can greatly outpace stream incision. Best-fit uplift histories were determined by visually comparing the model-predicted topography with those of the modern Sierra Nevada, including the extents of the Chagoopa and Boreal Plateaux and the major river knickpoints.

Figures 4.5b–4.5d illustrate results of the best-fit sediment-flux-driven model experiment, with $n = 1$ and $K_s = 2.4 \times 10^{-4}\,(\text{m kyr})^{-1/2}$. In this experiment, a 1-km pulse of late Cretaceous surface uplift was imposed from time $t = 0$ to 1 Myr (out of a total duration of 60 Myr) and a second 0.5-km pulse of late Cenozoic surface uplift was imposed from $t = 50$ to 50.5 Myr. Model experiments with $n = 2$ were also performed and yielded a similar best-fit uplift history, but the knickpoint shapes in that model were a poor match to the observed knickpoint shapes along the North Fork Kern River. Figures 4.5b–4.5d present color maps of the model topography at $t = 20$, 40, and 60 Myr. Figure 4.5e illustrates the modern topography of the southern Sierra Nevada with the same color scale for comparison. Figure 4.6a plots channel incision rate at three points along the Kings and South Fork Kings Rivers as a function of time. Location 2 is of particular significance since that point coincides with the South Fork Kings River where Stock *et al.* (2004, 2005) measured maximum incision rates of 0.3 mm/yr. Each point along the Kings River experiences two pulses of incision as knickpoints propagate upstream as topographic waves. Figure 4.6a indicates that incision rates decrease with increasing distance upstream and that knickpoint propagation is slower following the first phase of uplift compared to the second phase of uplift. Both of these behaviors can be associated with the "tools" effect of the sediment-flux-driven model. First, larger drainages have more cutting tools to wear down their beds, thereby increasing incision rates and knickpoint migration rates. Second, knickpoint propagation occurs slowly following the first pulse of uplift because of the lack of cutting tools supplied by the low-relief upland plateau. Figure 4.5b, in particular, illustrates that canyon cutting has affected only a very small percentage of the landscape even as late as 20 Myr following the initial uplift. In contrast, knickpoint and escarpment retreat following the second phase of uplift is enhanced by the upstream sediment supply associated with the first phase of uplift. As a result, the second knickpoint travels almost as far upstream as the first knickpoint in less than 20% of the time.

The results of Figures 4.5b–4.5d and Figures 4.6a–4.6b also provide several consistency checks that aid in confidence building. First, the maximum incision rate observed by Stock *et al.* (2004, 2005) from 3 to 1.5 Ma along the South Fork Kings

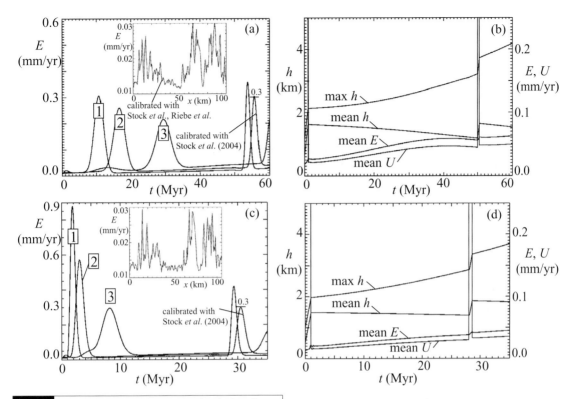

Fig 4.6 Plots of erosion rate versus time for the best-fit sediment-flux-driven model (a) and (b) and stream-power model (c) and (d). (a) Plot of erosion rate versus time for three locations along the Kings and South Fork Kings River (locations in Figure 4.5), illustrating the passage of two knickpoints corresponding to the two uplift pulses. The late-Cenozoic pulse of incision at location 2 provides a forward-model calibration based upon the cosmogenically derived incision rates (Stock et al., 2004, 2005) at this location. The maximum value of the erosion rate at this location is matched to the observed rate of 0.3 mm/yr by varying the value of K_s. Inset into (a) is the upland erosion rate along the transect located in Figure 4.5b, illustrating minimum erosion rates of 0.01 mm/yr and maximum values of 0.03 mm/yr, consistent with basin-averaged cosmogenic erosion rates of ≈ 0.02 mm/yr. (b) Maximum and mean elevations of the model, illustrating the importance of isostatically-driven uplift in driving peak uplift. (c) and (d) Plots of erosion rate and elevation versus time for the stream-power model, analogous to (a) and (b). Modified from Pelletier et al. (2007c). Reproduced with permission of Elsevier Limited.

rates before and after knickpoint passage. The inset graph in Figure 4.5d plots the erosion rate of the upland plateau along a linear transect shown in Figure 4.5b. The hillslope erosion rate E_h was prescribed to be 0.01 mm/yr based on measured cosmogenic erosion rates (Small et al., 1997; Stock et al., 2004, 2005). The plot in Figure 4.6a shows a range of erosion rates on the upland surface from a minimum of 0.01 mm/yr (on hillslopes) to a maximum value of 0.03 (in channels). Erosion rates in upland plateau channels are controlled by both E_h and K_s. The fact that the model predicts an average erosion rate of ≈ 0.02 mm/yr, consistent with basin-averaged rates measured cosmogenically (Riebe et al., 2000, 2001), provides additional confidence in this approach.

Figure 4.6b plots maximum and mean elevation values as a function of time in the model. Mean surface elevation increases during active uplift but otherwise decreases. Maximum elevation continually increases through time to a final value of over 4 km as a result of isostatic uplift of Boreal Plateau remnants driven by canyon cutting downstream. Mean erosion and uplift rates are also plotted in Figure 4.5e, showing

River was preceded and followed by much lower rates of incision (≈ 0.02–0.05 mm/yr) from 5 to 3 Ma and 1.5 Ma to present. The model predicts the same order-of-magnitude decrease in incision

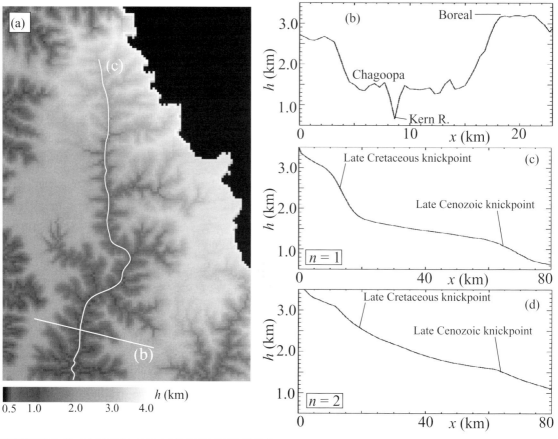

Fig 4.7 Best-fit sediment-flux-driven model results for the North Fork Kern River basin (i.e. 1 km of late Cretaceous uplift and 0.5 km of late-Cenozoic uplift), illustrating broadly similar features to those of the actual North Fork Kern River basin illustrated in Chapter 1. Modified from Pelletier *et al.* (2007c). Reproduced with permission of Elsevier Limited.

a gradual increase in mean erosion rate over a 40 Myr period to a maximum value of approximately 0.05 mm/yr. Isostasy replaces 80% of that erosion as rock uplift, as prescribed by Eq. (4.27). Figure 4.7 illustrates the ability of the model to predict details of the Sierra Nevada topography using the North Fork Kern River as an example. The model predicts the elevations and extents of the Chagoopa and Boreal Plateaux as well as the approximate locations and shapes of the two major knickpoints along the North Fork Kern River.

The best-fit results of the stream-power model are shown in Figures 4.5f–4.5h and 4.6c–4.6d, with $n = 1$ and $K_w = 8 \times 10^{-5}\,\mathrm{kyr}^{-1}$. Uplift in

this model occurs as two pulses: a 1-km pulse in the late Eocene ($t = 35\,\mathrm{Ma}$) and a 0.5-km pulse in the late Miocene ($t = 7\,\mathrm{Ma}$). Figure 4.6c illustrates the knickpoints as they pass points along the Kings and South Fork Kings River. In this model, knickpoint migration occurs at a similar rate in the first and second pulses of uplift, as illustrated by the equal durations between knickpoint passage following the first and second pulses. The rates of vertical incision are lower in the second phase because the knickpoint has a gentler grade. Model results are broadly comparable to those of the sediment-flux-driven model, except that the propagation of the initial knickpoint requires only about half as long as the sediment-flux-driven model to reach its present location.

The results of this model application support two possible uplift histories corresponding to the best-fit results of the sediment-flux-driven and stream-power models. Only one of these histories is consistent with the conclusion of high (> 2.2 km) Sierra Nevada in early Eocene time (Poage and Chamberlain, 2002; Mulch et al., 2006), however. That constraint, taken together with the fact that sediment abrasion is the dominant erosional process in massive granitic rocks (Whipple et al., 2000) support the greater applicability of the sediment-flux-driven model in this case.

The results of this section are consistent with the basic geomorphic interpretation of Clark et al. (2005) that the upland plateaux and associated river knickpoints of the southern Sierra Nevada likely record two episodes of range-wide surface uplift totaling approximately 1.5 km. The results presented here, however, suggest that the initial 1-km surface uplift phase occurred in late Cretaceous time, not late Cenozoic time. In the sediment-flux-driven model, the slow geomorphic response to the initial uplift phase is caused by the lack of cutting tools supplied by the slowly-eroding upland Boreal Plateau. If the behavior of the sediment-flux-driven model is correct, the model results suggest that 32 Myr (i.e. the time since onset of Sierra Nevadan uplift, according to Clark et al. (2005)) does not afford enough time to propagate the upland knickpoint to its present location. The self-consistency of the model results provide confidence in this interpretation. The model correctly reproduces details of the modern topography of the range, including the elevations and extents of the Chagoopa and Boreal Plateaux and the elevations and shapes of the major river knickpoints.

4.5 | The erosional decay of ancient orogens

One of the great paradoxes in geomorphology concerns the "persistence" of ancient orogens. Analyses of modern sediment-load data imply that mountain belts should erode to nearly sea level over time scales of tens of millions of years. Several mountain belts that last experienced active tectonic uplift in the Paleozoic (e.g. Appalachians, Urals) still have mean elevations at or near 1 km in elevation. In this section we explore this question using 1D and 2D modeling of bedrock channel and coupled bedrock-alluvial channel modeling using the stream power and sediment-flux-driven erosion models.

Analyses of sediment-load data have shown that erosion rates are approximately proportional to the mean local relief or mean elevation of a drainage basin. In tectonically inactive areas, for example, Pinet and Souriau (1988) obtained

$$E_{av} = 0.61 \times 10^{-7} H \qquad (4.28)$$

where E_{av} is the mean erosion rate in m/yr and H is the mean drainage basin elevation in m. Similar studies have documented approximately linear correlations between mean erosion rates and mean local relief, relief ratio, and basin slope both within and between mountain belts (e.g. Ruxton and MacDougall, 1967; Ahnert, 1970; Summerfield and Hulton, 1994; Ludwig and Probst, 1998). Equation (4.28) can be expressed as a differential equation for mountain-belt topography following the cessation of tectonic uplift:

$$\frac{\partial H}{\partial t} = -\frac{1}{\tau_d} H \qquad (4.29)$$

with $\tau_d = 16$ Myr using the correlation coefficient in Eq. (4.28). Equation (4.29) has the solution $H = H_0 e^{-t/\tau_d}$, where H_0 is the mean basin elevation immediately following uplift. Isostatic rebound will increase τ_d by a factor of $\rho_c/(\rho_m - \rho_c)$, where ρ_m and ρ_c are the densities of the mantle and crust, thereby giving $\tau_d \approx 50$–70 Myr. This value is about five times smaller than the age of Paleozoic orogens, several of which (e.g. Appalachians, Urals) still stand to well over 1 km in peak elevation. This is the paradox of persistent mountain belts.

One possible reason for persistent mountain belts is the role that piedmonts play in raising the effective base level for erosion. As a mountain belt ages, sediment can be deposited at the footslope of the mountain, causing the base level of the bedrock portion of the mountain belt to increase (Baldwin et al., 2003; Pelletier, 2004c). To explore this effect we need to develop models for the coupled evolution of bedrock channels and alluvial piedmonts. Here we consider a model that couples the stream power model (Eq. (4.4)) to a simple model for alluvial-channel

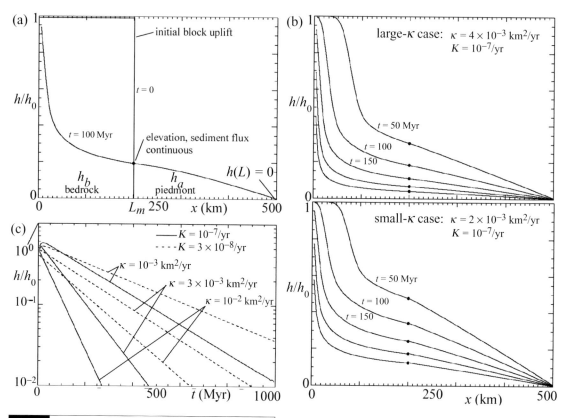

Fig 4.8 (a) Model geometry and key variables. (b) Plots of elevation vs. downstream distance for two values of κ. Each plot shows a temporal sequence from $t = 50$ Myr–250 Myr in 50 Myr intervals. (c) Plots of mean basin elevation vs. time for $L_m = 100$ km, $L = 500$ km, and a range of values of K and κ. From Pelletier (2004c).

evolution based on the diffusion equation (Paola et al., 1992):

$$\frac{\partial h}{\partial t} = \kappa \frac{\partial^2 h}{\partial x^2} \qquad (4.30)$$

where κ is the piedmont diffusivity. The diffusion equation is a highly simplified model of piedmont evolution, but enables us to make preliminary conclusions regarding the role of piedmont aggradation on the time scale of mountain-belt denudation.

The model geometry and boundary conditions are illustrated in Figure 4.8a. h_a and h_b refer to the alluvial and bedrock portions of the profile. Tectonic uplift is assumed to occur as a rigid block between $x = 0$ and $x = L_m$ prior to $t = 0$. h_0 is the maximum elevation immediately following the cessation of active uplift. This boundary

condition is applied a small distance L_h from the divide (i.e. the hillslope length) to give $h_b(L_h, 0) = h_0$. Sea level is assumed to be constant, giving $h_a(L, t) = 0$. The two remaining boundary conditions are continuity of elevation and sediment flux at the mountain front, $x = L_m$:

$$h_b(L_m) = h_a(L_m), \int_{L_h}^{L_m} A_r \frac{\partial h_b}{\partial t} dx = \kappa \left[\frac{\partial h_a}{\partial x} \right]_{x=L_m}$$

$$(4.31)$$

Equations (4.12) and (4.30) were solved using upwind differencing and FTCS techniques for the bedrock and alluvial portions of the basin, respectively. Codes for modeling coupled bedrock-alluvial channel evolution are given in Appendix 3. The effect of the piedmont diffusivity value on mountain-belt denudation is illustrated in Figure 4.8b. In these two experiments, the model parameters are identical ($K = 10^{-7}$ yr^{-1}, $L_m = 200$ km, $L = 500$ km) except for the values of κ, which differ by a factor of 2. A smaller value of κ results in a steeper piedmont, lower bedrock relief, and a smaller denudation rate. After 250 Myr, the small-κ case has a mean elevation about twice as large as the large-κ case. Varying the piedmont length L for a fixed value of κ has a similar effect.

Figure 4.8c illustrates the decline in mean basin elevation with time on a semi-log plot for a piedmont of fixed length ($L = 500$ km) and a range of values of K and κ. Following an initial rise to a maximum value, the decay in mean basin elevation is exponential in each case, but varies as a function of both the bedrock and piedmont parameters, illustrating their dual control on the tempo of denudation.

In nature, values of K, κ, L_m, and L vary from one mountain belt to another. Persistent mountain belts can be expected to have extreme values of one or more of these parameters, resulting in large values of τ_d compared to the global average. L and L_m can be determined from topographic and geologic maps. Values of K and κ are not well constrained, but qualitative data on climate, rock type, and sediment texture can be used to vary those parameters around representative values.

The effects of unusually resistant bedrock, for example, are illustrated in the Appalachian-type example of Figure 4.9a. Hack (1957, 1979) argued that the resistant quartzite of the Appalachian highlands was a key controlling factor in its evolution. In this example, the values of L_m and L were set to 100 km and 500 km. The values $K = 5 \times 10^{-8}$ yr^{-1} and $\kappa = 6 \times 10^{-4}$ km^2/yr led to profiles most similar to those observed (Figure 4.9a). In contrast, Figure 4.8b illustrates the Ural-type example (i.e. an unusually broad piedmont). Here, L_m and L were set to 600 km and 2000 km. The value of K most consistent with observed profiles in the Urals ($K = 10^{-7}$ yr^{-1}, $\kappa = 2 \times 10^{-2}$; Figure 4.9b) is a factor of two greater than the value for the Appalachians. The Ural example shows that a broad piedmont can diminish bedrock relief, reducing the average basin denudation rate compared to a narrower piedmont of the same slope. The eastern Appalachian piedmont is relatively modest in size, so resistant bedrock is the most likely explanation for its persistence and strongly concave headwater profiles. The values of K and κ in Figure 4.9 are not unique; a range of values is consistent with the observed profiles.

The model equations can also be solved analytically for the decay phase of mountain-belt evolution using separation of variables. Power-law and sinusoidal solutions are obtained for the bedrock and alluvial components, and a transcendental equation is obtained for τ_d:

$$h_b(x, t) = h_0 \left(\frac{L_h}{x} \right)^{A_r/(K\tau_d)} e^{-t/\tau_d} \tag{4.32}$$

$$h_a(x, t) = h_0 \left(\frac{L_h}{L_m} \right)^{A_r/(K\tau_d)} \frac{\sin\left(\frac{L-x}{\sqrt{\kappa\tau_d}} \right)}{\sin\left(\frac{L-L_m}{\sqrt{\kappa\tau_d}} \right)} e^{-t/\tau_d} \tag{4.33}$$

$$\frac{\sqrt{\kappa\tau_b}}{L_m} \left(1 - \frac{A_r}{K\tau_d} \right) = \tan\left(\frac{L - L_m}{\sqrt{\kappa\tau_b}} \right) \tag{4.34}$$

Equation (4.34) cannot be further reduced and must be solved numerically for specific values of the model parameters. As an example, for the model parameters of the Appalachian-type case of Figure 4.9a, Eq. (4.34) gives $\tau_b = 186$ Myr. In the limit of no piedmont (i.e. $L_m \to L$), Eq. (4.34) has the appropriate limiting behavior $\tau_b \to A_r/K$.

Shepard (1985) found actual bedrock profiles in nature to be best fit by a power law when slope was plotted versus downstream distance, consistent with Eq. (4.32). The power-law exponent in the model is a function of the bedrock erodibility (more resistant bedrock leads to more concave profiles) but the exponent is also a function of the basin diffusivity, κ (through τ_d), illustrating the explicit coupling of the bedrock and piedmont profile shapes. This result suggests that piedmont deposition must be included to correctly model bedrock drainage basin evolution. The sine function is a poor approximation to many observed piedmont profiles, but this discrepancy is largely the result of neglecting downstream fining. A more precise piedmont form could be achieved by making κ a function of downstream distance.

Montgomery and Brandon (2002) recently questioned the linear relationship between mean erosion rate and mean elevation (or mean local relief) that lies at the heart of the persistent mountain belt paradox. Their study is best known for documenting a rapid increase in erosion rates in terrain with a mean local relief greater than 1 km. However, these authors also expanded upon earlier data sets that established a linear relationship between mean erosion rate and mean local relief. Montgomery and Brandon (2002) found that the expanded data set was best

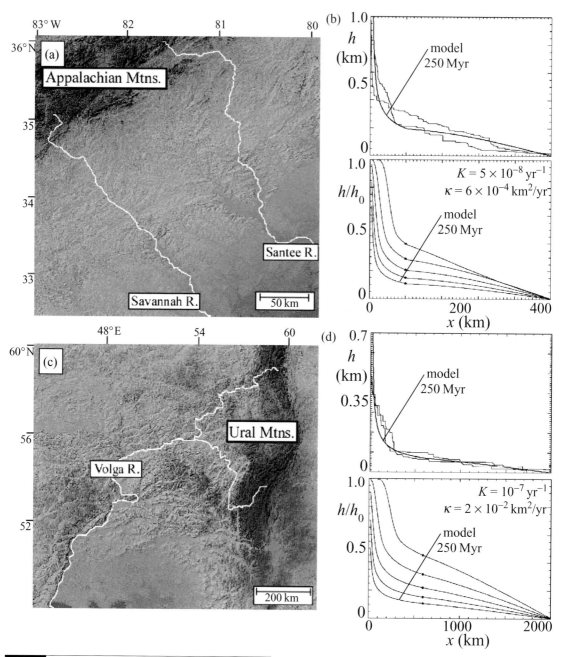

Fig 4.9 (a) and (b) Appalachian-type example. (b) Longitudinal profiles of channels in the Santee and Savannah drainage basins (location map in (a)), plotted with a model profile ($K = 5 \times 10^{-8}$ yr^{-1}, $\kappa = 6 \times 10^{-4}$ km^2/yr) at $t = 250$ Myr for comparison. Evolution of the model profile in intervals of 50 Myr is also shown. (c) and (d) Urals-type example. (d) Profiles in the Volga drainage basin (location map in (c)), plotted with a model profile ($K = 10^{-7}$ yr^{-1}, $\kappa = 2 \times 10^{-2}$ km^2/yr) for comparison. From Pelletier (2004c).

represented by a power law:

$$E_{av} = 1.4 \times 10^{-6} R_z^{1.8} \tag{4.35}$$

where E_{av} is in units of mm/yr and R_z is the mean local relief (defined as the average difference between the maximum and minimum elevation over a 10 km radius) in m. This result necessitates a reassessment of the persistent mountain-belt paradox.

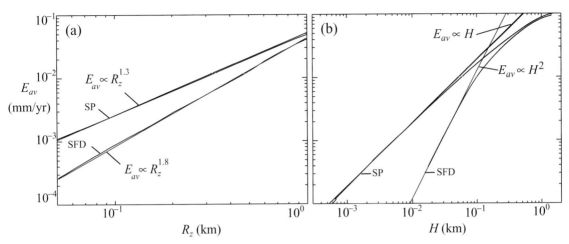

Fig 4.10 (a) Plots of mean erosion rate vs. mean local relief following uplift for the stream-power and sediment-flux-driven models. The stream-power model closely approximates a power-law relationship with an exponent of 1.3 (plots closely overlap), while the sediment-flux-driven model also follows a power law but with an exponent of 1.8. (b) Plots of mean erosion rate versus mean elevation following uplift for the stream-power and sediment-flux-driven models. Following a transient phase, erosion rates in the stream-power model are proportional to mean elevation H while in the sediment-flux-driven model they are proportional to the square of H.

In order to determine the relationships between erosion rate, mean local relief, mean elevation, and specific models of bedrock channel erosion, we must work with a model designed to use the stream-power model and a sediment-flux-driven model interchangeably. The model we will consider is similar to that of Figure 4.4 but it assumes that each pixel in the model is above the threshold for channelization. As such, it does not include hillslope processes. The stream-power version of the model uses Eq. (4.4) while the sediment-flux-driven version uses Eq. (4.25). The values of m and n used in the model are 0.5 and 1.0, respectively (Kirby and Whipple, 2001). Isostatic rebound in the model is estimated using Eq. (4.26). Sediment flux Q_s in Eq. (4.25) is computed in the model by downslope routing of the local erosion rate computed during the previous time step, multiplied by the pixel area.

Model runs reported here use a square domain of width 128 km subject to uniform vertical uplift of 0.05 mm/yr for the first 100 Myr of the simulation (sufficient to develop steady state), followed by a long period of erosional decay and isostatic response to erosion. The values of $K = 2 \times 10^{-3}\,\mathrm{kyr}^{-1}$ and $K_s = 4 \times 10^{-2}\,(\mathrm{m\,kyr})^{-1/2}$ used in these runs predict identical steady state conditions at the cessation of tectonic uplift at 100 Myr. Figure 4.10a plots the relationship between mean erosion rate E_{av} and mean local relief for the two models following tectonic uplift. The sediment-flux-driven model closely follows the power-law trend documented by Montgomery and Brandon (2002) with an exponent of 1.8 (model results and power-law trend strongly overlap). The stream-power model also follows a power-law trend but with a smaller exponent of 1.3. Figure 4.10b plots the relationships between mean erosion rate and mean elevation for the stream-power (SP) and sediment-flux-driven (SFD) models. Following an initial transient phase, mean erosion rates in the stream-power model are proportional to mean elevation while in the sediment-flux-driven model they are proportional to the square of mean elevation. The fact that the trends between mean erosion rate and mean local relief differ from those between mean erosion rate and mean elevation indicates that declines in mountain-belt relief and mean elevation are not identical. For the purposes of modeling erosional decay it is mean elevation, not relief, that is most significant because mean erosion rate is the derivative of mean elevation with respect to time, while erosion rates and relief are not as simply related. Nevertheless, observed relationships between

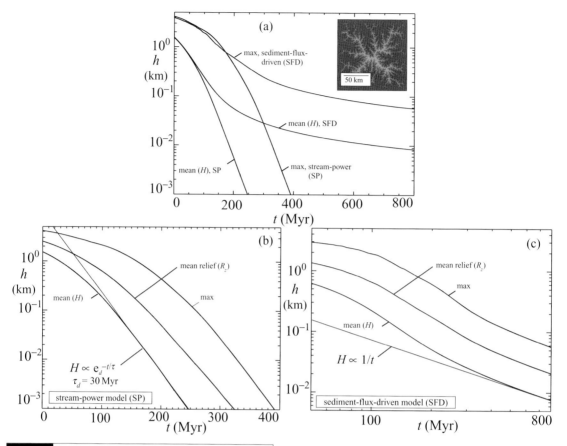

Fig 4.11 (a) Semi-log plots of maximum and mean elevation versus time following the cessation of tectonic uplift for the stream-power and sediment-flux-driven model of a uniformly uplifted square domain of width 128 km. The bedrock erodibilities of each model have been scaled to predict the identical steady-state landscapes. The sediment-flux-driven model exhibits persistent mountain belts relative to the exponential decay of the stream-power model. (b) Semi-log plots of maximum and mean elevations and mean local relief versus time following uplift for the stream-power model (same model output as in (a)). Also plotted is an exponential decay model with a time constant $\tau_d = 30$ Myr. (c) Log–log plot of maximum and mean elevations and mean local relief versus time for the sediment-flux-driven model, illustrating the asymptotic approach to $1/t$ behavior at large time scales.

mean erosion rates and mean relief are still very useful for distinguishing between different models of topographic evolution.

Data plotted in Figure 4.10b suggest that topographic decay in the sediment-flux-driven model is described at large times by

$$\frac{\partial H}{\partial t} = -cH^2 \tag{4.36}$$

where c is a constant. Equation (4.36) has the solution

$$H = \left(ct + \frac{1}{H_0}\right)^{-1} \tag{4.37}$$

where H_0 is the initial mean elevation following tectonic uplift.

The inverse power-law dependence in Eq. (4.37) predicts a fundamentally different history of mountain-belt decay than the exponential model that follows from Eq. (4.29). The inverse power law has a broad tail in which topography decays much more slowly than predicted by the exponential model at long time scales. Figure 4.11 compares the topographic decay of the stream-power and sediment-flux-driven models. Figure 4.11a plots the mean and maximum elevations and mean local relief of each model on a semi-log scale as a function of time following tectonic uplift. For large times, the mean elevation in the stream-power model decays as e^{-t/τ_d} with $\tau_d \approx 30$ Myr. The sediment-flux-driven model, however, decays much more slowly with time. One can think of the inverse power-law model as an exponential model in which the effective time scale (i.e. τ_d in Eq. (4.29)) is continually increasing as the topography decays. As such, the inverse-power-law model suggests that there is no single time scale for mountain belt decay, but rather a spectrum of time scales with increasing values for older mountain belts. The maximum and mean elevations in the stream-power and sediment-flux-driven models are plotted in Figures 4.11b and 4.11c on semi-log and log–log plots, respectively, with the exponential and power-law asymptotic trends of each model also shown.

Erosion in natural bedrock channels likely falls between the two end-member models of stream-power and sediment-flux-driven erosion, but, as previously noted, field studies suggest that the sediment-flux-driven model is more appropriate for massive bedrock lithologies common in ancient mountain belts. The results described in this section illustrate that the stream-power model in its basic form and a sediment-flux-driven model predict power-law relationships between mean erosion rate and mean local relief with exponents of 1.3 and 1.8, respectively. The results of the sediment-flux-driven model are more consistent with the analysis of Montgomery and Brandon (2002). The sediment flux-driven model predicts that mountain-belt topography decays proportionally to $1/t$, resulting in persistent mountain belts relative to the exponential-decay model. Conceptually, mountain belts persist in the sediment-flux-driven model because the cutting tools responsible for abrading channel beds decrease over time, thereby decreasing channel incision rates in a positive feedback. This mechanism provides a new hypothesis for mountain belt persistence in massive bedrock lithologies dominated by the saltation-abrasion process.

Exercises

4.1 The migration of radionuclides into a soil by diffusive processes was considered in Chapter 2. Modify the model to include advective transport. This ratio of the diffusivity to the advection velocity is often called the dispersivity. Given a dispersivity of 1 cm and a diffusivity of $\kappa = 0.1\ \mathrm{cm^2/yr}$, model the relative concentration of an abrupt pulse of radionuclides into a semi-infinite soil profile with time. Plot the concentration profiles for $t = 10$, 100, and 1000 yr following the fallout event.

4.2 Assume that the precipitation in a mountain range increases linearly with elevation. In such a case, the stream-power model for a semi-circular basin in a humid environment (Eq. (4.12)) becomes

$$\frac{\partial h}{\partial t} = cx \left(1 + \frac{h}{h_0}\right) \frac{\partial h}{\partial x} \qquad (E4.1)$$

where h_0 is a characteristic length scale. Using the method of characteristics, plot the evolution of knickpoints in this system. How do the knickpoints change in shape and speed as they propagate up into the headwaters?

4.3 Extract a fluvial channel profile from a topographic map or a DEM starting at the channel head. Using Excel, evolve the channel forward in time according to Eq. (4.12). Neglect uplift. Choose a value of c that produces realistic erosion rates (e.g. 1 m/kyr for very steep terrain).

4.4 Construct a simple model of bank retreat assuming that the rate of bank retreat is inversely proportional to the channel width w:

$$\frac{\partial h}{\partial t} = \frac{c}{w}\frac{\partial h}{\partial x} \qquad (E4.2)$$

with $c = 0.001\ \mathrm{yr^{-1}}$. Assume a channel with an initial width of 100 m. As the banks retreat and the channel widens, update the value of w in Eq. (E4.2). How long is required before the channel doubles in width?

4.5 Cliff retreat can be modeled with the advection equation. Consider a cliff and talus slope as illustrated in Figure 4.12. Initially, the cliff

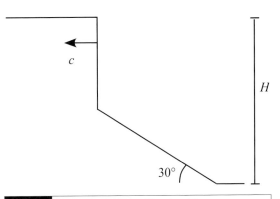

Fig 4.12 Schematic diagram of Exercise 4.5.

and talus slope both have height $H/2$. Assume that the cliff retreat can be modeled advectively with retreat rate $c = 1 \, \text{m/kyr}$. Model the evolution of the cliff and talus slope through time. Impose conservation of mass at the base of the cliff (i.e. the volume of rock removed from the slope must be deposited on the slope).

Assume that the rock and talus slope have equal densities.

4.6 Repeat the above for a case in which the channel crosses a major lithologic boundary (e.g. the Kaibab limestone at the rim of the Grand Canyon). Choose a smaller value of c to represent the more resistant unit.

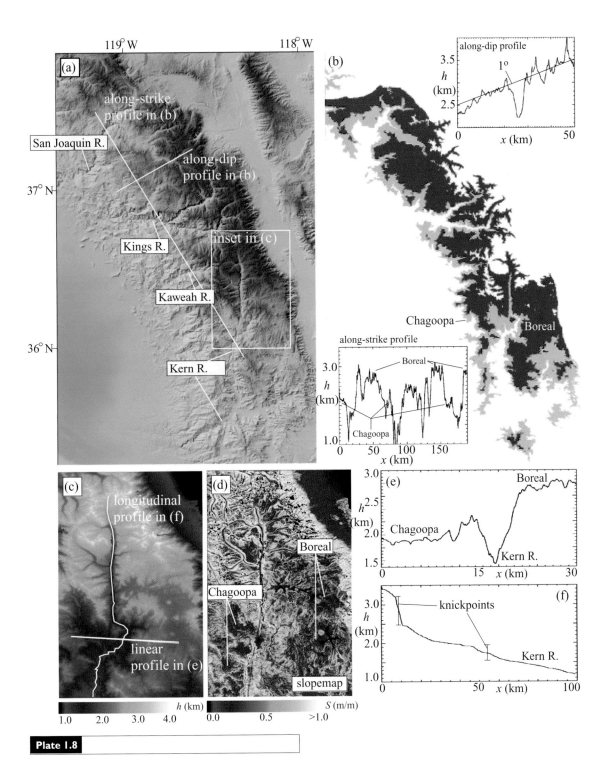

(a)

119° W   118° W

along-strike profile in (b)

San Joaquin R.

along-dip profile in (b)

37° N

Kings R.

inset in (c)

Kaweah R.

36° N

Kern R.

(b)

along-dip profile

h (km)

3.5

1°

2.5

0   50 x (km)

Chagoopa

Boreal

along-strike profile

3.0

Boreal

h (km)

2.0

Chagoopa

1.0

0   50   100   150 x (km)

(c)

longitudinal profile in (f)

linear profile in (e)

(d)

Boreal

Chagoopa

slopemap

h (km)

1.0   2.0   3.0   4.0

S (m/m)

0.0   0.5   >1.0

(e)

3.0

Boreal

h (km)

2.7

2.5

Chagoopa

2.0

1.5

Kern R.

0   15   30 x (km)

(f)

3.0

knickpoints

h (km)

2.0

1.0

Kern R.

0   50 x (km)   100

Plate 1.8

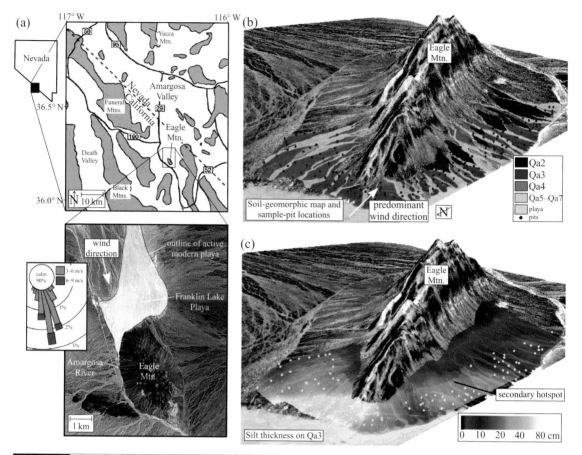

(a)

117° W · Yucca Mtn. · 95 · 60 · Nevada · Amargosa Valley · 116° W · Funeral Mtns. · Nevada California · 29 · 36.5° N · 190 · Eagle Mtn. · Death Valley · Black Mtns. · N 10 km · 52

wind direction · outline of active modern playa · calm 90% · 3–6 m/s · 6–9 m/s · 1% · 2% · 3% · Franklin Lake Playa · Amargosa River · Eagle Mtn. · 1 km

(b) Eagle Mtn. · Qa2 · Qa3 · Qa4 · Qa5–Qa7 · playa · pits · Soil-geomorphic map and sample-pit locations · predominant wind direction · N

(c) Eagle Mtn. · secondary hotspot · Silt thickness on Qa3 · 0 10 20 40 80 cm

Plate 2.18

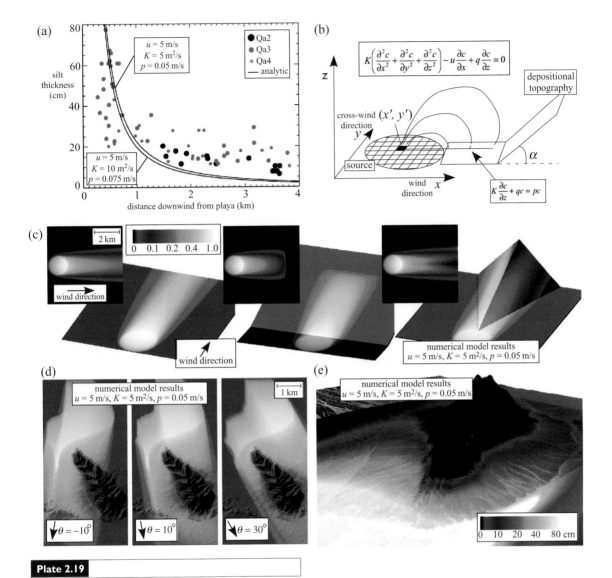

(a)

silt thickness (cm)

$u = 5$ m/s
$K = 5$ m^2/s
$p = 0.05$ m/s

- Qa2
- Qa3
- Qa4
— analytic

$u = 5$ m/s
$K = 10$ m^2/s
$p = 0.075$ m/s

distance downwind from playa (km)

(b)

$$K\left(\frac{\partial^2 c}{\partial x^2} + \frac{\partial^2 c}{\partial y^2} + \frac{\partial^2 c}{\partial z^2}\right) - u\frac{\partial c}{\partial x} + q\frac{\partial c}{\partial z} = 0$$

z

cross-wind (x', y')
direction
y

source

wind direction x

depositional topography

α

$K\dfrac{\partial c}{\partial z} + qc = pc$

(c)

2 km

0 0.1 0.2 0.4 1.0

wind direction

wind direction

numerical model results
$u = 5$ m/s, $K = 5$ m^2/s, $p = 0.05$ m/s

(d)

numerical model results
$u = 5$ m/s, $K = 5$ m^2/s, $p = 0.05$ m/s

1 km

$\theta = -10°$ $\theta = 10°$ $\theta = 30°$

(e)

numerical model results
$u = 5$ m/s, $K = 5$ m^2/s, $p = 0.05$ m/s

0 10 20 40 80 cm

Plate 2.19

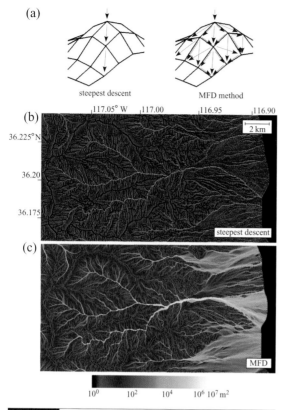

(a)

steepest descent   MFD method

(b)

117.05° W   117.00   116.95   116.90

2 km

36.225° N

36.20

36.175

steepest descent

(c)

MFD

10^0 10^2 10^4 10^6 10^7 m²

Plate 3.1

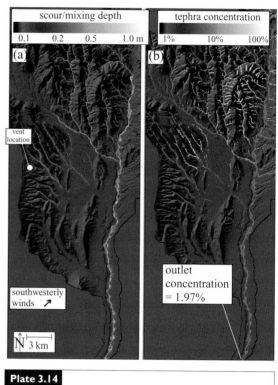

scour/mixing depth

0.1   0.2   0.5   1.0 m

tephra concentration

1%   10%   100%

(a)

(b)

vent location

southwesterly winds ↗

outlet concentration = 1.97%

N 3 km

Plate 3.14

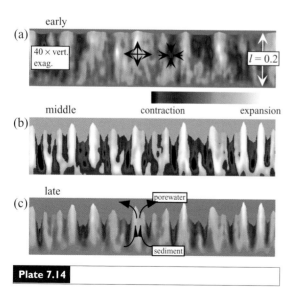

early

(a)

40 × vert. exag.

$I = 0.2$

contraction   expansion

middle

(b)

late

(c)

porewater

sediment

Plate 7.14

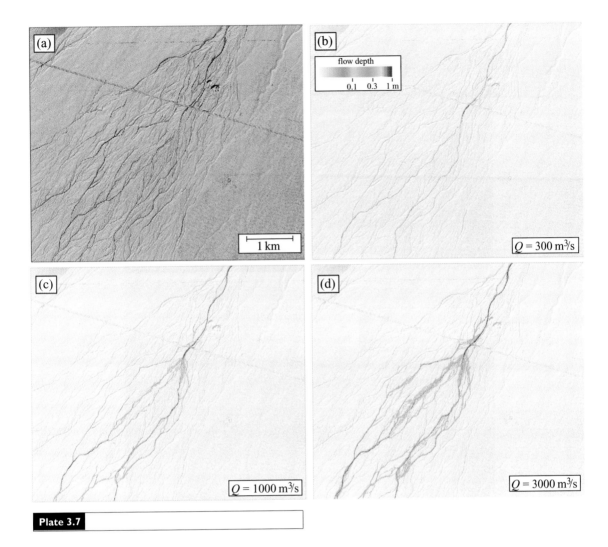

(a)

1 km

(b)

flow depth

0.1 0.3 1 m

$Q = 300$ m^3/s

(c)

$Q = 1000$ m^3/s

(d)

$Q = 3000$ m^3/s

Plate 3.7

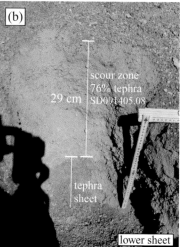
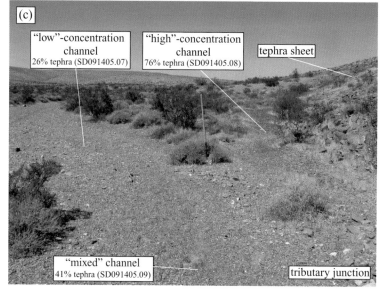

Plate 3.11

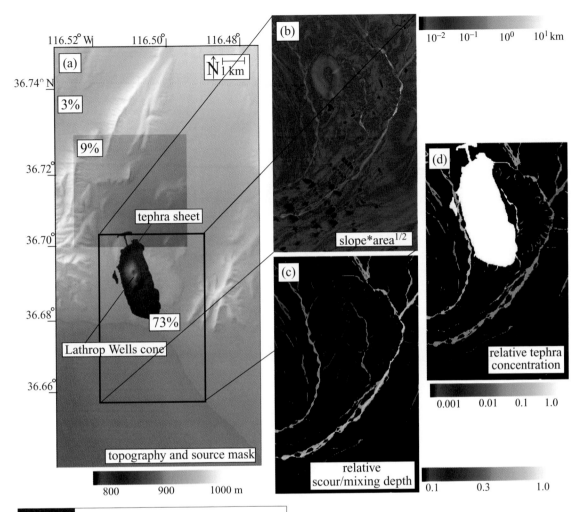

116.52° W   116.50°   116.48°

(a)

N
1 km

36.74° N

3%

9%

36.72°

tephra sheet

36.70°

36.68°

73%

Lathrop Wells cone

36.66°

topography and source mask

800   900   1000 m

(b)

10^{-2} 10^{-1} 10^{0} 10^{1} km

slope*area$^{1/2}$

(c)

relative
scour/mixing depth

(d)

relative tephra
concentration

0.001   0.01   0.1   1.0

0.1   0.3   1.0

Plate 3.12

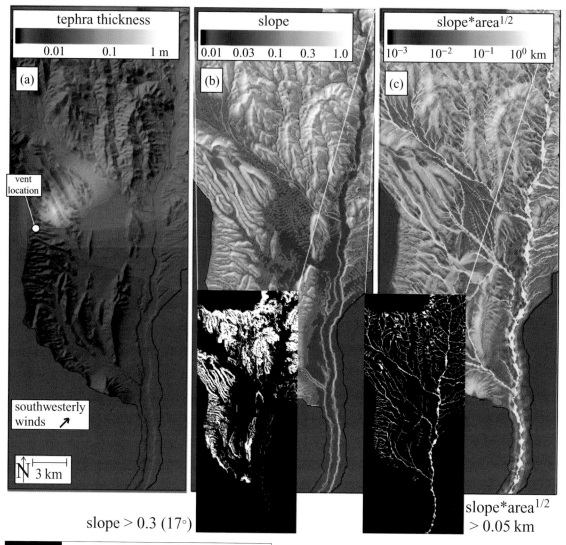

tephra thickness

0.01   0.1   1 m

(a)

vent location

southwesterly winds ↗

N 3 km

slope

0.01 0.03 0.1 0.3 1.0

(b)

slope > 0.3 (17°)

slope*area$^{1/2}$

10^{-3} 10^{-2} 10^{-1} 10^0 km

(c)

slope*area$^{1/2}$
> 0.05 km

Plate 3.13

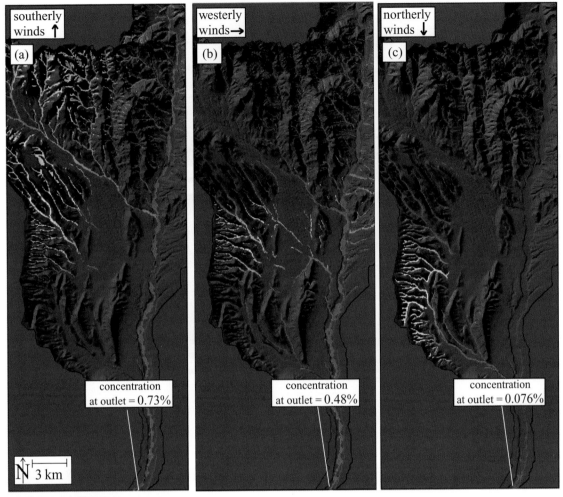

southerly winds ↑

(a)

westerly winds →

(b)

northerly winds ↓

(c)

concentration at outlet = 0.73%

concentration at outlet = 0.48%

concentration at outlet = 0.076%

N 3 km

tephra concentration

1%   10%   100%

Plate 3.15

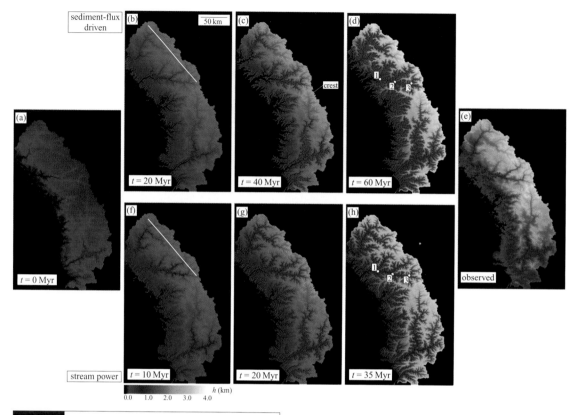

sediment-flux driven

(a) $t = 0$ Myr

(b) 50 km $t = 20$ Myr

(c) crest $t = 40$ Myr

(d) $t = 60$ Myr

(e) observed

(f) $t = 10$ Myr

(g) $t = 20$ Myr

(h) $t = 35$ Myr

stream power

h (km)
0.0 1.0 2.0 3.0 4.0

Plate 4.5

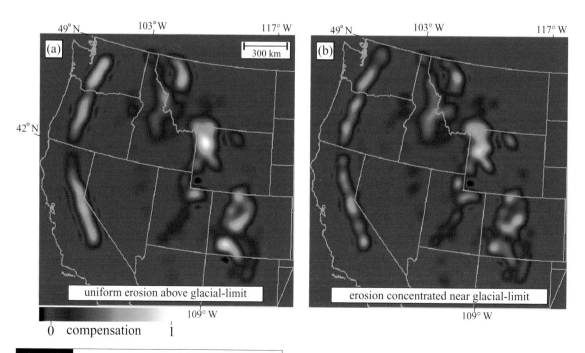

(a) uniform erosion above glacial-limit

(b) erosion concentrated near glacial-limit

49° N   103° W   117° W   300 km

42° N

109° W   109° W

0 compensation 1

Plate 5.11

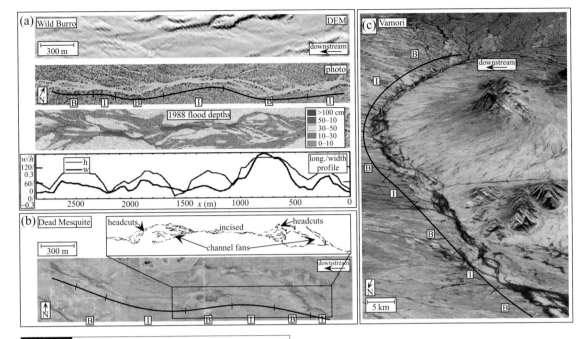

Plate 7.9

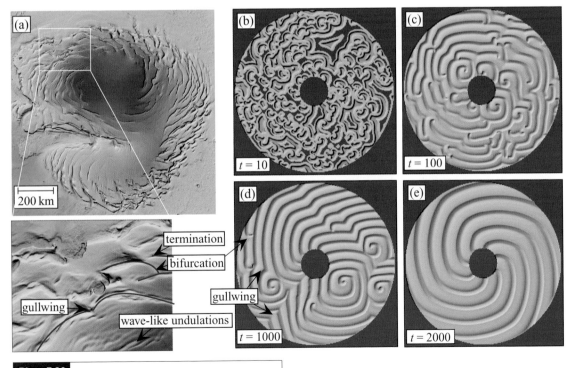

Plate 7.20

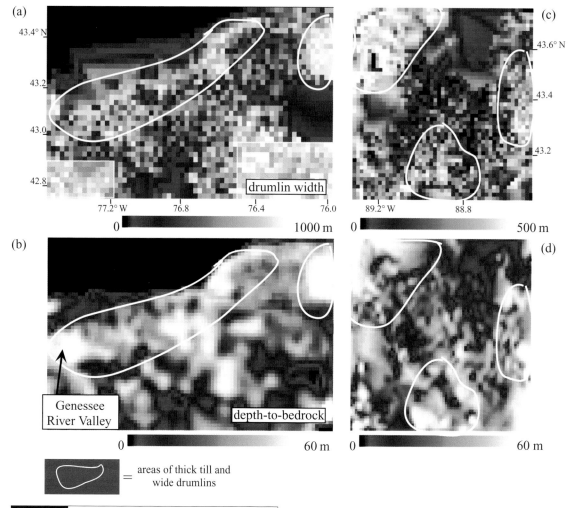

(a)

43.4° N

43.2

43.0

42.8

77.2° W   76.8   76.4   76.0

drumlin width

0 ▮▮▮▮▮▮▮ 1000 m

(c)

L

43.6° N

43.4

43.2

89.2° W   88.8

0 ▮▮▮▮▮▮▮ 500 m

(b)

Genessee
River Valley

depth-to-bedrock

0 ▮▮▮▮▮▮▮ 60 m

(d)

0 ▮▮▮▮▮▮▮ 60 m

= areas of thick till and
    wide drumlins

Plate 7.17

Chapter 5

Flexural isostasy

5.1 | Introduction

Flexural isostasy is the deflection of Earth's lithosphere in response to topographic loading and unloading. When a topographic load is generated by motion along a thrust fault, for example, the lithosphere subsides beneath the load. The width of this zone of subsidence varies from place to place depending on the thickness of the lithosphere, but it is generally within the range of 100 to 300 km. Conversely, a reduction in topographic load causes the lithosphere to rebound, driving rock uplift. Flexural-isostatic uplift in response to erosion replaces approximately 80% of the eroded rock mass, thereby lengthening the time scale of mountain-belt denudation by a factor of approximately five because erosion must remove all of the rock that makes up the topographic load *and* the crustal root beneath it in order to erode the mountain down to sea level. Given the ubiquity of erosion in mountain belts, it is reasonable to assume that flexural-isostasy plays a key role in nearly all examples of large-scale landform evolution. Flexural isostasy also plays an important role in the evolution of ice sheets because the topographic load of the ice sheet causes lithospheric subsidence, thereby influencing rates of accumulation and ablation on the ice sheet. In this chapter, we will discuss three broadly-applicable methods (series and integral solutions, Fourier filtering, and the Alternating-Direction Implicit (ADI) method) for solving the flexural-isostatic equation in geomorphic applications.

Isostasy refers to the buoyant force created when crustal rock displaces mantle rock. Isostatic balance requires that, over geologic time scales, the weight of the overlying rock must be uniform at any given depth in the mantle. Because the mantle acts as a fluid over long time scales, any lateral pressure gradient will initiate mantle flows to restore equilibrium. Isostasy simply says that the hydrostatic pressure gradient in the mantle must be zero, otherwise the mantle would correct the imbalance by flowing. Isostatic balance requires that the hydrostatic force produced by the topographic load, $\rho_c gh$ (where ρ_c is the density of the crust, g is the acceleration due to gravity, and h is the elevation of the mountain belt) be equal to the buoyancy force $(\rho_m - \rho_c)gw$ (where ρ_m is the density of the mantle and w is the depth of the crustal root) produced by the displacement of low-density crustal rocks with higher-density mantle rocks (Figure 5.1):

$$\rho_c gh = (\rho_m - \rho_c)gw \qquad (5.1)$$

Solving for w, the thickness of the crustal root, gives:

$$w = h\frac{\rho_c}{\rho_m - \rho_c} \qquad (5.2)$$

For typical crust and mantle densities, e.g. $\rho_c = 2.7\,\text{g/cm}^3$ and $\rho_m = 3.3\,\text{g/cm}^3$, the depth of the crustal root is approximately five times the height of the topographic load.

Flexure refers to the forces and displacements involved in bending the elastic lithosphere. Flexure affects isostasy because small-scale variations in the topographic load (e.g. peaks and valleys)

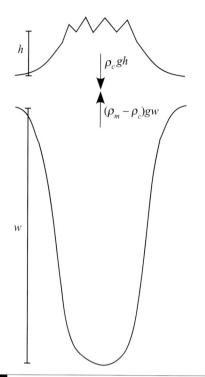

Fig 5.1 Isostatic balance requires that a topographic load, given by $\rho_c g h$, be balanced by a much larger crustal root, which extends to a depth given by Eq. (5.2).

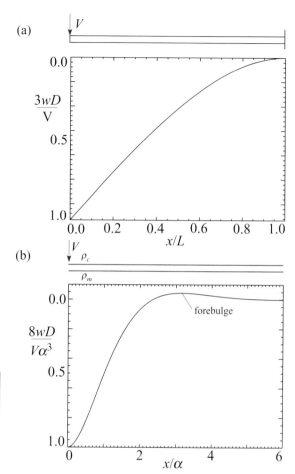

Fig 5.2 (a) Nondimensional deflection of an elastic beam of rigidity D and length L subject to an applied force V at the end of the beam. (b) Nondimensional deflection of an elastic beam overlying an inviscid fluid of different density, subject to a line load at $x = 0$. The deflection of the elastic beam in this case has a characteristic length scale given by α.

can be supported by the rigidity of the lithosphere. It is only at length scales larger than the flexural wavelength (i.e. $\approx$ 100–300 km) that isostatic balance is achieved. The equations used to describe flexure were originally developed in the mechanical engineering literature to describe the response of elastic beams and plates to applied forces. The deflection of a diving board under the weight of a diver is a simple example of 2D flexure. If a diver stands still at the end of the board, all of the forces and torques on the diving board must be in balance, otherwise the board would accelerate. A force-balance analysis of the diving board indicates that the fourth derivative of the displacement w is proportional to the applied load $q(x)$ (Turcotte and Schubert, 2002):

$$D \frac{d^4 w}{dx^4} = q(x) \qquad (5.3)$$

where D is a coefficient of proportionality that defines the flexural rigidity of the board. The solution to Eq. (5.3) corresponding to the weight V

of a diver at the end of a diving board ($x = 0$) of length L can be obtained by integrating four times to obtain the cubic function (Figure 5.2a):

$$w(x) = \frac{V(L - x)^3}{3D} \qquad (5.4)$$

The flexural rigidity of the lithosphere is, in turn, controlled by the elastic thickness T_e as well as the elastic properties of rock by the relationship

$$D = \frac{E T_e^3}{12(1 - v^2)} \qquad (5.5)$$

where E is the elastic modulus and v is the Poisson ratio of the rocks in the crust. Equation (5.5) indicates that the flexural rigidity is very sensitive to the elastic thickness T_e. One common technique for mapping T_e in a region involves comparing the topography and gravity fields at different length scales. At small length scales, topography and gravity are poorly correlated, while at large scales they have a high degree of coherence. Local values of T_e can be inferred by mapping the smallest length scales at which topography and gravity are strongly correlated.

In this chapter our primary focus is on vertical (i.e. topographic) loads. In some cases, such as when a subducting slab exerts a horizontal frictional force on the overriding plate, horizontal forces play an important role in lithospheric flexure. Such cases are rare, however, in geomorphic applications. We will also ignore the time-dependent response of the mantle to loading, which can be important for some problems (e.g. flexural isostatic response to Quaternary ice sheets). Once a load is applied, there is necessarily some time delay before the crust can respond by uplift or subsidence. This time scale is generally on the order of 10^4 yr, and generally longer for narrower loads. The transient response of the lithosphere to loading and unloading can generally be neglected over time scales of interest in mountain building.

The equation for the 1D flexural-isostatic displacement of a uniformly rigid lithosphere to vertical loading and unloading is:

$$D\frac{d^4w}{dx^4} + (\rho_m - \rho_c)gw = q(x) \tag{5.6}$$

This equation combines flexural bending stresses (first term on left side), the upward buoyancy force exerted on the bottom of the lithosphere (second term on left side), and the applied load $q(x)$ (right side).

5.2 | Methods for 1D problems

5.2.1 Displacement under line loading

In some cases the topographic load is narrow enough compared to the flexural wavelength of the lithosphere that the load can be approximated as acting at a single point along a 1D profile. Within a 1D model, such a load is referred to as a "line" load. A narrow mountain range that extends for thousands of kilometers along-strike is one example of a model well approximated by line loading. In such cases, the term $q(x)$ on the right side of Eq. (5.6) is equal to zero except at $x = 0$ where it is equal to some prescribed value V. In order to solve Eq. (5.6) under a line load it is easiest to set $q(x)$ equal to zero for all x and then introduce the load V as a boundary condition for $x = 0$. With $q(x) = 0$, Eq. (5.6) becomes

$$D\frac{d^4w}{dx^4} + (\rho_m - \rho_c)gw = 0 \tag{5.7}$$

The general solution to Eq. (5.7) is obtained by integration:

$$w(x) = e^{x/\alpha}\left(c_1\cos\frac{x}{\alpha} + c_2\sin\frac{x}{\alpha}\right)$$
$$+ e^{-x/\alpha}\left(c_3\cos\frac{x}{\alpha} + c_4\sin\frac{x}{\alpha}\right) \tag{5.8}$$

where c_1 through c_4 are integration constants, and

$$\alpha = \left[\frac{4D}{(\rho_m - \rho_c)g}\right]^{1/4} \tag{5.9}$$

We can use the symmetry of the problem to specify the boundary conditions. The value of w must go to zero as x goes to ∞ and dw/dx must go to zero at $x = 0$. This implies that c_1 and c_2 are zero and $c_3 = c_4$ to give

$$w(x) = c_3 e^{-x/\alpha}\left(\cos\frac{x}{\alpha} + \sin\frac{x}{\alpha}\right) \tag{5.10}$$

The value of c_3 is proportional to the applied load V. From Eq. (5.3) and Eq. (5.10) we have

$$\frac{1}{2}V = D\left.\frac{d^3w}{dx^3}\right|_{x=0} = \frac{4Dc_3}{\alpha^3} \tag{5.11}$$

We used $\frac{1}{2}V$ for the load in Eq. (5.11) because we are solving for only half of the profile (i.e. the other half of the profile supports the other half of the load). Substituting Eq. (5.11) into Eq. (5.10) gives

$$w(x) = \frac{V\alpha^3}{8D}e^{-x/\alpha}\left(\cos\frac{x}{\alpha} + \sin\frac{x}{\alpha}\right) \tag{5.12}$$

for $x \geq 0$ (Figure 5.2b).

In the case of deflection of a diving board, we found that the solution is characterized by

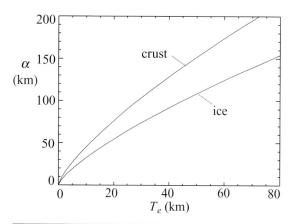

Fig 5.3 Plot of flexural parameter α as a function of elastic thickness T_e for crustal and ice loads, assuming $E = 70$ GPa and $\nu = 0.25$.

a cubic polynomial with no characteristic length scale. In the case of flexural-isostasy, however, the deflection is characterized by the length scale α. Also, the zone of depression is accompanied by a smaller zone of uplift flanking the margin of the load. This uplift zone is the forebulge. The location of the forebulge is determined by differentiating w with respect to x and setting the result equal to zero:

$$\frac{dw}{dx} = -\frac{2w_0}{\alpha} e^{-x/\alpha} \sin \frac{x}{\alpha} = 0 \qquad (5.13)$$

Therefore,

$$x_b = \alpha \sin^{-1}(0) = \pi \alpha \qquad (5.14)$$

The amplitude of the forebulge can be found by substituting x_b into Eq. (5.12) to obtain

$$w_b = -w_0 e^{-\pi} = -0.0432 w_0 \qquad (5.15)$$

Therefore, the height of the forebulge is equal to about 4% of the maximum deflection of the lithosphere beneath the load. This calculation assumes that the lithosphere is entirely rigid. In some cases, the loading causes the lithosphere to break under the stresses of the load. In these cases, the sin term in Eq. (5.12) becomes zero and the width of the zone of subsidence becomes narrower (Turcotte and Schubert, 2002).

The relationship between α and T_e is plotted in Figure 5.3 for loading by crustal rocks (i.e. $\Delta\rho \approx 600$ kg/m^3) and ice (i.e. $\Delta\rho \approx 2300$ kg/m^3), assuming $E = 70$ GPa and $\nu = 0.25$. For an elastic

thickness of 40 km, for example, the flexural parameter under crustal loading is ≈ 130 km, while for ice loading it is approximately 90 km.

5.2.2 Series solutions and Fourier filtering

The flexure equation is a linear equation. As such, we can use Fourier series and superposition to develop general solutions for the deflection of the lithosphere under any topographic load, given the solution for deflection under a periodic load. Let's assume a periodic topography with amplitude h_0 and wavelength λ:

$$h(x) = h_0 \sin\left(\frac{2\pi x}{\lambda}\right) \qquad (5.16)$$

The vertical load corresponding to this topography is given by

$$q_a(x) = \rho_c g h_0 \sin\left(\frac{2\pi x}{\lambda}\right) \qquad (5.17)$$

and hence the flexural-isostatic equation we must solve is

$$D\frac{d^4 w}{dx^4} + (\rho_m - \rho_c) g w = q_a(x) \qquad (5.18)$$

The displacement of the lithosphere will follow the same periodic form as the loading – the *magnitude* of the deflection will depend sensitively on the wavelength of the load and flexural parameter of the lithosphere, but there will be no offset or shift between the load and the response. Therefore, we look for solutions of the form

$$w = w_0 \sin\left(\frac{2\pi x}{\lambda}\right) \qquad (5.19)$$

where w_0 is yet to be determined. Substituting Eq. (5.19) into Eq. (5.18) gives

$$w_0 = \frac{h_0}{\frac{\rho_m}{\rho_c} - 1 + \frac{D}{\rho_c g}\left(\frac{2\pi}{\lambda}\right)^4} \qquad (5.20)$$

For small wavelengths, i.e. where

$$\lambda \ll 2\pi \left(\frac{D}{\rho_c g}\right)^{\frac{1}{4}} \qquad (5.21)$$

the denominator in Eq. (5.20) is very large and hence $w_0 \ll h_0$. This means that small-wavelength topographic variations cause very little displacement of the crust and we say that the crust is *rigid* for loads of this scale. At large scales, where

$$\lambda \gg 2\pi \left(\frac{D}{\rho_c g} \right)^{\frac{1}{4}} \tag{5.22}$$

Eq. (5.20) gives

$$w_0 = h_0 \frac{\rho_c}{\rho_m - \rho_c} \tag{5.23}$$

which is equivalent to the case of pure isostatic balance given by Eq. (5.2). In the large-scale limit, therefore, flexure plays no role and the topographic load is said to be fully compensated. At smaller scales, where the deflection is smaller than the Airy isostatic limit, the topography is said to be only partially compensated.

The solution for periodic topography is useful for determining the degree of compensation of a topographic load. The real power of the periodic solution, however, is that it can be used in conjunction with the Fourier series approach to determine the response of the lithosphere to any load using superposition. If the topographic load within a domain of length L is expressed as a Fourier series:

$$h(x) = \sum_{n=1}^{\infty} a_n \sin \left(\frac{\pi x}{L} \right) \tag{5.24}$$

with coefficients

$$a_n = \frac{2}{L} \int_0^L h(x') \sin \left(\frac{n\pi x}{L} \right) dx' \tag{5.25}$$

then, according to Eq. (5.20), the deflection caused by the load is given by

$$w(x) = \sum_{n=1}^{\infty} a_n \frac{\sin \left(\frac{n\pi x}{L} \right)}{\frac{\rho_m}{\rho_c} - 1 + \frac{D}{\rho_c g} \left(\frac{n\pi}{L} \right)^4} \tag{5.26}$$

Equation (5.26) can be used to obtain analytic expressions for the flexural-isostatic displacement underneath loads with simple (rectangular, triangular, etc.) loads.

Most real topography, however, does not lend itself to simple analytic forms. Equation (5.26) can be implemented numerically for *any* topographic load, however, using the Fourier filtering technique. In this technique, the Fast Fourier Transform (FFT) is used to quickly calculate the Fourier coefficients a_n for an input load function. Then, each coefficient is multiplied by the filter

(so called because it filters out low wavenumbers) before the inverse Fourier transform is applied to bring the function back to real space.

Appendix 4 provides code that implements the Fourier filtering technique for any input topographic load. Figure 5.4 illustrates example output of that program for an east–west profile through the central Andes. The east–west topographic profile of the central Andes is characterized by a high-elevation plateau (the Altiplano-Puna basin), high peaks on the margins of the Altiplano-Puna, and an active fold thrust belt to the east of the Altiplano-Puna. In this example, the topographic load of the Andes is considered to be all topography above 1 km in elevation. Figure 5.4b plots the modeled deflection of the crust for $\rho_c = 2.7\,\text{g/cm}^3$, $\rho_m = 3.3\,\text{g/cm}^3$, and three hypothetical values of α. The depth of total deflection beneath the Andes ($\approx 15\,\text{km}$) depends linearly on the relative density contrast between the crust and mantle, but only weakly on the flexural parameter for wide topographic loads. Low values of the flexural parameter predict narrow foreland sedimentary basins, while higher values predict broad basins (Figure 5.4b–5.4c). The height of the flexural forebulge (Figure 5.4c) also increases with higher values of the flexural parameter. Gravity modeling has been used to constrain the width of the Andean foreland basin by taking advantage of the density contrast between the basin infill and the crust below. Those constraints, coupled with forward modeling of flexural profiles, suggest that $\alpha = 100\,\text{km}$ is a good estimate for the flexural parameter of the central Andes.

5.3 Methods for 2D problems

5.3.1 Integral method

Thus far we have considered solutions in 1D only. These solutions are very useful for long mountain chains or other landforms where the loading is relatively constant in one direction. In many cases, however, (e.g. an isolated seamount or curving mountain belt) a full 2D solution is needed. In such cases, 2D methods analogous to those we developed for 1D are available.

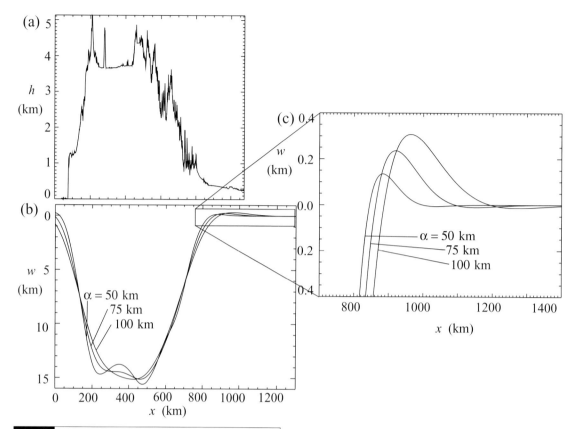

Fig 5.4 Flexural model of the response of the lithosphere to the distributed load of the central Andes. (a) Topographic profile of the central Andes. (b) Deflection of the lithosphere beneath the central Andes for three hypothetical values of $\alpha = 50, 75, 100$ km. (c) Close-up plot of the flexural profile of the foreland basin and forebulge, illustrating the dependence of basin width and forebulge amplitude on α.

First we consider the solution to the 2D flexural-isostatic equation:

$$D\nabla^4 w + \Delta\rho g w = q \tag{5.27}$$

corresponding to a point load V at $r = 0$. The solution is the Kelvin function kei:

$$w(r) = \frac{V\alpha^2}{\pi D}\text{kei}\left(\frac{\sqrt{2}r}{\alpha}\right) \tag{5.28}$$

The shape of the kei function is broadly similar to Eq. (5.12), i.e. it has a deep zone of subsidence beneath the load and a smaller forebulge flanking the load.

Solutions corresponding to a spatially distributed load can be obtained by representing the load as a collection of many point sources and integrating the response to each point. For a spatially distributed topographic load given by $h(x, y)$, the deflection is given by:

$$w(r) = \frac{\rho_c}{\rho_m - \rho_c}\int_0^\infty \int_0^\infty$$
$$\times \text{kei}\left(\frac{\sqrt{2}}{\alpha}\sqrt{(x - x')^2 + (y - y')^2}\right)$$
$$\times h(x', y')dx'dy' \tag{5.29}$$

This approach is broadly analogous to the series solution method of Eq. (5.26).

Few analytic solutions can be obtained using Eq. (5.29), both because topographic loads rarely conform to analytic expressions and because the kei function is difficult to integrate. Appendix 4 provides code for series approximations to the kei

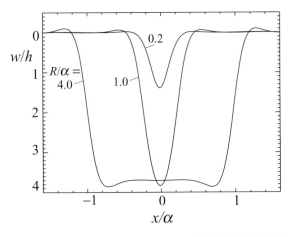

Fig 5.5 Deflection of disk-shaped loads for various values of R/α.

function that can be used to integrate Eq. (5.29) numerically for any topographic load.

One analytic expression is available, however, that nicely illustrates the response of the lithosphere to loads of different size. The response of the lithosphere to a disk-shaped load depends on the ratio of the disk radius R to the flexural parameter α. The solution is given by (Hetenyi, 1979):

$$w(r) = \frac{h\rho_c}{\rho_m - \rho_c} \left[\frac{R}{\alpha}\mathrm{ker}'\left(\frac{R}{\alpha}\right)\mathrm{ber}\left(\frac{r}{\alpha}\right) \right.$$
$$\left. - \frac{R}{\alpha}\mathrm{kei}'\left(\frac{R}{\alpha}\right)\mathrm{bei}\left(\frac{r}{\alpha}\right) + 1 \right] \quad (5.30)$$

for $r < R$ and

$$w(r) = \frac{h\rho_c}{\rho_m - \rho_c} \left[\frac{R}{\alpha}\mathrm{ber}'\left(\frac{R}{\alpha}\right)\mathrm{ker}\left(\frac{r}{\alpha}\right) \right.$$
$$\left. - \frac{R}{\alpha}\mathrm{bei}'\left(\frac{R}{\alpha}\right)\mathrm{kei}\left(\frac{r}{\alpha}\right) \right] \quad (5.31)$$

for $r > R$.

Figure 5.5 plots Eq. (5.31) for loads of different radii, expressed as a ratio of the flexural parameter α. For cases in which the load radius is much less than the flexural parameter, the load is only partially compensated. For cases in which the radius is equal to the flexural parameter, the maximum deflection is equal to the Airy isostatic limit (i.e. Eq. (5.2)), but the maximum deflection is attained only beneath the center of the load. For load radii much larger than the flexu-

ral parameter, the deflection follows the disk-like shape of the load except at the edges where flexural bending stresses limit the curvature of the deflection profile.

5.3.2 Fourier filtering
In the previous section we showed how the deflection beneath spatially distributed loads could be calculated by dividing the load into many small point sources and summing the contributions of each point. This is a powerful approach, but it can be computationally challenging when calculating the response at high spatial resolution over a large region. Such calculations can require evaluating the kei function 10^4–10^6 times for each of 10^4–10^6 points. The Fourier-filtering method provides a faster way by taking advantage of the computational speed of the Fast Fourier Transform algorithm.

In order to implement the Fourier filtering algorithm in 2D, the FFT of the 2D load $h(x, y)$ over a domain of width L_x and length L_y is computed to obtain a set of Fourier coefficients $a_{n,m}$. Each Fourier coefficient is then multiplied by the factor

$$\left(\frac{\rho_m}{\rho_c} - 1 + \frac{D}{\rho_c g}\left(\pi\sqrt{\left(\frac{n}{L_x}\right)^2 + \left(\frac{m}{L_y}\right)^2}\right)^4 \right)^{-1} \quad (5.32)$$

Then, the inverse Fourier transform is applied to the filtered data set.

Appendix 4 provides a code for implementing the Fourier filtering method in 2D. Figure 5.6 illustrates example output from this program for the central Andes for $\alpha = 50$, 75, and 100 km. Figures 5.6b–5.6d are grayscale plots of the absolute value of the deflection on a logarithmic scale. As the value of α increases, the pattern of deflection becomes smoother. In each grayscale image the forebulge is represented by the first ring around the central zone of subsidence. Subsequent rings are the backbulge and other smaller zones of alternating uplift and subsidence with increasing distance from the load. Figure 5.6e illustrates the variations in forebulge amplitude in the vicinity of the Bolivian Orocline using a contour map. The forebulge amplitude varies from 160 m to 400 m

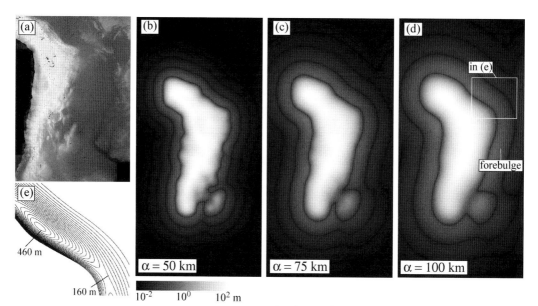

Fig 5.6 2D flexural-isostatic response of the lithosphere to the topographic load of the central Andes (a) for $\alpha = 50$ (b), 75 (c), and 100 km (d). In (b)–(d), the grayscale images represent the absolute value of the deflection on a logarithmic scale. This color scale clearly shows that the width of the forebulge increases with increasing α. (e) A contour map of the forebulge in the vicinity of the Bolivian Orocline. This example shows that the height of the forebulge changes dramatically (from a minimum of 160 m to a maximum of 400 m) in the vicinity of a curved load.

within a distance of ≈ 100 km, illustrating the importance of 2D effects in the vicinity of a bend in the topographic load.

5.4 | Modeling of foreland basin geometry

Flexure plays an important role in large-scale landform evolution because, as erosion and deposition change the topographic load, flexural isostasy causes uplift and subsidence responses that vary as a function of spatial scale and lithospheric rigidity. In this section we explore the interaction between flexural isostasy and sedimentation in a foreland basin adjacent to a migrating thrust belt. Foreland basins provide a potentially useful record of tectonics and erosion of the adjacent range, but our ability to interpret that record uniquely is limited. Foreland

basin environments are also attractive environments for studying the linkages of climate, tectonics, and erosion, because the tectonic parameters controlling foreland basin geometry can be unusually well constrained. Specifically, thrustbelt migration rates can be determined using a mass-balance framework that relates crustal shortening to thrust-belt migration (DeCelles and DeCelles, 2001) and basin subsidence is often well characterized using flexural modeling, gravity measurements, and seismic stratigraphy (e.g. Watts *et al.*, 1995). Sediment transport in fluvial channels has a linear relationship with channel slope, resulting in a diffusion model for the basin topographic profile (Begin *et al.*, 1981; Paola *et al.*, 1992).

Foreland basin evolution can be modeled as a combination of thrust-belt migration, flexural subsidence, and diffusive fluvial sediment transport (Flemings and Jordan, 1989). Modeling this system requires solving the diffusion equation with a moving boundary (the thrust front, where sediment enters the basin) and moving sink term (the flexural depression). Typically, moving boundaries and sinks in diffusion problems require numerical approaches. However, in this case the problem can be solved analytically by working in the frame of reference of the migrating thrust front and assuming steady-state conditions. Conceptually, basins are in steady state when sediment progradation occurs at the same rate as thrust-belt migration. Basins naturally evolve toward this condition because when sediment progradation lags behind thrust-belt migration at any point along its profile, the basin steepens locally until the two are in balance. Conversely, if sediment progradation outstrips thrust-belt migration, slope decreases and progradation slows. This approach is analogous to that of the Kenyon and Turcotte model of deltaic sedimentation considered in Chapter 2. In this case, however, we will also consider the pattern of flexural subsidence that controls accommodation space in the basin.

Basin evolution in the model is governed by the diffusion equation:

$$\frac{\partial h}{\partial t} = \kappa \frac{\partial^2 h}{\partial x^2} \tag{5.33}$$

where h is the elevation, t is time, κ is the diffusivity, and x is the distance from the thrust front. Boundary conditions are needed at both ends of the basin to solve Eq. (5.33). At the thrust front, a constant sediment flux Q_s (in m^2/yr) enters the

where $w(x)$ is the flexural profile, given by Turcotte and Schubert (2002) as

$$w(x) = -w_0 e^{-x/\alpha} \cos \frac{x}{\alpha} \tag{5.36}$$

where w_0 is the basin depth beneath the thrust front and α is the flexural parameter. Equation (5.36) corresponds to the flexural profile corresponding to a line load. This is only an approximation to the Andes, where loading is spatially distributed.

Combining Eqs. (5.33) and (5.35) and assuming steady-state conditions gives the time-independent equation:

$$\kappa \frac{d^2 h}{dx^2} + F \frac{dh}{dx} = F \frac{dw}{dx} \tag{5.37}$$

which can be written as a first-order differential equation for the slope $S = dh/dx$ with solution

$$S(x) = \frac{-e^{-Fx/\kappa}}{\kappa} \left(Q_s - F \int_0^x e^{-Fx'/\kappa} \frac{dw}{dx'} dx' \right) \tag{5.38}$$

Substituting Eq. (5.36) into Eq. (5.38) gives

$$S(x) = \frac{-e^{-Fx/\kappa}}{\kappa} \left(Q_s - \frac{F}{(\alpha F/\kappa)^2 + 2\alpha F/\kappa + 2} \left(e^{-(F/\kappa + 1/\alpha)x} \left(\frac{\alpha F}{\kappa} \left(\cos \frac{x}{\alpha} - \sin \frac{x}{\alpha} \right) - 2\sin \frac{x}{\alpha} \right) - \frac{\alpha F}{\kappa} \right) \right) \tag{5.39}$$

basin, providing a flux boundary condition

$$\left. \frac{\partial h}{\partial x} \right|_{x=0} = \frac{Q_s}{\kappa} \tag{5.34}$$

In this 2D framework, the sediment flux is given by $E L_d$, where E is the basin-averaged erosion rate and L_d is the upstream drainage basin length. At the basin outlet, a constant base-level elevation boundary condition is prescribed. Two cases can be considered. For an infinite basin, $h = 0$ as $x = \infty$. For a basin of length L_b and base-level elevation h_b (determined by sea level or a valley-floor channel), the boundary condition is $h(L_b) = h_b$. In the moving reference frame, the foreland basin moves towards the thrust front and enters the flexural depression with velocity F, the thrust-migration velocity. This motion is represented by an advection equation:

$$\frac{\partial h}{\partial t} = F \left(\frac{\partial w}{\partial x} - \frac{\partial h}{\partial x} \right) \tag{5.35}$$

The basin topographic profile is obtained by integrating Eq. (5.39) analytically or numerically and enforcing the basin-outlet boundary condition to constrain the integration constant.

Model solutions are plotted in Figure 5.7, representing the effects of varying Q_s on basin profiles around a reference case with parameter values $F = 10\,\mathrm{mm/yr}$, $\alpha = 150\,\mathrm{km}$, $w_0 = 5\,\mathrm{km}$, $\kappa = 5000\,\mathrm{m}^2/\mathrm{yr}$, $Q_s = 10\,\mathrm{m}^2/\mathrm{yr}$, and assuming an infinite basin. Figure 5.7a illustrates three important points. First, the basin profile is highly sensitive to incoming sediment supply: the topographic relief of the basin doubles when Q_s is increased by 50% from 8 to 12 m^2/yr. This suggests that basin profiles may provide useful constraints on upstream sediment supply through forward modeling and comparison with observed profiles. Second, as sediment supply decreases, the basin undergoes a transition from overfilled to underfilled (closed) basins. As such, the model provides quantitative criteria for topographic closure in

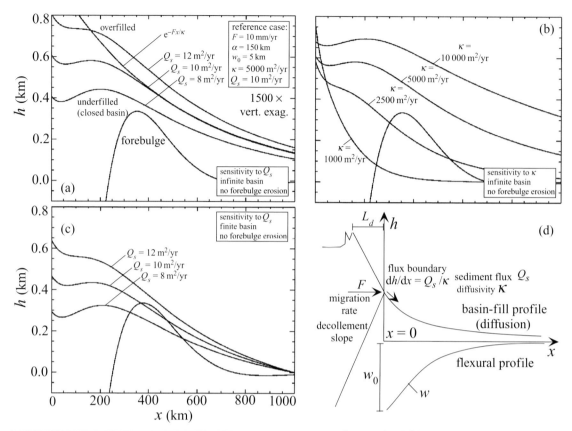

Fig 5.7 (a) and (b): Model solutions for the topographic profile of an infinite basin assuming no forebulge erosion, showing sensitivity to variations in Q_s and κ, respectively, around a reference solution with $F = 10$ mm/yr, $\alpha = 150$ km, $w_0 = 5$ km, $\kappa = 5000$ m²/yr, and $Q_s = 10$ m²/yr. (c) Solutions for a finite basin of length $L_b = 1000$ km for the same model parameters as in (a). (d) Schematic diagram of the model. Modified from Pelletier (2007b).

foreland basins. Third, the distal basin profile is approximated by a simple exponential function with length scale κ/F. This result suggests that distal basin profiles can be used to uniquely and easily constrain κ values if thrust-belt migration rates are known.

Figure 5.7b illustrates the sensitivity of model solutions to variations in κ around the reference value of $\kappa = 5000$ m²/yr. Diffusivity values are controlled by upstream basin length, precipitation, and sediment texture (with longer basins, higher precipitation rates, and finer textures promoting greater diffusivity values) (Paola *et al.*, 1992). Low diffusivity values (e.g. $\kappa = 1000$ m²/yr)

cause sediment backfilling at the thrust front to produce steep, short basins. Surprisingly, higher diffusivities (e.g. more humid conditions) promote basin closure if sediment supply is held constant. Naively, we might expect more humid conditions to result in steeper, more overfilled basins. However, to understand the influence of climate on basin geometry we must consider the effects of climate on sediment supply (i.e. weathering) and basin transport rates independently. Greater sediment supply puts more total sediment into the basin and therefore promotes steeper and more overfilled basins. Diffusivity values, on the other hand, control how uniformly that sediment is spread across the basin. Higher diffusivity values promote closed basins because a larger fraction of the total sediment is transported past the foredeep, leaving behind a depression if the ratio Q_s/κ is sufficiently small. These results emphasize that the key to understanding basin geometry is not simply the sediment supply to the basin, but the ratio of sediment supply to transport capacity (e.g. as emphasized in the geomorphic context by Bull (1979)). Figure 5.7c illustrates model solutions for a finite basin using the same model parameters as in Figure 5.7a. Shorter basins have lower basin relief, all else being equal.

Equation (5.40) assumes no forebulge erosion (i.e. that the forebulge is buried or, if exposed, that the erosion rate is low). In some cases, it may be more appropriate to assume that the forebulge erodes completely to sea level. In this alternative end-member scenario, Eq. (5.36) is amended to $w(x) = 0$ for $x > 3\pi\alpha/4$ (i.e. the point at which the forebulge rises above sea level). In this case, the slope is given by Eq. (5.36) for $x < 3\pi\alpha/4$, and by

$$S(x) = \frac{-e^{-Fx/\kappa}}{\kappa}\left(Q_s - \frac{F}{(\alpha F/\kappa)^2 + 2\alpha F/\kappa + 2}\left(1.414e^{-\left(\frac{F}{\kappa}+\frac{1}{\alpha}\right)\frac{3\pi\alpha}{4}}\left(1+\frac{\alpha F}{\kappa}\right)+\frac{\alpha F}{\kappa}\right)\right) \quad (5.40)$$

for $x > 3\pi\alpha/4$. Model solutions assuming forebulge erosion are given in Figure 5.8.

This model framework also provides a simple means to estimate the basin sediment delivery ratio (the ratio of the sediment yield to erosion rate). The sediment flux at the entrance to the foreland basin is Q_s. For a basin of length L_b, the flux leaving the basin is computed by evaluating Eq. (5.40) at $x = L_b$ and multiplying by κ. For basins that have a base-level control located beyond the forebulge distance, Eq. (5.40) reduces to a simple exponential function with length scale κ/F. In the reference case, for example, a 500-km wide basin has a sediment delivery ratio of $1/e = 0.37$, and a 1000-km wide basin has a value of $1/e^2 = 0.14$. This latter value is broadly consistent with measured sediment delivery ratios for large drainage basins in areas of active tectonism (Schumm, 1977).

As a test of the model, we consider the modern foreland basin of the central Andes. The latest phase of Andean deformation responsible for Eastern Cordilleran and Subandean uplift is generally considered to be late Miocene in age (Gubbels et al., 1993). The foreland basin is approximately 3.5–4.5 km thick beneath the thrust front based on seismic reflection data (Horton and DeCelles, 1997). DeCelles and DeCelles (2001) estimated the thrust-migration rate to be approximately 10 mm/yr for the central Andes based

on crustal shortening rates. Flexural modeling coupled with geophysical and geomorphic observations constrain the flexural parameter to be approximately 150 km (resulting in a foredeep approximately 250 km wide and a forebulge at a distance 400 km from the thrust front) (Coudert et al., 1995; Watts et al., 1995; Horton and DeCelles, 1997; Ussami et al., 1999). These data provide relatively firm constraints on three of the five model parameters: $F = 10$ mm/yr, $\alpha = 150$ km, $w_0 = 4$ km.

Figure 5.9a illustrates a map of the topography of the central Andes and the adjacent foreland basin from 17° S to 31° S. This region corresponds to a humid-to-arid transition in the Andes based upon mean winter precipitation values above 1500 m elevation. Associated with this climatic transition is a geomorphic transition from strongly overfilled (700 m in relief) to weakly overfilled (100 m in relief) basins. Closed basins (e.g. L. Ambargasta and L. Mar Chiquita) also occur near the southernmost transect, but they are located in wedgetop depozones of a broken foreland and hence are not well described by the model.

Three topographic profiles were extracted from the Shuttle Radar Topographic Mission (SRTM) data and are plotted in Figure 5.9b. To model these profiles, it is most accurate to use finite-basin solutions constrained by values of h_b and L_b for each profile. The base level elevation h_b of the Andean foreland is controlled by Rio Paraguay, which ranges in elevation from 80 to 10 m (decreasing from north to south) at distances L_b varying from 350 to 720 km from the thrust front. Model results in Figure 5.7a suggest that the distal basin profile is an exponential function with length scale κ/F. The inset plot in Figure 5.9b showing $h$$h_b$ (note vertical logarithmic scale) as a function of x confirms this prediction for the central Andean foreland. Best-fit exponential profiles (straight lines on this logarithmic scale) provide estimates of $\kappa = 1500$ m²/yr for the northern profile and $\kappa = 4000$ m²/yr for

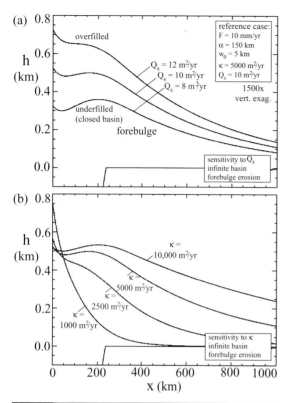

Fig 5.8 (a) and (b) Model solutions for the topographic profile of an infinite basin assuming complete forebulge erosion, showing sensitivity to variations in Q_s and κ, respectively, for the same model parameters as Figures 5.7a and 5.7b. These results show that model results are broadly similar to those obtained by assuming no forebulge erosion; the only significant difference is a minor reduction in basin relief when forebulge erosion is assumed.

the two southern profiles using $F = 10$ mm/yr. Given observed or inferred values for h_b, L_b, and κ, as well as estimated values for F, α, and w_0 from the published literature, a family of solutions corresponding to a range of Q_s values can be generated for each profile and compared to the observed data. Figure 5.9c illustrates the observed profiles plotted with their corresponding best-fit profiles (κ values for the two southernmost profiles were also varied to find the optimal fit for both κ and Q_s). Model solutions match the observed data very closely for the strongly overfilled profile, and they match the first-order trends of the weakly overfilled profile. Discrepancies in the southernmost profiles may be the

result of assuming uniform κ values. Downstream fining occurs in all basins, and therefore κ values are likely to increase somewhat downstream, resulting in gentler distal-basin slopes compared to the model solutions of Figure 5.9b.

5.5 | Flexural-isostatic response to glacial erosion in the western US

The pace of alpine glacial erosion increased significantly starting 2–4 Myr ago when global cooling initiated the Plio-Quaternary era of large continental ice sheet advances and retreats (Zhang *et al.*, 2001). This erosion must have triggered significant flexural-isostatic unloading of glacially carved mountain belts. Since glacial erosion is strongly focused in the landscape, scouring out valley floors where ice flow is focused but eroding the highest points in the landscapes relatively slowly, it is likely that glaciated mountain belts have increased in relief during the Plio-Quaternary era due to isostatic rock uplift, even in areas that are tectonically inactive. Determining precisely how much glacial erosion has taken place is difficult to do in most cases. Nevertheless, it is still useful to ask how much rock uplift occurred for a given amount of glacial erosion. By using the geographic extent of glacial cover as a proxy for the spatial distribution of erosion, we can use the 2D flexural-isostatic model to compute the ratio of late Cenozoic erosionally driven rock uplift to glacial erosion. In this section we compute that ratio for the western United States, following the approach of Pelletier (2004a). Areas where this ratio is large have likely undergone the greatest relief production and should be the focus of future efforts to identify signatures of erosionally driven rock uplift and relief production. In this application we will use the Alternating-Direction-Implicit technique, taking into account spatial variations in flexural rigidity.

The ratio of rock uplift to erosion depends on the area and shape of the eroded region and on the local flexural wavelength of the lithosphere. If the flexural wavelength is large compared to

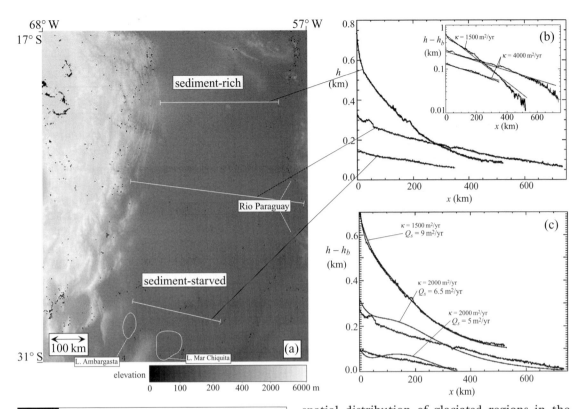

Fig 5.9 (a) Grayscale map of topography in the central Andes indicating sediment-rich, strongly overfilled basins in the northern central Andes (near 20° S) and sediment-starved, underfilled and weakly overfilled basins (near 30° S) corresponding to a humid-to-arid transition. (b) Basin topographic profiles extracted from SRTM data. Also plotted using logarithmic scale (with base-level elevation subtracted) to illustrate exponential trend and associated best-fit κ values. (c) Best-fit model profiles with associated Q_s values. Sediment supply decreases from north to south with precipitation rates and as deformation becomes more spatially distributed. Note that 100 m has been added to the northernmost profile so that the plots can more easily be distinguished. Modified from Pelletier (2007). Reproduced with permission of the Geological Society of America.

the width of the glaciated region, uplift will be limited by flexural bending stresses. In contrast, if the flexural wavelength and the width of the eroded region are comparable, bending stresses will be smaller and uplift will approach the Airy isostatic limit in which 80% of the eroded material is replaced through rock uplift.

To map the compensation resulting from glacial erosion, we first must determine the spatial distribution of glaciated regions in the western United States. Porter *et al.* (1983) mapped areas of Pliocene–Quaternary glaciation in the western United States, based on the elevations of the lowest cirques. Their map identifies the approximate local elevation of vigorous glacial erosion. To construct a map of the extent of late Cenozoic glacial erosion, the contours of glacial-limit elevation from Porter *et al.* (1983) were first digitized, projected to Lambert Equal-Area, and then interpolated to obtain a grid of glacial-limit elevations. Second, a 1 km resolution DEM (digital elevation model) of the western United States was rectified to the glacial-limit-elevation grid. Third, the glacial-limit-elevation and DEM grids were compared pixel-by-pixel to determine whether local elevations were above or below the glacial-limit elevation. Pixels with elevations above the glacial limit were considered areas of glacial erosion and were given a value of 1. Conversely, pixels with elevations below the glacial limit were considered not to have been subjected to glacial erosion and were given a value of 0. The resulting binary grid is a map of relative glacial erosion; an underlying assumption is that

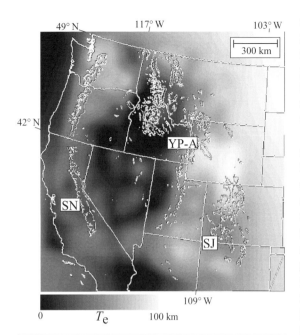

Fig 5.10 Grayscale map of elastic thickness in the western United States. Data from Lowry *et al.* (2000) and Watts (2001). Boundaries of glaciated areas are shown in white. Labeled ranges are: SN (Sierra Nevada), SJ (San Juan), and YP-A (Yellowstone Plateau-Absaroka). Modified from Pelletier (2004a).

erosion is uniform within the glaciated regions. The boundaries of the glaciated regions obtained in this way are mapped in white in Figure 5.10 (which also shows state boundaries for reference). Also shown in Figure 5.10 is a grayscale map of the elastic thickness in the western United States. Ranges along the Pacific coast were not included in Porter *et al.* (1983) and have also been excluded from this analysis. In addition, the northern Cascades were excluded because they were completely covered by the Cordilleran Ice Sheet and hence subject to a different style of glaciation than other ranges of the western United States.

The deflection of a thin elastic plate with variable elastic thickness under a vertical load is given by

$$\nabla^2(D(x, y)\nabla^2 w) + \Delta\rho g w = q(x, y) \qquad (5.41)$$

where w is the deflection, $D(x, y)$ is the flexural rigidity, $\Delta\rho$ is the density contrast between the crust and the mantle, g is the acceleration due

to gravity, and $q(x, y)$ is the vertical load (Watts, 2001). In order to account for spatial variations in elastic thickness we retained $D(x, y)$ inside the parentheses in Eq. (5.41). This results in additional terms involving gradients of D that are not present when D is assumed to be spatially uniform.

The binary grids of glacial erosion were directly input into the flexure equation as the loading term, so that $q(x, y) = 1$ where erosion occurred and $q(x, y) = 0$ where erosion did not occur. The solution to the flexure equation with these loading terms is the ratio of the positive deflection of the lithosphere to the depth of glacial erosion. This is the compensation resulting from glacial erosion. Here we use the alternating difference implicit (ADI) technique to solve Eq. (5.41), following the work of van Wees and Cloetingh (1994). Previous techniques we have discussed (e.g. Fourier filtering) cannot be used when D varies spatially.

For the analysis, we will use Lowry *et al.*'s (2000) data set for elastic thickness (T_e) in the western United States. These data are based on the coherence method and provide a direct measure of lithospheric strength that includes the weakening effects of faults. These data have the highest resolution available for the region, but do not completely cover the western United States. To cover the entire area, the Lowry *et al.* data was augmented with the global compilation of Watts (2001), and the data sets joined smoothly with a low-pass filter. The resulting T_e map was rectified to the glacial erosion maps and is shown as a grayscale map in Figure 5.10. T_e values were converted to a grid of flexural rigidity, $D(x, y)$, for use in Eq. (5.41), by means of the relationship

$$D = \frac{E T_e^3}{12(1 - \nu^2)} \qquad (5.42)$$

assuming $E = 70$ GPa and $\nu = 0.25$.

A map of glacial-erosion compensation in the western United States is given in Figure 5.11a. Also shown in Figure 5.11b is the compensation determined by assuming that glacial erosion was concentrated near the equilibrium line, not uniformly within the glacial region as in Figure 5.11a. Although compensation is strictly defined to be between 0 and 1, we have mapped

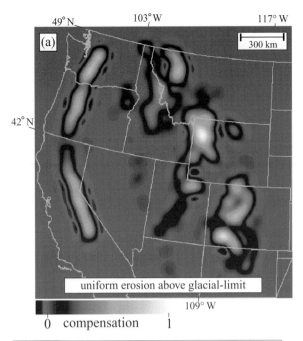

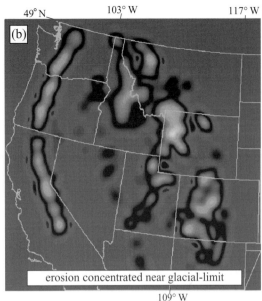

the full solution to the flexure equation in Figure 5.11, including areas of minor subsidence (in black). All of the major ranges in the western United States have compensation values greater than or equal to 0.4, indicating that nearly half of all glacially eroded mass is returned to the region's ranges as erosionally driven rock uplift. The pattern of compensation varies little between Figures 5.11a and 5.11b because flexure is insensitive to small-scale variations in loading.

The greatest difference between Figures 5.11a and 5.11b is the mean values: values of compensation obtained by assuming concentrated erosion near the glacial limit are 20% lower than for uniform erosion. These lower values reflect the additional bending stresses present in the case of localized erosion; i.e., for the same eroded mass, a circular area (i.e., uniform erosion within the range) results in less intense bending stresses than a "ring-shaped" area (i.e., localized erosion near the glacial limit).

The highest compensation values in the western United States are found in the Yellowstone Plateau–Absaroka Range. The high values in this region reflect the extensiveness of the glaciated area, which is nearly 200 km in width. In addition, however, nearby ranges are also being eroded, reducing the bending stresses that would otherwise be concentrated at the edges of the Yellowstone area, limiting its potential uplift. Instead, nearby ranges, including the Beartooth, Gallatin, Wyoming, and Wind River Ranges, accommodate the bending stresses to produce a "bulls-eye" over the central part of the area – the Yellowstone Plateau–Absaroka Range.

Exercises

5.1 Calculate the degree of isostatic compensation for continental topography with a wavelength of 20 km and an elastic thickness of 5 km, assuming $E = 70$ GPa, $nu = 0.25$, $\rho_m = 3300$ kg/m^3, and $\rho_c = 2700$ kg/m^3.

5.2 In addition to driving rock uplift, the isostatic response to erosion also causes subsidence in some locations. Consider a 1D model of periodic line unloading (Figure 5.12). Each line unloading might represent glacial erosion concentrated in the high peaks of a Basin and Range-style topography. Calculate the distance between unloading, L, that produces the largest value of subsidence in spaces

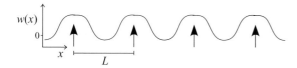

Fig 5.12 Schematic diagram of the 1D periodic line loading calculation used in Exercise 5.2.

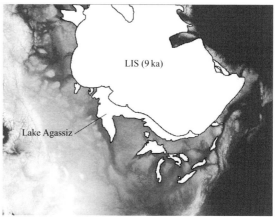

Fig 5.13 Outline of the Laurentide Ice Sheet at 9 ka, showing the extents of proglacial lakes at the ice sheet margins.

between the locations where unloading is concentrated.

5.3 Calculate the lithospheric deflection associated with a narrow seamount of radius 10 km and height 1 km using the Kelvin function solution.

5.4 Use the Fourier series approach to construct an expression for the flexural response of the lithosphere to a periodic topographic load of triangular mountain ranges with width L and height H.

5.5 Proglacial lakes formed at the margins of the Laurentide Ice Sheet. The largest of these, Lake Agassiz, had a width of several hundred kilometers (e.g. Figure 5.13). Model the 1D bathymetric profile of Lake Agassiz perpendicular to the ice margin assuming that the lake formed as a flexural depression. Assume that the proglacial topography was initially flat and at sea level, and that the ice sheet acts as a rectangular load with height 2 km. How wide is the lake according to

this model, assuming that the lake fills up to sea level?

5.6 Following the Andean example, download topographic data for a portion of the Main Central Thrust of the Himalaya. Using published estimates for the elastic thickness of the lithosphere, model the 1D flexural profile and determine the approximate location of the Himalayan forebulge in India.

Chapter 6

Non-Newtonian flow equations

6.1 | Introduction

Chapter 1 introduced the concept of non-Newtonian fluids and highlighted their importance for our understanding of the mechanics of glaciers. In this chapter we explore methods for modeling non-Newtonian fluids with applications to ice sheets, glaciers, lava flows, and thrust sheets. In most cases we will make the simplifying assumption that the rheology of the flow is constant in time. In nature, however, many types of non-Newtonian flows are complicated by the fact that the rheology of the flow depends on parameters that evolve with the flow itself. The viscosity of lava, for example, depends on its temperature, which, in turn, depends on the flow behavior (e.g. slower moving lava will cool more quickly, increasing its viscosity and further slowing flow in a positive feedback). Here we focus primarily on modeling flows with a constant, prescribed rheology. In cases where the rheology depends strongly on the state of the flow the methods introduced in this chapter will serve as the foundation for more complex models which fully couple the rheology with the flow.

6.2 | Modeling non-Newtonian and perfectly plastic flows

First we consider the forces acting on a fluid spreading over a flat surface (Figure 6.1). We begin with a 1D example, so the fluid is assumed to extend infinitely far into and out of the cross section. The horizontal force F exerted on the left side of a thin vertical slice of the fluid of width Δx is equal to the total hydrostatic pressure in the fluid column. The hydrostatic pressure at a depth y below the surface is given by $\rho g y$. The total hydrostatic pressure is calculated by integrating $\rho g y$ over the fluid column from $y = 0$ to $y = h$:

$$F(x) = \int_0^h \rho g y \, dy = \frac{\rho g h^2}{2} \tag{6.1}$$

On the right side of the vertical slice (i.e. at position $x + \Delta x$), the pressure is lower because the flow is thinner:

$$F(x + \Delta x) = \frac{\rho g (h - \Delta h)^2}{2} = \frac{\rho g}{2}(h^2 - 2h \Delta h$$
$$+ (\Delta h)^2) \approx \frac{\rho g}{2}(h^2 - 2h \Delta h) \tag{6.2}$$

The $(\Delta h)^2$ term in Eq. (6.2) is much smaller than the $2h \Delta h$ term for flows with small (i.e. < 0.1) surface slopes, and hence the second-order term can be neglected in such cases. This "small-angle" approximation is commonly used when modeling ice sheets, glaciers, and lava flows on Earth. The difference between the horizontal force exerted on the opposite sides of the fluid slice must be equal to the shear stress τ_0 exerted over the width of the slice, Δx:

$$-\tau_0 \Delta = F(x + \Delta x) - F(x) = \rho g h \Delta h \tag{6.3}$$

Rearranging Eq. (6.3) gives

$$\tau_0 = -\rho g h S \tag{6.4}$$

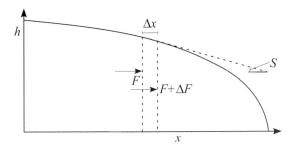

Fig 6.1 Schematic diagram of the forces acting on a gravity flow atop a flat surface.

where S is equal to the local ice-surface gradient, $\Delta h / \Delta x$. Field studies of modern ice sheets and glaciers have found the basal shear stress calculated with Eq. (6.4) to be typically between 0.5 and 1.5 bars (Nye, 1952), a surprisingly narrow range considering the spatial variability of observed basal conditions, including gradients in temperature, meltwater content, basal debris, till rheology, and other variables.

6.2.1 Analytic solutions

Equation (6.4) is a general result for any "slowly" moving gravity flow (i.e. slow enough that accelerations can be neglected). In cases where the fluid moves rapidly when a threshold shear stress is exceeded, Eq. (6.4) can be used to model the flow geometry by treating τ_0 as a constant. Taking the limit as Δx goes to zero and expressing Eq. (6.4) as a differential equation gives

$$h\frac{dh}{dx} = \frac{\tau_0}{\rho g} \tag{6.5}$$

Multiplying both sides of Eq. (6.5) by dx and integrating yields

$$h^2 = \frac{\tau_0}{\rho g}x + C \tag{6.6}$$

The integration constant C is equal to the terminus position. If the flow has a half-width of L, for example, then $C = L$, and Eq. (6.6) becomes

$$h = \sqrt{\frac{2\tau_0}{\rho g}(L - x)} \tag{6.7}$$

Figure 6.2a illustrates several examples of this simple parabolic ice-sheet model solution. Equation (6.7) is usually referred to as a "perfectly plastic" model for ice sheets. The ratio $\tau_0/\rho g$ on

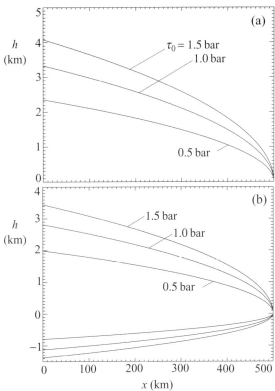

Fig 6.2 Plots of ice sheet profiles predicted by the perfectly plastic model with (b) and without (a) isostatic adjustment included, for a range of threshold basal shear stresses.

the right side of Eq. (6.7) has units of length and is often written as h_0 (not to be confused with an initial, or maximum thickness). For an ice sheet with $\tau_0 = 1\,\mathrm{bar}$ ($10^5\,\mathrm{Pa}$), $\rho = 920\,\mathrm{kg/m^3}$, and $g = 9.8\,\mathrm{m^2/s}$, $h_0 \approx 12\,\mathrm{m}$. It should be noted that this approach breaks down near the divide, where additional stress components become important.

The parabolic solution of Eq. (6.7) is usually associated with perfectly-plastic deformation of near-basal ice. Perfectly plastic deformation, however, is an idealization of just one of the complex basal flow mechanisms we recognize today, including debris-controlled frictional sliding and till deformation. So, how can Eq. (6.7) provide even an approximate model for real ice sheets and glaciers? The reason is that many deformation mechanisms mimic the threshold-controlled flow of the perfectly plastic model. Many flow mechanisms are threshold controlled and result in low effective friction when a threshold shear stress is exceeded. Threshold behavior, for example, characterizes the plastic deformation of subglacial till (e.g. Tulaczyk et al., 2000) and the shear stress necessary to overcome kinetic friction during basal sliding (e.g. Llibustry, 1979). The perfectly plastic model, therefore, can be generally applied to model the geometry of ice sheets and glaciers that move at low effective friction when a threshold shear stress is exceeded, whether relative motion occurs internally or by sliding over deforming till or directly over the bed. To emphasize this broader applicability, we will also use the term "threshold-sliding" to refer to the perfectly plastic model.

The results of Figure 6.2a are unrealistic in at least one major respect: the subsidence of the crust beneath the ice sheet was neglected. For ice sheets larger than a few hundred kilometers, the weight of the overlying ice will cause significant bedrock subsidence beneath the ice sheet in order to achieve isostatic equilibrium. This effect can be quantified with a simple adjustment to the threshold-sliding model. If the bed elevation is given by $b(x)$, then isostatic equilibrium requires that

$$\rho_i z = -\rho_m b \tag{6.8}$$

where ρ_i is the density of ice, $z = b + h$ is the ice-surface elevation, and ρ_m is the density of the mantle. Substituting $z = b + h$, Eq. (6.8) becomes

$$b = -\frac{\rho_i}{\rho_m - \rho_i} z = -\delta z \tag{6.9}$$

where

$$\delta = \frac{\rho_i}{\rho_m - \rho_i} \tag{6.10}$$

The ice thickness can be written as

$$h = (1 + \delta) z \tag{6.11}$$

Substituting Eq. (6.11) into Eq. (6.7) yields the following equations for the ice-sheet elevation and thickness:

$$z = \sqrt{2 \frac{\tau_0}{\rho g (1 + \delta)} (L - x)} \tag{6.12}$$

$$h = \sqrt{2 \frac{\tau_0 (1 + \delta)}{\rho g} (L - x)} \tag{6.13}$$

Plots of Eqs. (6.12) and (6.13) are shown in Figure 6.2b for the same cases as Figure 6.2a with $\delta = 0.4$. Equation (6.13) indicates that the volume of an ice sheet increases when isostatic effects are taken into account. In essence, the "hole" created by the weight of the ice provides an additional confining force that enables the ice to be thicker than it would be on a flat bed.

Equation (6.4) can be generalized to cases with variable bed topography. In such cases, the form of Eq. (6.4) does not change, but the surface slope S is no longer simply equal to dh/dx. Instead, S is equal to $dz/dx = dh/dx + db/dx$ where $b(x)$ is the bed elevation. Equation (6.4) then becomes

$$h \left(\frac{dh}{dx} + \frac{db}{dx} \right) = \frac{\tau_0}{\rho g} \tag{6.14}$$

Equation (6.14) can be used to derive an analytical expression for the approximate shapes of alpine glaciers with a uniform bed slope. In such cases, $db/dx = S_b$ and Eq. (6.14) becomes

$$h \left(\frac{dh}{dx} + S_b \right) = h_0 \tag{6.15}$$

where $h_0 = \tau_0 / \rho g$. Dividing both sides of Eq. (6.15) by h gives

$$\frac{dh}{dx} = S_b - \frac{h_0}{h} \tag{6.16}$$

which integrates to

$$x - C = \frac{h}{S_b} + \frac{h_0}{S_b^2} \ln(h_0 - S_b h) \tag{6.17}$$

where C is the integration constant. Figure 6.3a illustrates solutions to Eq. (6.17) for $C = 0$, $S_b = 0.05$ and 0.1, and $h_0 = 12$ m (i.e. $\tau_0 = 1$ bar). In order to better visualize the results, it is helpful to plot $z = -S_b x + h$ instead of simply h. Figure 6.3b presents these results. The value of C can be varied to shift the glacier terminus to the right or left, but otherwise the value of C does not affect the solution.

Another solution relevant to alpine glaciers involves the variations in thickness caused by minor variations in valley slope. If the bed topography is slowly-varying and the region close to the glacier terminus is not considered, the dh/dx term can be neglected relative to the db/dx term

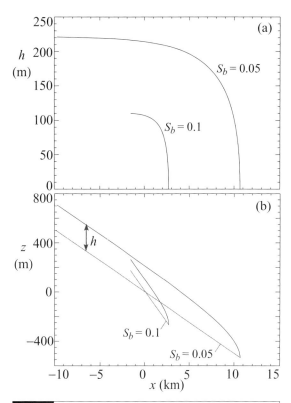

Fig 6.3 Plots of (a) thickness and (b) ice-surface elevation for $S_b = 0.05$ and 0.1 for $h_0 = 12$ m (i.e. $\tau_0 = 1$ bar).

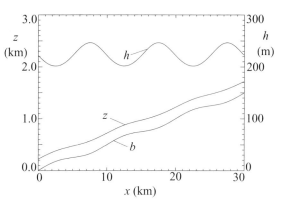

Fig 6.4 Plots of ice thickness variations in an alpine glacier with slowly varying bed topography, according to Eq. (6.18), with $h_0 = 12$ m.

to give

$$h = h_0 \left(\frac{db}{dx} \right)^{-1} \tag{6.18}$$

Figure 6.4 plots solutions to Eq. (6.18) for $h_0 = 12$ m, a mean valley slope of 0.05, and an idealized sinusoidal bed topography. The ice thickness is inversely proportional to the bed slope, thickening in areas of gentle slopes and thinning in areas of steep slopes.

When considering the geometry of ice sheets and glaciers within the framework of the perfectly plastic model, it is not necessary to prescribe the accumulation or ablation rates because the shape of the ice sheet or glacier is controlled entirely by the threshold shear stress τ_0 (given standard values for ice density and gravity). Changes in the spatial distribution of accumulation may cause the ice terminus to migrate. As such, the position of the terminus is controlled by mass balance. If the terminus position is pre-scribed, however, the geometry of the ice sheet or glacier does not explicitly depend on mass balance. When considering ice bodies obeying Glen's Flow Law, however, it is necessary to prescribe a mass balance rate M. Here we will assume M to be uniform and constant, but in general it will vary in both space and time.

The starting point for modeling ice-sheet or glacier geometries with a power-law fluid rheology is the relationship between shear stress and strain rate, given in Chapter 1 for the case of simple shear:

$$\frac{du}{dy} = A\tau^n \tag{6.19}$$

where u is the fluid velocity, y is the depth, τ is the shear stress, and A and n are rheological parameters. Substituting Eq. (6.4) into Eq. (6.19) and integrating from 0 to h gives the velocity profile

$$u(y) = -\frac{A}{2(n+1)}(\rho g S)^n \left[(h-y)^{n+1} - h^{n+1} \right] \tag{6.20}$$

Vialov (1958) was the first to obtain an analytic solution for the profile of an ice sheet on a flat bed evolving according to a power-law fluid rheology. Vialov assumed a radially symmetric ice sheet with radius R. Equation (6.20) must be integrated from 0 to h to obtain the total horizontal flux of ice:

$$U = -Ch^4 \left| \frac{dh}{dr} \right| \tag{6.21}$$

where C is a single constant that combines parameters for viscosity, density, and gravity. Conservation of mass is given by

$$\frac{\partial h}{\partial t} = -\nabla \cdot (HU) + M = -\frac{\partial(rhU)}{rdr} + M \quad (6.22)$$

For an ice sheet in steady state,

$$d(rhU) = Mrdr \quad (6.23)$$

which integrates to

$$U = \frac{M}{2}\frac{r}{h} \quad (6.24)$$

Combining Eqs. (6.21) and (6.24) yields

$$h^{1+2/n}dh = -\left(\frac{M}{2C}\right)^{1/n}r^{1/n}dr \quad (6.25)$$

which can be integrated to obtain

$$\frac{n}{2n+2}(h^{2+2/n}-h_c^{2+2/n}) = -\frac{n}{n+1}\left(\frac{M}{2C}\right)^{1/n}r^{1+1/n} \quad (6.26)$$

where h_c is the thickness at the center where $r = 0$. Equation (6.26) can be rearranged to give

$$h^{2+2/n} - h_c^{2+2/n} = -2\left(\frac{M}{2C}\right)^{1/n}r^{1+1/n} \quad (6.27)$$

The value of h_c can be constrained by setting a boundary condition of vanishing ice thickness at the margin, i.e. $h(R) = 0$, to give

$$h_c^{2+2/n} = 2\left(\frac{M}{2C}\right)^{1/n}R^{1+1/n} \quad (6.28)$$

Substituting Eq. (6.28) into Eq. (6.27) gives the full solution

$$h = 2^{n/(2n+2)}\left(\frac{M}{2C}\right)^{1/(2n+2)}$$
$$\left[R^{1+1/n} - r^{1-1/n}\right]^{n/(2n+2)} \quad (6.29)$$

For $n = 3$, Eq. (6.29) becomes

$$h = 2^{3/8}\left(\frac{M}{2C}\right)^{1/8}\left[R^{4/3} - r^{4/3}\right]^{3/8} \quad (6.30)$$

It is difficult to compare the Vialov solution precisely to the perfectly plastic solution because one was derived for a 1D ice-sheet profile and the other for a radially symmetric ice sheet. Nevertheless, Figure 6.5 compares solutions obtained for the two models. The ratio M/C acts as a free parameter in the Vialov solution that relates to the maximum height h_c. The solution plotted in

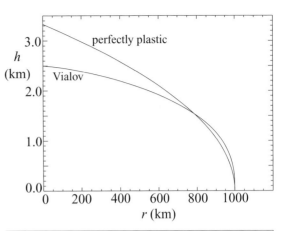

Fig 6.5 Comparison of the perfectly plastic model solution (with $\tau_0 = 1$ bar) to the Vialov solution with Glen's Flow Law with a maximum thickness of 2.5 km.

Figure 6.5 prescribes $h_c = 2.5$ km. The Vialov solution predicts gentler slopes in the ice sheet interior and steeper slopes near the margin relative to the perfectly plastic model.

6.2.2 Numerical solutions

Thus far we have considered only "one-sided" ice sheets and glaciers, i.e. those with elevations that decrease monotonically from a known divide position or increase from a known terminus position. Equation (6.14) is limited to one-sided ice sheets because $dh/dx + db/dx$ must be positive throughout the domain, increasing from the margin (at $x = 0$) to the divide (at $x = L$). In the more general case that includes variable bed topography, only the positions of the two margins are known and the topography on both sides of the ice sheet must be solved for. In such cases, the divide position cannot be prescribed but must be solved for along with the geometry. For such cases, Eq. (6.14) must be modified to include both positive and negative ice-surface slopes to obtain

$$\left|\frac{dh}{dx} + \frac{db}{dx}\right| = \frac{\tau}{\rho g h} \quad (6.31)$$

The boundary conditions for Eq. (6.31) are $h = 0$ at the two ice margins.

Equation (6.31) can be solved with standard methods for boundary-value problems. The equation must first be modified to remove the absolute-value sign by squaring both sides:

$$\left(\frac{dh}{dx} + \frac{db}{dx}\right)^2 = \left(\frac{\tau_0}{\rho g h}\right)^2 \qquad (6.32)$$

and then by differentiating both sides of Eq. (6.32) to obtain:

$$\frac{d^2h}{dx^2} + \frac{d^2b}{dx^2} = -\frac{\tau_0^2}{\rho^2 g^2} \frac{1}{h^3 \left(\frac{dh}{dx} + \frac{db}{dx}\right)} \qquad (6.33)$$

Equation (6.33) is nonlinear and singular at divides (i.e. where $dh/dx + db/dx = 0$). As such, care must be taken to solve it. One approach is to use the relaxation method with a Newton iteration step (Press *et al.*, 1992). In this approach, Eq. (6.33) is discretized as

$$h_{i+1} - 2h_i + h_{i-1} + b_{i+1} - 2b_i + b_{i-1}$$
$$= -(\Delta x)^2 \frac{\tau_0^2}{\rho^2 g^2} \frac{2\Delta x}{h_i^3 (h_{i+1} - h_{i-1} + b_{i+1} - b_{i-1})} \qquad (6.34)$$

where Δx is the resolution cell size of the grid. This equation is solved iteratively with

$$h_i^{new} = h_i^{old} - \frac{L_i(h_i^{old})}{\partial L_i(h_i^{old})/\partial h_i} \qquad (6.35)$$

where

$$L(h_i) = h_{i+1} - 2h_i + h_{i-1} + b_{i+1} - 2b_i + b_{i-1}$$
$$+ \Delta^2 \frac{\tau^2}{\rho^2 g^2} \frac{2\Delta}{h_i^3 (h_{i+1} - h_{i-1} + b_{i+1} - b_{i-1})} \qquad (6.36)$$

Figures 6.6 and 6.7 illustrate example solutions for 1D ice sheets of different sizes and with varying bed topography. In the simple case of a flat bed, the "two-sided" solution predicts gentler slopes near the divide compared to the "one-sided" parabolic solution. In the case of tilted beds, the divide position is solved for simultaneously with the ice geometry using the "two-sided" model. The divide position is located farther up-slope on steeper beds.

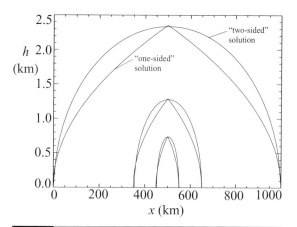

Fig 6.6 Comparison of analytic and numerical perfectly plastic solutions for an ice sheet over a flat bed with $\tau_0 = 1$ bar and $L = 100$, 300, and 1000 km.

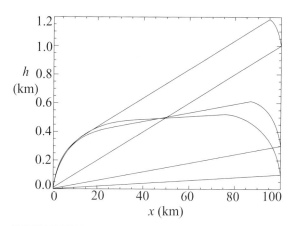

Fig 6.7 Solutions for ice flows on a tilted bed with $\tau_0 = 1$ bar, $L = 100$ km, and bed slopes of 0.01, 0.003, and 0.001. As part of this model solution, the divide position is solved for simultaneously with the ice thickness.

6.3 | Modeling flows with temperature-dependent viscosity

Lava is the classic geological example of a non-Newtonian fluid with a temperature-dependent viscosity. As lava cools, it becomes stiffer. The temperature-dependent viscosity of lava is partly responsible for the rich dynamics of lava flows. Pillow lavas, for example, form underwater through a periodic surge-type mechanism in which pressure builds up beneath a cool, stiff flow margin until the margin fails, causing a rapid surge of hot, rapidly flowing lava out in front of the flow. The surging lava then slows rapidly as the lava cools in contact with the water. Slow velocities are again maintained until sufficient pressure builds up behind the flow margin to cause another surge.

Lava flows in the geologic record often occur as low-relief deposits with a relatively flat top. When exhumed by uplift and erosion, such deposits are often more resistant than the surrounding rocks and create a topographic "inversion" in which the resistant lava flow comprises the highest outcrops in the landscape, forming a mesa. Such lava-flow morphologies are not consistent with the domal morphology of isoviscous gravity flows. Motivated by the observation that lava-flow deposits often lack a domal morphology analogous to ice sheets, Bercovici (1994) investigated a radially-symmetric 1D model of the coupled evolution of mass and heat flow in a fluid spreading by gravity on a flat plane. In this model, strongly temperature-dependent viscosities cause most of the relief of the gravity flow to be accommodated along a cool, stiff gravity-flow margin, while the hot, central portion of the flow is able to flow at very low gradients. Hence, the temperature-dependent viscosity of lava is responsible for the wide, tabular nature of some lava flows according to this model. In this section we review Bercovici's model as an example of modeling the coupled evolution of fluid flow and rheology in gravity flows.

Bercovici (1994) assumed that the lava viscosity, η, was inversely proportional to its temperature, T:

$$\eta(T) = \frac{\eta_h \eta_c}{\eta_h + (\eta_c - \eta_h)T} \tag{6.37}$$

where η_h and η_c are the viscosities of the hottest and coldest parts of the lava, respectively. His model also assumed that cooling takes place along the upper and lower boundaries of the flow where the temperature is kept at a constant reference value $T = 0$. This thermal boundary condition, together with the assumption of a uniform heat flux with depth z, is consistent with a parabolic vertical temperature distribution function that goes to zero at $z = 0$ and $z = H$:

$$T = 6T_{av}\frac{z}{H}\left(1 - \frac{z}{H}\right) \tag{6.38}$$

where $T_{av} = (1/H)\int_0^H T\,dz$ is the depth-averaged temperature. The flow itself is governed by the radially symmetric Stokes equation in which the pressure gradient is balanced by viscous stresses:

$$\Delta\rho g\frac{\partial H}{\partial r} = \frac{\partial}{\partial z}\left(\eta\frac{\partial v_r}{\partial z}\right) \tag{6.39}$$

where g is gravity, v_r is the radial velocity, and $\Delta\rho$ is the density contrast between the fluid and its surrounding medium (e.g. air or water). The vertically-averaged radial velocity is given by

$$U = \frac{1}{H}\int_0^H v_r\,dz = -\frac{\Delta\rho g H^2}{3\eta_h\eta_c}$$
$$(\eta_h + (\eta_c - \eta_h)T_{av})\frac{\partial H}{\partial r} \tag{6.40}$$

Conservation of mass for this system can be written as

$$\frac{\partial H}{\partial t} + \frac{1}{r}\frac{\partial(rUH)}{\partial r} = 0 \tag{6.41}$$

and heat transport is given by

$$\frac{\partial T_{av}}{\partial t} + U\frac{\partial T_{av}}{\partial r} = -\frac{12\kappa}{H^2}T_{av} \tag{6.42}$$

where κ is the thermal diffusivity. Substituting Eq. (6.40) into Eqs. (6.41) and (6.42), Bercovici obtained

$$\frac{\partial H}{\partial t} = \frac{1}{r}\frac{\partial}{\partial r}\left[r(1 + vT_{av})H^3\frac{\partial h}{\partial r}\right] \tag{6.43}$$

$$\frac{\partial T_{av}}{\partial t} = (1 + vT_{av})H^2\frac{\partial h}{\partial r}\frac{\partial T_{av}}{\partial r} - \frac{T_{av}}{H^2} \tag{6.44}$$

where H has been nondimensionalized by $(3\eta_c Q/2\pi\Delta\rho g)$ (where Q is the flux of lava to the surface), t by $H_0^2/12\kappa$, and r by $\sqrt{(\Delta\rho g H_0^5/12C\kappa v_c)}$. The only free parameter in Eqs. (6.43) and (6.44) is the relative viscosity contrast $v = (\eta_c - \eta_h)/\eta_h$.

Here we consider the case of a constant rate effusion, Q to the surface through a cylindrical conduit of radius r_i (Figure 6.8). This case corresponds to a constant-flux boundary condition at $r = r_i$:

$$r(1 + vT_{av})H^3\frac{\partial H}{\partial r} = 1 \text{ at } r = r_i \tag{6.45}$$

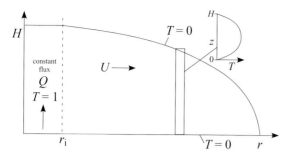

Fig 6.8 Schematic diagram of the lava flow model with coupled mass and heat flow.

Also, the lava is assumed to erupt onto the surface with its maximum temperature $T_{av} = 1$ and $r = r_i$.

Figure 6.9 illustrates the flow-thickness and temperature profiles of a spreading lava flow for a case of (a) no viscosity contrast and (b) a large viscosity contrast. For the case of isoviscous flow, a domal shape develops in which the surface slope is steepest at the margin, but varies continuously from the margin to the conduit radius, defined to be at $r = 1$. This solution is broadly similar to the parabolic geometry of a perfectly plastic ice sheet on a flat bed. In the case of a strong viscosity contrast, a more tabular flow develops. In this case, relief is accommodated along the cool, stiff flow margin, while the hot, fluid central portion of the flow lacks the stiffness to support significant surface slopes.

6.4 | Modeling of threshold-sliding ice sheets and glaciers over complex 3D topography

Analytic solutions are generally restricted to 1D ice sheets and glaciers with relatively simple bed geometries. Real ice sheets and glaciers, of course, are more complex. In this section we describe a method for modeling the 2D geometry of an ice sheet or glacier above complex topography. The solution method relies upon a cellular algorithm that iteratively builds up discrete packets of ice until a threshold condition is achieved. The model is illustrated using applications to modern Greenland and Antarctic ice sheets and the for-

mer Laurentide Ice Sheet and alpine glaciers of the Alaskan Brooks Range.

Before considering the extension of Eq. (6.33) to 2D, we will investigate the spatial variability of basal shear stress beneath modern ice sheets. Basal shear stresses are not uniform. However, by characterizing the spatial variability observed in real ice sheets, we can extend the threshold-sliding model to include the observed variations as a further development of the model. Recently compiled datasets for the ice-surface and bed topography of Greenland (Bamber *et al.*, 2001) and East Antarctica (Lythe *et al.*, 2001) are available for this purpose. Figures 6.10a and 6.10b are grayscale images of the basal shear stresses in Greenland and East Antarctica calculated from Eq. (6.4). The brightness scale for these images is given in the lower left of the figure. The scale varies from 0 (black) to 5 bars (white), with typical values in the range of 0.5 to 3 bars. The average basal shear stress for Greenland and East Antarctica, calculated from these maps, are almost identical: 1.41 bars for Greenland and 1.39 for East Antarctica. These ice sheets also share a common pattern of spatial variation.

Figures 6.10a and 6.10b clearly show that basal shear stresses increase with distance from divides, with values close to 0.5 bars near divides to an average of 3 bars near margins. Notably, the bed-elevation data for East Antarctica are interpolated from a much lower density of observations than for Greenland, especially south of 87°. Nevertheless, in both Greenland and East Antarctica, where the data are most complete, this pattern is clear.

Data from example profiles are plotted in Figure 6.10c. Although there is significant scatter, the basal shear stresses increase from 0.5 to an average of 3 bars in each of the profiles. The scatter observed in these data is partly a function of the high resolution of the source data. At these small scales (i.e. 5 km), variations in ice-surface slope occur that are not significant at the larger scales of greatest interest in ice-sheet reconstructions.

One means of incorporating the observed spatial variability of basal shear stresses is to consider them to be a function of ice-surface slope. In Figure 6.11a and 6.11b, the basal shear stresses

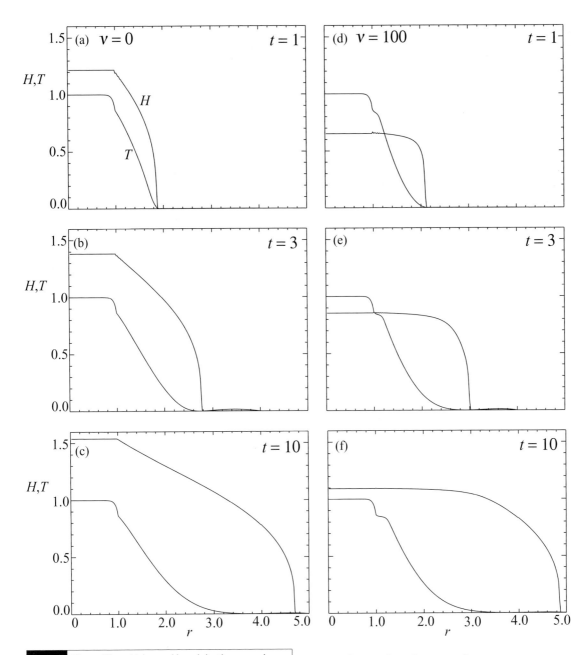

Fig 6.9 Plots of flow thickness, H, and depth-averaged temperature, T, as a function of time, t, for a flow with (a)–(c) constant viscosity, and (d)–(f) temperature-dependent viscosity with $\nu = 100$.

of Greenland and East Antarctica are plotted as a function of ice-surface slope on logarithmic scales. These plots illustrate that basal shear stresses increase from values as low as 0.5 bars at low slope values to higher, more variable val-

ues as slope values increase. Least-squares power-function fits to these data are indicated by the solid lines in Figure 6.11. The least-squares fit for Greenland is $\tau = 15S^{0.55}$, while the fit for East Antarctica is slightly higher at $\tau = 18S^{0.55}$. Certainly a higher-order fit would characterize the data more precisely (because both data sets have a significant convex curvature in their dependence on ice-surface slope), but a power-law function of slope provides a good first-order approximation.

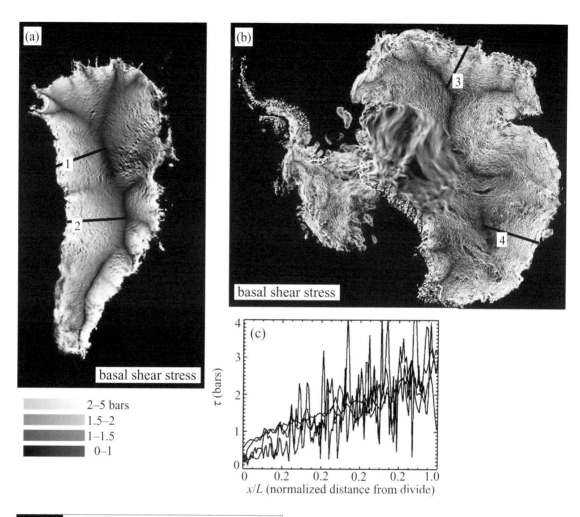

Fig 6.10 Basal shear stresses beneath Greenland and East Antarctica, calculated from Eq. (6.4). Legend gives brightness scale. (a) Greenland, calculated using datasets of Bamber et al. (2001), (b) East Antarctica, calculated using datasets of Lythe et al. (2001). (c) Profiles of τ versus distance from the divide for profiles 1–4. In these profiles, τ varies from about 0.5 bars at divides to an average of 3 bars near the margins. Basal shear stresses are also more spatially variable near margins, as indicated by the large scatter on the right side of the plot.

Our approach is to apply the empirical results of Figure 6.10 directly to the threshold-sliding model by replacing τ with $\tau(S)$ in Eq. (6.31). In 2D, Eq. (6.31) applies along the direction of flow lines. Since flow lines are parallel to the local ice-surface gradient, the 2D version of Eq. (6.33) is obtained by replacing $S = dh/dx + db/dx$ with

$S = \nabla h + \nabla b$ to obtain

$$|\nabla h + \nabla b| = \frac{\tau(S)}{\rho g h} \tag{6.46}$$

where

$$\tau(S) = 15 S^{0.55} \tag{6.47}$$

to incorporate the spatial variations in basal shear stress observed in Figure 6.10. Equation (6.46) then becomes

$$|\nabla h + \nabla b|^{0.45} = \frac{15}{\rho g h} \tag{6.48}$$

Equation (6.48) can be transformed to a second-order differential equation analogous to Eq. (6.33). The resulting equation cannot be solved by relaxation methods, however, because the slopes vary both in direction and magnitude. This

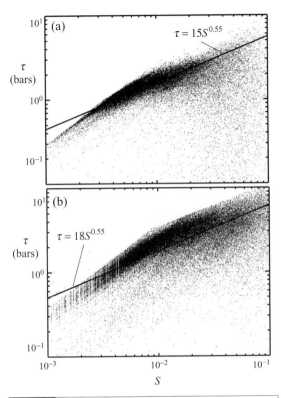

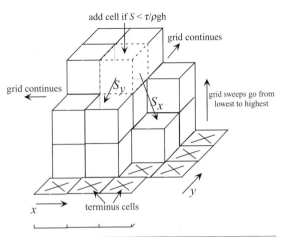

Fig 6.12 Illustration of the "sandpile" method for solving Eq. (6.46). At each iteration of the algorithm, a unit of ice is added to each grid point within the area of ice coverage if the addition does not violate the condition that $S > \tau/\rho g h$ (where S is the ice-surface slope, equal to $\sqrt{S_x^2 + S_y^2}$). The ordering of the grid points is important in this algorithm. In order to avoid oversteepenings, the sweep through the grid should be from lowest to highest elevations.

Fig 6.11 Relationship of basal shear stress, τ, to ice-surface slope for (a) Greenland, and (b) East Antarctica. Each point represents a pixel from Figure 6.10. Solid lines are least-squares power-law fits.

additional degree-of-freedom renders the relaxation method unstable in 2D.

Reeh (1982, 1984) developed a method for applying the perfectly-plastic model to 2D ice sheets by solving the 1D model on a curvilinear coordinate system that follows flow lines. Equation (6.48) can be solved more simply, however, using a simple algorithm based on the accumulation of discrete "blocks" of ice on a grid. This method mimics the accumulation of ice thickness and slope until a threshold is reached. The algorithm is illustrated schematically in Figure 6.12. Before the algorithm begins, grid points within the ice margin are identified. These are the points where ice will accumulate; other grid points will remain ice free. The fundamental action item for each allowed grid point is simple: add a discrete unit of ice thickness if the resulting surface slope is less than a threshold value, given by $\tau(S)/\rho g h$. The ice-surface slope is calcu-

lated as $S = (S_x^2 + S_y^2)^{1/2}$, where S_x is the downhill slope in the x-direction and S_y is the downhill slope in the y-direction. During each "sweep" of the grid, the algorithm attempts to add ice to all of the allowed grid points. The grid is swept repeatedly until no additional blocks can be added. The end result is an exact solution to Eq. (6.46) for any bed topography. τ can be taken to be uniform in this model or it can be a function of the ice-surface slope S. In addition to mimicking the accumulation of ice in a thickening ice sheet, the method is also analogous to the thickening of a sandpile until a threshold angle of repose is reached. A fundamental difference between this model and a real sandpile, however, is that the angle of repose is not uniform but is lower in areas of thicker ice.

Two additional aspects need to be specified to fully define the algorithm. First, in what order should blocks be added to the grid? The order matters because the addition of a block at one grid point may increase the slope of one of its neighboring grid points to an oversteepened condition (i.e. one with a larger basal shear stress than the threshold value). To avoid this problem, it is best to move through the grid in order of ascending elevations. To do this, a table is created at the start of each grid sweep that indexes grid points from lowest to highest elevation. Efficient methods based on the `quicksort` algorithm are available for this purpose (e.g. Press *et al.*, 1992). Oversteepenings do not occur with this ordering because slopes are dependent only on the values of *downslope* points. Therefore, adding ice to *higher* elevations does not change the slopes of grid points that have already been swept, and oversteepenings are avoided. The reconstructions in this chapter were obtained using this ordering scheme. Another ordering scheme that works is to proceed from the outermost to the innermost pixels.

Second, how thick should the discrete "unit" of ice be for each iteration? In order to achieve a smooth, precise ice-surface topography, the unit should generally be less than or equal to 1 m thick. However, for an ice sheet that is several kilometers thick, this would require adding blocks of ice to each grid point thousands of times over, resulting in an inefficient method. The optimal approach is to start with a coarse block size (e.g. 100 m) and reduce the size by a constant factor (e.g. 3) whenever no more blocks can be added to the grid. The model gradually reduces the block size until a prescribed minimum thickness (e.g. 10 cm) is reached. This approach builds up a coarse solution quickly in the early phase of the run, gradually smoothing the rough edges as the block size is reduced.

For many glacial-geologic applications, ice velocities are an important variable to constrain and map. In glacial bedrock erosion, for example, basal-sliding velocities are likely to be the key variable controlling bedrock-erosion rates (Andrews, 1972; Hallet, 1979, 1996). Therefore, a better understanding of the formation of finger lakes, U-shaped valleys, cirques, and other glacial-erosional landforms will likely require a method for calculating the velocity field of former ice-sheets and glaciers in 2D.

Budd and Warner (1996) developed an algorithm for computing the depth-averaged balance velocities of ice sheets. These authors illustrated their method by computing a map of balance velocities for East Antarctica. This important work quickly inspired calculations of balance veloci-

ties in Greenland (Bamber *et al.*, 2000) and Austfonna, in eastern Svalbard (Dowdeswell *et al.*, 1999). These studies have generally emphasized the important of the distance from the ice margin in controlling the spatial distribution of flow. The interiors of thick ice sheets are generally characterized by relatively slow, distributed flow while outlet glaciers and ice streams are characterized by much faster and more focused flow. The flow-distribution algorithm at the heart of Budd and Warner (1996) is similar to fluvial flow-distribution algorithms (e.g. Murray and Paola, 1994; Coulthard *et al.*, 2002), suggesting that glacial drainage networks may be modeled using similar approaches developed for fluvial drainage networks (e.g. Willgoose *et al.*, 1991).

Balance velocities are those required to maintain a time-independent ice-sheet topography. The analogy of an ice sheet or glacier as a conveyor belt is useful here; balance velocities are those that transport exactly the right amount of ice at each point to prevent thickening or thinning of the ice at any point. For threshold-sliding ice sheets and glaciers, balance velocities are equivalent to instantaneous velocities because in a threshold-sliding ice sheet or glacier the entire ice mass continuously adjusts to a change in accumulation or ablation. Although Budd and Warner's method computes *depth-averaged* velocities rather than *basal* velocities, the two are nearly identical for threshold-sliding ice sheet and glaciers in which the ice-surface slope is much less than one. Nye (1951) provided equations for both depth-averaged and basal velocities in threshold-sliding ice sheets and glaciers. He found that the magnitude of the correction term relating depth-averaged velocities to basal velocities is on the order of the ice-surface slope, and hence is very small for most applications.

In the Budd and Warner algorithm, the ice-surface topography, ice thickness, and rates of accumulation and ablation are used as input. Local accumulation rates are added to each grid point and incoming ice is distributed to the downslope nearest neighbors. The amount of ice distributed to each downslope grid point is proportional to the ice-surface slope in the direction of that grid point. The input and output through each grid point is calculated using a highest-to-lowest ordering scheme. This scheme ensures that all of the contributing discharge from upstream sources is accumulated at each grid point before it is distributed to downslope neighbors. In the final step of the algorithm, the local ice discharge is divided by the local ice thickness to obtain the depth-averaged velocity for each grid point.

As a complement to our reconstructions of ice-surface geometry, we have used the Budd and Warner method to calculate relative depth-averaged velocities for each of our reconstructions. For these calculations we have assumed that accumulation is uniform and that there is no ablation. Although these assumptions are clearly invalid, the resulting maps are useful because flow is so strongly controlled by the distance from divides. Therefore, even if accumulation rates differ by a factor of ten or more from one side of an ice sheet to the other, the basic pattern of flow is essentially unchanged. This insensitivity is analogous to the spatial variations in discharge within a river basin. Discharge in a river network is predominantly a function of drainage area; variations in precipitation and runoff within a basin play a much less significant role. The flow maps we have created also serve to indicate the extent to which the ice flow is controlled by the subglacial topography. For ice sheets in which the thickness is much greater that the subglacial topographic relief, ice flow is only weakly controlled by the topography. In contrast, if the subglacial-topographic relief is comparable to the ice thickness, flow can be strongly focused into troughs and depressions beneath the ice.

In this section we present reconstructions of the ice-surface topography and flow of the modern Greenland and Antarctic ice sheets using the sandpile and Budd and Warner (1996) algorithms. Greenland is perhaps the most important case study in this chapter because the bed topography is both well constrained and has a significant influence on the morphology of the ice sheet. The principal divide in Greenland is offset from central Greenland by approximately 100 kilometers. This asymmetric profile reflects a bed-topographic control associated with higher bed elevations in eastern Greenland.

Figure 6.13 illustrates the bed and ice-surface topography for modern Greenland (Bamber et al., 2001). Areas below sea level are indicated in black. This and all other reconstructions in this chapter require three inputs: a DEM of bed topography, a "mask" grid that defines the areas of the ice margin, and a function defining the basal shear stress, such as $\tau = 15S^{0.55}$. Appendix 5 provides a C code that implements the sandpile and balance velocity method for the Greenland and Antarctic ice sheets.

The bed topography is input directly into the model. The ice-thickness data, derived from the difference between the ice-surface and bed topography, is used to provide a binary "mask" grid defining the grid points that are allowed to accumulate ice in the sandpile model. The mask grid has values of 1 where the ice thickness is greater than 0 (i.e. areas covered by ice), and values of 0 where the thickness is 0 (i.e. no ice coverage). For this reconstruction we used the shear-stress relationship $\tau = 15S^{0.55}$ observed in Figure 6.11.

Figure 6.13c is a shaded-relief and contour map of the solution to Equation (6.46) obtained with the sandpile model. Figure 6.13c has been constructed with the same vertical exaggeration (30×) as Figure 6.13b to provide a direct, side-by-side comparison. The similarity of the location and shape of the contours indicate that the overall solution is in good agreement with the observed topography. The location of the major divide, offset from center by approximately 100 km, is also reproduced in the model. One major difference between the observed and modeled topography, however, is the steepness of divides in the model solution. The model topography has a significantly more angular appearance than the observed topography. This angularity can be traced to the poorness of fit between the power-function and the data of Figure 6.11a. In the lower-left corner of the plot, representing areas of low slope, the data fall far below the power function, indicating that the threshold basal shear stress near divides is significantly lower than the values predicted by the power function. As a result, the model solution overestimates the ice-surface slopes near divides by as much as a factor of 2. This does not introduce significant errors in the *elevations* of these regions because the slopes in

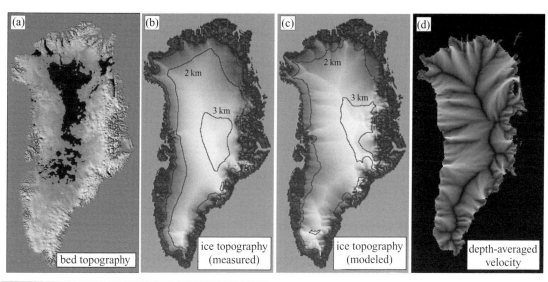

Fig 6.13 Reconstruction of the modern Greenland Ice Sheet using Eq. (6.46) and the observed correlation $\tau = 15S^{0.55}$ from Figure 6.11. Bed and ice-surface topography are illustrated with shaded-relief (30 × vertical exaggeration) grayscale maps. (a) Input bed topography from Bamber *et al.* (2001). Darkest areas are below sea level. (b) Ice topography observed from radar interferometry and given by Bamber *et al.* (2001). Contours are 1 km in spacing (1 km contour is too close to the margin to be easily seen). (c) Numerical reconstruction of the ice-surface topography (with same grayscale and shading as in (b)). The divide position and elevations are a good match to (B) except that the divides are too peaked. (d) Grayscale map of depth-averaged velocities computed using the balance-velocity algorithm of Budd and Warner (1996) assuming uniform accumulation. The grayscale map is scaled logarithmically.

either case are small. The flow map for Greenland is indicated in Figure 6.13d and is similar to the map provided by Bamber *et al.* (2000b). The grayscale map is scaled logarithmically in this and all subsequent flow maps to illustrate the internal flow distribution. A linear map, in contrast, would show significant velocities only at the outlet areas.

Input data for the East Antarctic Ice Sheet reconstruction (Figure 6.14) includes the DEM of bed topography provided by Lythe *et al.* (2001). Lythe *et al.*'s ice-thickness data were also used to derive a mask of ice coverage. Rather than including all areas of finite ice thickness in our mask, however, we used a threshold ice thickness of 150 meters instead. This elevation was used to elim-

inate ice shelves and areas of sea ice from the reconstruction.

Here we used the shear–stress relationship $\tau = 18S^{0.55}$ observed in Figure 6.11b. For comparison, we also performed a reconstruction using the relationship of $\tau = 15S^{0.55}$ observed for Greenland. Using $\tau = 18S^{0.55}$ results in a slightly better match to the modern ice-surface topography than we obtain by using the Greenland relationship, which yields a maximum elevation of just under 4 km.

These reconstructions illustrate that the sandpile model is capable of accurately predicting ice-sheet geometries if the threshold basal shear stress is well constrained. However, the principal objective of glaciological work within the context of geomorphology is to reconstruct former ice sheets and glaciers.

Reconstructions of the Laurentide Ice Sheet (LIS) have had a long history that includes many classic contributions. However, despite decades of effort there is still vigorous debate over the mean thickness of the LIS and the locations of its major domes. These questions bear on many important issues, including the magnitude of Late Quaternary sea-level change. Numerical reconstructions of the Laurentide Ice Sheet at the Last Glacial Maximum have generally reproduced a relatively thick ($\geq$ 2 km) ice sheet with a central east-west divide (e.g. Denton and Hughes, 1981). In contrast, Peltier's (1994) ICE4G relative sea-level (RSL)

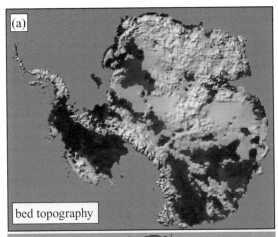

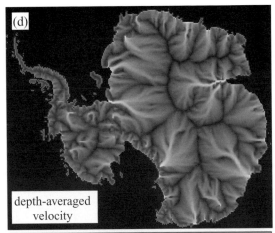

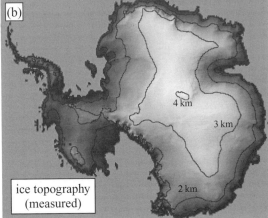

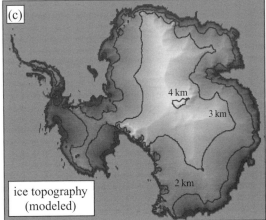

Fig 6.14 Reconstruction of the modern Greenland Ice Sheet from Eq. (6.46) with $\tau = 18S^{0.55}$. (a)–(d) are identical to the corresponding images for Greenland in Figure 6.13.

inversion reproduces a thinner ice sheet with a more complex ice-surface topography.

An inherent uncertainty in many ice-sheet reconstructions is the amount of isostatic rebound experienced by the bed between the time of glaciation and the present. This uncertainty is not significant for areas that have been unglaciated for several tens of thousands of years. In these cases, we can be sure that the modern topography includes no dynamic component: postglacial rebound has run its course. It is common in these cases to ignore the time delay of mantle response and add an instantaneous deflection amplitude proportional to the local ice load: for every kilometer of ice thickness, approximately 270 meters of ice are added to the thickness

to account for bed deflection (e.g. Tomkin and Braun, 2002). The ratio of 0.27 comes from the relative density of ice and mantle. This approach is straightforward to incorporate in the sandpile model by replacing h with $(1 + \rho_i/\rho_m)h$ in the threshold condition. It should be noted, however, that this approach is valid only for ice sheets larger than the flexural wavelength. Smaller ice sheets and glaciers may be partially or fully flexurally compensated.

Including the effects of isostasy is a much greater challenge for LGM ice sheet reconstructions because postglacial rebound has not yet run its course. It is difficult to estimate the amount of postglacial rebound that has taken place or to estimate the amount left to occur because the dynamics of postglacial rebound depend on the size and shape of the load through time. In particular, we cannot extrapolate the present rate of sea-level rise along an exponential curve

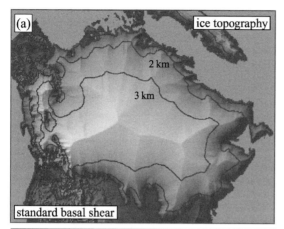

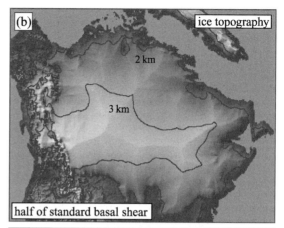

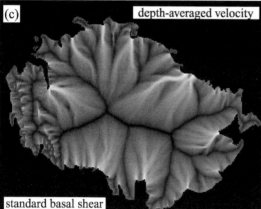

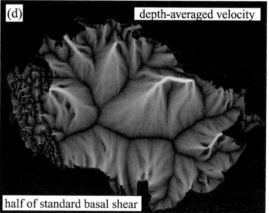

Fig 6.15 Reconstruction of the Laurentide Ice Sheet at Last Glacial Maximum (18 ka) assuming (a) $\tau = 18S^{0.55}$ (referred to as standard basal shear) and (b) $\tau = 9S^{0.55}$ (one-half the standard basal shear). Bed and ice-surface topography are illustrated with shaded-relief (100× vertical exaggeration) grayscale maps. (c) Grayscale map of depth-averaged velocities for the standard basal shear model assuming uniform accumulation.

because the Maxwell-relaxation time of the mantle depends on the wavelength of the load (e.g. Turcotte and Schubert, 2002). Therefore, postglacial rebound is not a simple function of the local ice thickness but is also controlled by the large-scale morphology of the ice sheet, particularly at long time scales.

Nevertheless, we can take a similar approach to including isostatic deflection that is often taken in cases with no dynamic topography. Rather than adding 27% of the ice thickness to account for deflection, we can add some fraction

between 0 and 27%. If we believe that postglacial rebound beneath the LIS is nearly complete, we should choose a value close to 0.27. In contrast, if little postglacial rebound has occurred, the value should be closer to zero. For our Laurentide reconstructions, we will assume a ratio of 0.2. This value assumes that the majority of postglacial rebound has taken place. The uncertainty of this value, which could plausibly vary between 15 and 25%, introduces a 10% uncertainty into determination of the LIS thickness.

Two reconstructions of the LIS and LGM are presented in Figure 6.15. Figures 6.15a and 6.15c illustrate the ice-surface topography and flow using the basal shear stress observed in East Antarctica ($\tau = 18S^{0.55}$). We refer to this shear–stress relationship as the "standard" basal shear model, because it characterizes the observed trends in the Greenland and Antarctic Ice Sheets (Figure 6.11). For comparison, we have also provided a second reconstruction which uses a shear stress equal to one half of the standard basal shear, or $\tau = 9S^{0.55}$.

The DEM of bed topography was constructed by projecting the ETOPO5 DEM (Loughridge, 1986) for North America to a Lambert Equal-Area projection. The ETOPO5 data include both bathymetry and topography data. The bathymetry is important in this case because the LIS covered many areas that are now submerged. The ice margins were digitized from Dyke and Prest (1987) by Eric Grimm of the Illinois State Museum (personal communication, 2002) and rectified to the DEM by the author. The bed and ice-surface topography are illustrated with shaded-relief ($100\times$ vertical exaggeration) grayscale maps in Figure 6.15. The bathymetry in these images has been removed so that the coastlines could be indicated instead. Use of the standard basal shear model yields an ice sheet with a central divide striking east-west. The average elevation and thickness of the ice sheet in Figure 6.15a (the standard basal shear case) are 2211 m and 2011 m. Decreasing the shear stress preserves the overall shape of the ice sheet but results in a migration of the divide to the southwest where the ice is propped up by the topographic front of the Canadian Rockies. The average elevation and thickness of Figure 6.15b are 1482 m and 1134 m, respectively. Although the *average* elevation and thickness of the standard reconstruction are a relatively modest 2.2 and 2.0 km, respectively, much of the central core of the ice sheet is quite high, with average elevations above 3 km and greater than 3.5 km in thickness. As a caveat, it should be noted that we did not remove the post-glacial sediment from Hudson Bay or the Great Lakes prior to this reconstruction. This will not affect the large-scale geometry of the ice sheet significantly. Certain areas, however, including the Great Lakes, will have ice overestimated thicknesses due to the postglacial sediment fill in the DEM. A more precise reconstruction would require that any post-glacial sediment be removed from these areas to reflect the topography that the ice sheet experienced.

The results of Figure 6.15 indicate that even though the thickness of the LIS is sensitive to the basal shear stress, the overall shape is not sensitive. The principal divide migrates to the west

and south between Figures 6.15a and 6.15b, but otherwise the flow pattern is unchanged.

Figures 6.16 and 6.17 illustrate the ice-surface topography and flow of the Laurentide Ice Sheet at 14, 13, 12, 11, 9, and 8 ka. These reconstructions do not include ice coverage between northern Canada and Greenland (Dyke et al., 2002), and are therefore only approximate in their northernmost regions.

The reconstructions of Figures 6.16a–6.16f are characterized by a thick central region that changes little as it shrinks in size. By 11 ka the ice sheet has shrunk to less than half its original area (including coverage in the Cordillera) but still maintains a significant central region with elevations above 3 km. Only by 9 ka has the overall shape of the ice sheet and the altitude of its central region changed significantly. At this point, the central region of the ice sheet has collapsed into three distinct domes. This collapse is followed by the rapid deglaciation of Hudson Bay between 9 and 8 ka.

An important caveat should be given regarding the reconstruction of the ice in the Canadian Cordillera. In alpine terrain it is important to use DEM data that enable the model to estimate bed slopes accurately. It is unlikely that the ETOPO5 dataset, with a resolution of 5 km, is an adequate representation of the bed topography in the Cordillera. Coarse DEMs such as ETOPO5 may underestimate bed slopes, especially in high-relief terrain where closely spaced ridges and valleys may not be resolved. This problem can be minimized by using DEMs of the highest-available resolution. Nevertheless, some size reduction may be necessary to maintain reasonable grid sizes and computing times. When size-reduction must be done, it should be done by subsampling the DEM rather than by averaging neighboring pixels. Subsampling will tend to preserve the small-scale topographic relief in the DEM better than averaging. In the Laurentide reconstruction of Figure 6.15, the thickness of the Cordilleran Ice Sheet may be overestimated because of the coarse scale of the DEM (5 km). Applications focused on extensive areas of high-relief terrain should be reproduced on multiple scales to verify that the results are independent of the DEM scale.

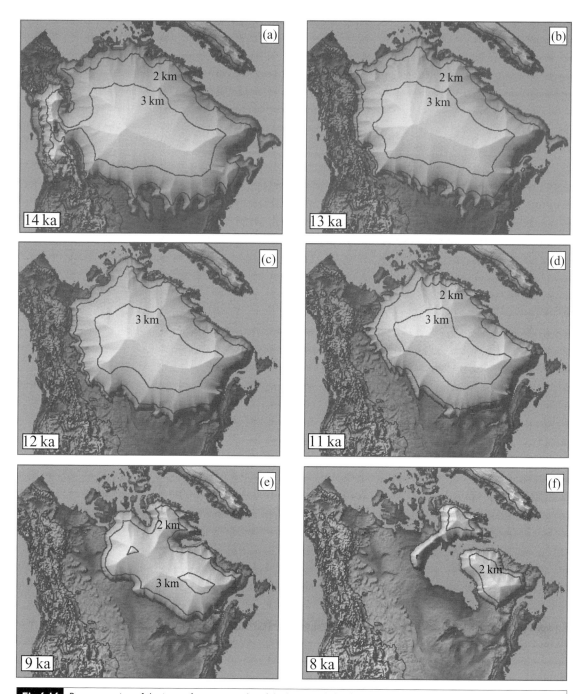

Fig 6.16 Reconstruction of the ice-surface topography of the Laurentide Ice Sheet from 14–8 ka assuming the standard basal shear model and using the margin positions of Dyke and Prest (1987). Bed and ice-surface topography are illustrated with shaded-relief (100× vertical exaggeration) grayscale maps. Ice coverage between Greenland and northern Canada has not been included in this reconstruction.

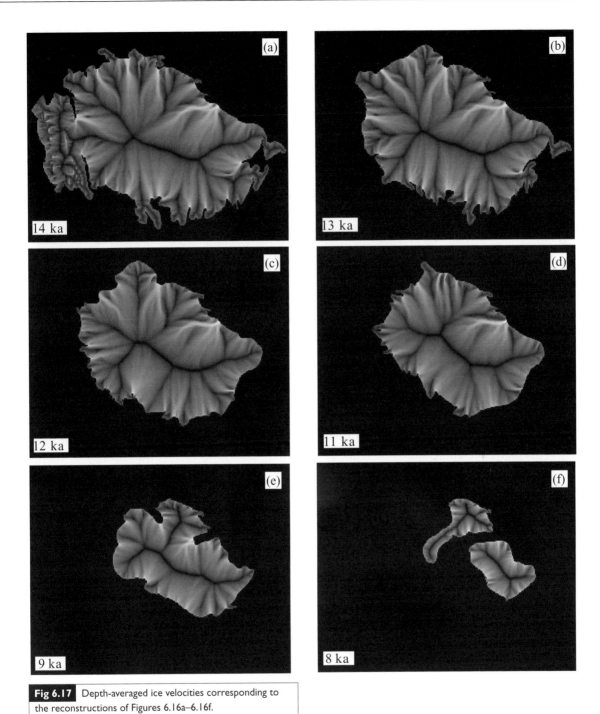

Fig 6.17 Depth-averaged ice velocities corresponding to the reconstructions of Figures 6.16a–6.16f.

Next we illustrate the model application to a high-resolution reconstruction of the Laurentide Ice Sheet in the Finger Lakes Region of New York State. This example will illustrate the application of the model to an ice-marginal environ-ment in which only a portion of the ice sheet can be modeled at one time. In addition, the Finger Lakes are an example where both the bed topography and the distance from the ice margin have an important influence on ice-surface topography and flow. In contrast, large ice sheets and alpine glaciers are predominantly influenced by only one of these variables. For example, the local topography and flow in large ice sheets is primarily a function of distance from the margin, while the thickness and flow in alpine glaciers are almost entirely controlled by bed topography.

The Finger Lakes Region has been the focus of many glacial-geologic studies, including glaciological reconstructions (Ridky and Bindschadler, 1990), stratigraphic studies of meltwater production and lake infilling (Mullins and Hinchey, 1989), and geomorphic studies of the Finger Lakes themselves (von Engeln, 1956). Geomorphically, the Finger Lakes Region is dominated by subparallel, glacially-scoured troughs with their southernmost extents in Seneca and Cayuga Lakes, the two largest of the Finger Lakes. The five largest troughs of the region comprise the Finger Lakes proper, but there are numerous other troughs cut into the Allegheny Plateau of smaller size that are not deep enough to be enclosed depressions. Figure 6.18a is a shaded-relief image of the topography of the region. The troughs vary in spacing from 10 to 30 km along strike, with the greatest spacing between Seneca and Cayuga Lakes. The southern tips of the Finger Lakes coincide with the Valley Heads Moraine (14 ka). As such, the scouring of the Finger Lakes Region most likely took place when the ice margin was coincident with this moraine, although several phases of glaciation may have contributed to their formation.

Little is known about the processes and dynamics of the scouring of glacial troughs in general or the Finger Lakes in particular. The regular spacing of the Finger Lakes suggests a positive-feedback or instability mechanism in which incipient depressions in the bed topography focus ice flow, resulting in enhanced deepening and focusing of ice flow. In order for this model to be valid, the ice-surface topography in the region must reflect variations in bed topography in order for ice flow to be focused into troughs. The first step towards testing this hypothesis is to reconstruct the ice-surface topography and flow in the region to determine whether the ice-flow patterns were likely to have been focused by the bed

topography. We will not model the full dynamics of Finger Lakes formation in this chapter but will rather determine what the ice-surface topography and flow looked like after the Finger Lakes had formed.

The reconstruction of this region requires the following inputs: bed topography (Figure 6.18a), a mask grid incorporating the position of the Valley Heads Moraine, and the threshold basal shear stress. The topography of the region was obtained by joining several 90-m resolution USGS DEMs. The position of the Valley Heads Moraine was used to create the mask grid by overlaying the surficial geologic map of the area (New York State Geological Survey, 1999) onto the bed topography to delineate the ice margin. At the spatial scale of the Finger Lakes, the relationship between basal shear stress and ice-surface slope observed in Greenland and East Antarctica is not useful because ice-surface slopes this close to the margin are uniformly steep. In addition, the average basal shear stress observed for modern ice-sheet margins (3 bars) may also be inapplicable because shear stresses near ice-sheet margins are spatially variable and may be much smaller than 3 bars locally. In lieu of a better constraint on basal shear stresses in the region, we have used the traditional end-member values of 0.5 and 1.5 bars in two alternative reconstructions. In addition, we have modified the ice-flow portion of the reconstruction so that all of the flow originates at the top boundary of the grid. This modification represents the incoming ice flow from the Laurentide Ice Sheet, which we will assume dominates over any locally-derived accumulation.

The results for the ice-surface topography and depth-averaged flow are given in Figures 6.18b–6.18c and 6.18d–6.18e for 1.5 and 0.5 bars, respectively. These values bracket an important transition in ice-sheet behavior. If the basal shear stress is 1.5 bars or larger, the topography and flow are only weakly dependent on bed topography. Figure 6.18b illustrates the ice-surface topography in this case. The reconstructed ice sheet in this figure is dominated by a forked central divide and distributary ice flow. In contrast, for basal shear stresses of 0.5 bars or smaller, the

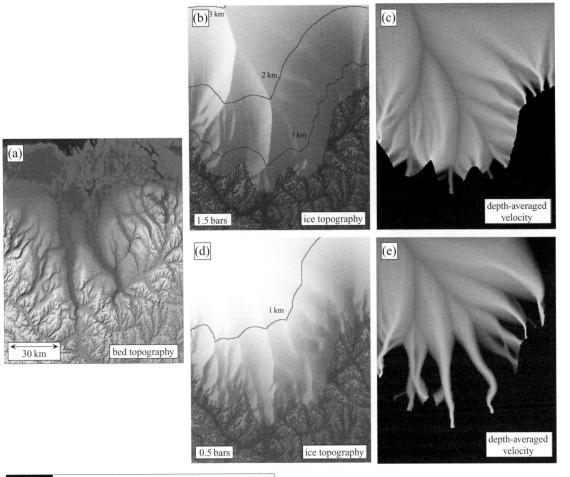

Fig 6.18 Reconstruction of the ice-surface topography and depth-averaged velocities in the Finger Lakes Region. Two reconstructions are presented with uniform basal shear stresses of (a) and (c) 1.5 bars and (d) and (e) 0.5 bars. (a) Shaded-relief image of bed topography. (b) Shaded-relief image of ice-surface topography and (c) depth-average velocity assuming 1.5 bars. For this case, representing a relatively rigid base, ice topography and flow are only weakly controlled by subglacial topography. (d) Shaded-relief image of ice-surface topography and (e) depth-average velocity assuming 0.5 bars.

ice-surface is strongly controlled by the subglacial topography. Ice flow is focused into the troughs of the Finger Lakes in this case. If the Finger Lakes were formed by focused ice flow, as the positive-feedback model requires, these reconstructions suggest that the basal shear stress was probably closer to 0.5 bars than to 1.5 bars. Future work will require that the postglacial sediment in the

Finger Lakes be removed to more accurately reflect the bed topography experienced by the ice sheet. Also, drumlin orientations can be used to provide a constraint on the flow pattern and, in conjunction with reconstructions of different shear stresses, provide an indirect constraint on the appropriate value of shear stress. Ridky and Bindschadler (1990), for example, used drumlin orientations to guide their 2D flow line reconstructions in this area.

One additional point should be made regarding this example. Although the solution of partial differential equations such as Eq. (6.46) typically require the application of boundary conditions at the edges of the grid, the sandpile algorithm does not require boundary conditions at grid boundaries in *upslope* directions. The reason

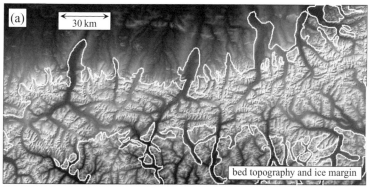

(a)

30 km

bed topography and ice margin

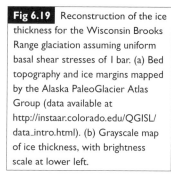

Fig 6.19 Reconstruction of the ice thickness for the Wisconsin Brooks Range glaciation assuming uniform basal shear stresses of 1 bar. (a) Bed topography and ice margins mapped by the Alaska PaleoGlacier Atlas Group (data available at http://instaar.colorado.edu/QGISL/data_intro.html). (b) Grayscale map of ice thickness, with brightness scale at lower left.

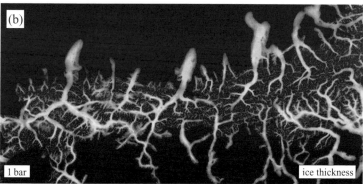

(b)

1 bar

ice thickness

300–400 m
200–300
100–200
0–100

boundary conditions are not required is that the algorithm refers only to downslope grid points. This is useful for applications in ice-marginal areas because only a portion of the ice sheet need be considered.

One of the benefits of the threshold-sliding model is that it can be broadly applied to the reconstruction of any ice sheet or glacier in the continuum from large ice sheets to alpine glaciers. As a robust and broadly applicable model, it can provide preliminary reconstructions for any area and requires the constraint of just a single input variable, the basal shear stress.

As an example of the model application in an alpine environment, we will reconstruct the Wisconsin glaciation of the central Brooks Range of Alaska using input data compiled by the Alaska PaleoGlacier Group (http://instaar.colorado.edu/QGISL/data_intro.html). This data includes a 100-m resolution DEM of bed topography and a vector dataset of Wisconsin ice

margins. We have assumed a uniform threshold shear stress of 1 bar for this reconstruction.

In the alpine environment, the bed-slope component of the basal shear stress is dominant. As such, variations in ice thickness are small and are associated with areas of bed curvature: thickening occurs in areas of bed concavity and thinning occurs in areas of bed convexity. Rapid changes in glacier thickness are, therefore, restricted to slope breaks and glacier snouts.

Figure 6.19b presents the results of our reconstruction based on the input data in Figure 6.19a. Ice is up to several hundred meters thick in glacial valleys and is generally thicker where bed slopes are gentler, except for areas close to glacier snouts where thicknesses taper. Modifying the assumed basal shear stress from 1 bar to a greater or lesser value leads to thicker or thinner glaciers but does not significantly change the relative thicknesses illustrated in Figure 6.19b.

In this reconstruction, steep areas in the highlands of the Brooks Range are predicted to have thin ice, as indicated by the dark tones in the grayscale map in Figure 6.19b. However, it is likely that below some minimum thickness, ice cannot form and locally the area will be ice free.

For example, in steep terrain the snow depth may be insufficient to insulate deeper layers from summer melting or it could be too thin to initiate compaction. To account for this, the alpine reconstructions could be post-processed to remove ice from areas that have insufficient thickness for ice formation. In such cases, the reconstruction should then be recomputed with the revised mask to provide an improved reconstruction.

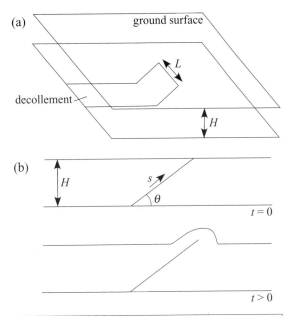

Fig 6.20 Schematic diagram of the 2D thrust sheet model.

6.5 | Thrust sheet mechanics

Deformation within Earth's crust takes place by brittle fracture and slip along faults and joints. As such, it is not appropriate, in detail, to model the crust as a fluid. At large scales, however, plastic flow models do a surprisingly good job at reproducing the observed geometries of structural folds. The reason why plastic flow models are successful is that they represent the collective effects of slip along a complex system of faults and joints. As a rock is stressed, it deforms very little until the yield stress of the rock is exceeded and a new fault or joint is formed. Subsequent deformation is accommodated by slip along faults and joints when the threshold frictional force is exceeded. Rock fracture and slip along a network of faults are both threshold-controlled motions. As such, a perfectly plastic or threshold-sliding model is often a good model for the large-scale deformation of the crust.

In this section we consider a simple example of thrust-sheet displacement using a perfectly plastic model. We assume the existence of a crustal ramp of angle θ and width L separating the ground surface from a decollement surface at depth H (Figure 6.20). Motion along the thrust fault takes place with slip rate s. The key questions we will ask are: (1) what is the 2D shape of the fold above the thrust fault, and (2) how does the shape of the fold depend on the yield stress and fault geometry? In this example we prescribe the tectonic geometry and slip rate, but it is also possible to prescribe the stress state along the fault and allow the slip rate to respond to both the tectonic stress and the overburden stress, allowing the fault to slow down as overburden accumulates.

The thickness of rock above the fault surface is governed by Eq. (6.46) with a uniform yield stress:

$$h|\nabla h + \nabla b| = \frac{\tau_0}{\rho g} \tag{6.49}$$

Equation (6.49) can be solved with the sandpile algorithm using a very similar approach to the one used for modeling 2D ice sheets over complex topography. The advance of the fault tip is modeled by adding mass to the system directly above the fault tip at a rate equal to the vertical component of slip. Appendix 5 provides a C code that implements this approach.

Figure 6.21 illustrates the fold evolution for a case with $\tau_0 = 20$ bars, $\rho = 2700 \text{ kg/m}^3$, $H = 3 \text{ km}$, $L = 300 \text{ km}$, $\theta = 2°$, and $s = 1 \text{ cm/yr}$ at three time increments $t = 10 \text{ Myr}$, 20 Myr, and 30 Myr following the initiation of thrusting. The

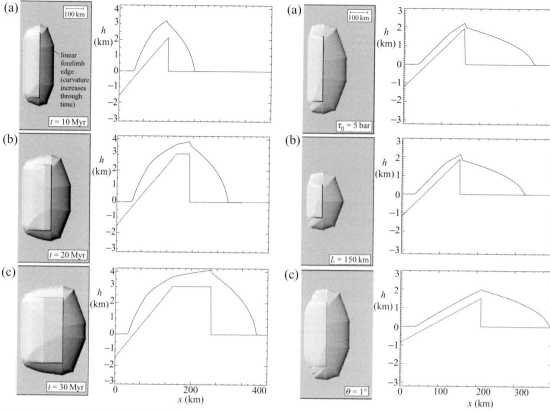

Fig 6.21 Shaded-relief images and cross-sectional plots of the fold topography developed over a thrust fault for a perfectly-plastic model of crustal deformation, assuming $\tau_0 = 20$ bar, $\rho = 2700$ kg/m^3, $H = 3$ km, $L = 300$ km, $\theta = 2°$, and s = 1 cm/yr at three time increments $t = 10$ Myr, 20 Myr, and 30 Myr following the initiation of thrusting.

Fig 6.22 Sensitivity of the thrust sheet system to variations in the (a) yield stress ($\tau_0 = 5$ bars compared to 20 bars for the reference case), (b) fault width ($L = 150$ km instead of 300 km), and (c) fault angle ($\theta = 1°$).

maximum structural relief in the model has also been restricted to be $2H$ (i.e. the fault becomes horizontal, forming a large syncline, once the fault tip has achieved an elevation equal to H above the ground surface). The shaded-relief images show the topography of the fold (given by the sum of the fault-surface elevation b and the rock thickness h). Early in the simulation, fault displacement creates an asymmetric fold characterized by a steeper backlimb than frontlimb. The forelimb progrades out in front of the fault to a maximum length controlled by the structural relief and the yield stress. Early in the simulation, the forelimb has a linear planform shape. Over time, the curvature migrates from the edge of the orogen into the center. Only when the amount of

total shortening equals the width of the thrust sheet does the fault achieve a true steady-state taper.

The sensitivity of fold geometry to model parameters is shown in Figure 6.22. Each panel shows the shaded relief and cross section of the fold at $t = 10$ Myr for a case with (a) a low yield stress ($\tau_0 = 5$ bars), (b) a low yield stress ($\tau_0 = 5$ bars) and a narrow fault ($L = 150$ km), and (c) a low yield stress ($\tau_0 = 5$ bars) and thrust angle ($\theta = 1°$). Lower yield stresses lead to thinner rock, as expected. Out in front of the fault, this thinning has the effect of creating a broader zone of deformation and hence a higher degree of orogenic curvature for faults with otherwise equal geometries and shortening amounts.

To date, most fluvial landform evolution models have assumed relatively simple tectonic forcings. When more realistic tectonic forcings have been considered (e.g. Willett *et al.*, 2000), the effects on mountain-belt geometry have been shown to be very important. Although simplified, the perfectly plastic model of topographic evolution above a crustal-scale fault provides a starting point to quantify rock uplift rates above prescribed structural elements. Coupling this model with bedrock erosion processes could help to further unravel the relationships between tectonics and erosion in thrust systems.

6.6 | Glacial erosion beneath ice sheets

Bedrock erosion beneath ice sheets is the dominant agent of Quaternary erosion and landform evolution in most of the Northern Hemisphere. Despite the dominance of subglacial erosion in many parts of the world, our understanding of subglacial erosional landforms and the feedback processes between ice flow, bedrock erosion, and bedrock topography beneath ice sheets is limited. For example, what processes and feedbacks control the morphology of overdeepenings beneath glaciers and ice sheets? How do these processes differ on the margins of ice sheets from their interiors? What processes control the size and distribution of regularly spaced glacial lakes such as the New York State Finger Lakes? Although considerable work has focused on modeling landform evolution in the alpine-glacial environment (e.g. Harbor, 1992; Braun *et al.*, 1999; MacGregor *et al.*, 2000), previous work has just touched on landform evolution beneath ice sheets. In addition to providing a better understanding of glacial-landform evolution, determining the controls on glacial-lake formation could help to constrain paleo ice-sheet thickness, a critical parameter that is difficult to determine using existing methods such as relative sea-level inversion (Peltier, 1994).

In this example application we use the threshold-sliding model as the first step in a glacial-landform evolution model for erosion ice sheets. Following the ice-surface reconstruction, basal-sliding velocities and bedrock erosion rates

are calculated based on the reconstructed ice-surface topography. Our approach is to model the geometry and flow in ice sheets and the resultant subglacial erosion iteratively through time. Figure 6.23 illustrates the basic steps of the model and the input and output data at each step using the New York State Finger Lakes as the example application. In this example, we use the *modern* Finger Lakes topography as input to determine the bedrock-erosion pattern that would occur if an ice sheet were to advance over the area today. Input data for step 1 of the model includes (1) the bedrock topography, (2) the position of the ice margin, and (3) the threshold basal shear stress for motion. These inputs are used to reconstruct the geometry of the ice sheet using the threshold-sliding model and the sandpile method. Following the ice-surface reconstruction, the lithospheric deflection under the ice load must be considered if the study area is of large extent. If the study area is less than $\approx 100\,\mathrm{km}$ in width, the lithosphere is assumed to be perfectly rigid. If the area is significantly larger than $100\,\mathrm{km}$ the deflection of the lithosphere should be determined using the 2D flexure equation and estimates for the local elastic thickness of the crust. If deflection is computed, the model returns to step 1 where the ice-surface topography is recomputed taking into account the lithospheric deflection in the bed topography. Although $100\,\mathrm{km}$ is given as a rule-of-thumb, the precise threshold length scale where flexural compensation becomes significant depends on the local elastic thickness.

In step 2, basal-flow velocities are determined with the balance-velocity method of Budd and Warner (1996). The balance-velocity method requires that the pattern of accumulation and ablation be specified. For many paleo ice sheets, this pattern may not be well constrained. For these cases, the accumulation/ablation pattern may be modeled using an equilibrium line altitude (ELA) and an accumulation/ablation gradient. For example, assuming an ELA of 1 km and an accumulation gradient of 0.1 m/yr per km, the ablation rate at an elevation of 500 m is -0.5 m/yr. In step 3, the basal velocities are used to determine the bedrock-erosion rate using Hallet's model. Hallet's model estimates erosion rates as a power

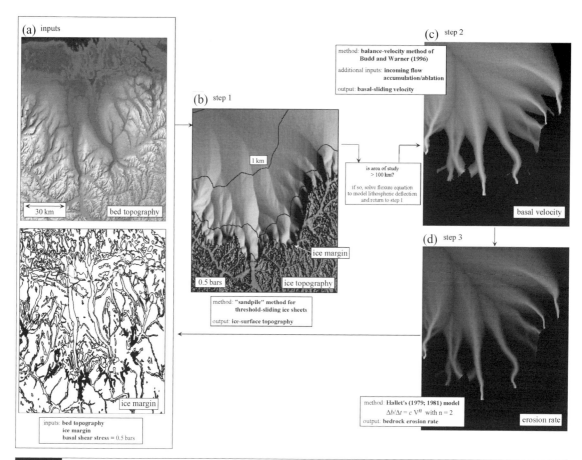

Fig 6.23 A flow chart illustrating the model inputs and outputs with erosion of the modern New York State Finger Lakes as the example application. (a) Inputs to the model include the initial bedrock topography (from US Geological Survey DEMs), the ice-margin position (given by the position of the Valley Heads Moraine (New York State Geological Survey, 1999)), and the threshold basal shear stress (assumed to be 0.5 bars). (b) In step 1 these inputs are used to reconstruct the ice-surface topography. Following this step, the deflection of the lithosphere under the ice load should be determined if the study area is equal to or larger than 100 km in extent. For this example we have neglected flexural subsidence. (c) In step 2 the ice-surface topography and pattern of accumulation and ablation are used with the balance-velocity method of Budd and Warner (1996) to determine the basal-sliding velocities. (d) In step 3, Hallet's model, which relates the bedrock erosion rate to a power function of the sliding velocity, is used to determine the spatial pattern of erosion. The model then returns to step 1 where the erosion pattern is used to determine a new bed topography for the next iteration.

function of the basal-sliding velocity. In step 4, the pattern of erosion rate is used, together with the model time step, to modify the bedrock topography according to $\Delta b = E \, \Delta t$, where b is the bedrock topography, E is the erosion rate, and Δt is the time step. The new bedrock topography, reflecting the bedrock scour during the time step Δt, is then used as input for the next model iteration, which begins back at step 1.

Any study of glacial erosion must contend with two complicating factors. First, the initial, preglacial surface is generally unknown. This complication applies to individual glacial advances as well as to erosion during the full duration of the Quaternary. As a specific example, it is commonly assumed that the ice flow responsible for the carving of the New York State Finger Lakes took advantage of preexisting fluvial valleys of the Allegheny Plateau formed by north-flowing rivers (von Engeln, 1965). It is difficult to test this hypothesis, however, since we have no constraints on the depth of pre-glacial fluvial valleys. If valleys were deep, it may still be possible that ice flow and bedrock erosion were controlled principally by the ice-surface topography of the Ontario Lobe rather than by the underlying bed topography. Numerical modeling of glacial erosion is ideal for coping with this uncertainty because multiple working hypotheses can be tested and evaluated for their impact on model results. For example, a suite of initial topographies can be used to determine a suite of corresponding model results to assess the sensitivity of the model results to the initial topography. Results that are insensitive to variations in initial topography can be considered to be robust.

A second complicating factor in studying erosion beneath ice sheets is the difficulty of separating variations in the intensity of glacial erosion from variations in the erodibility of bedrock. In some cases, structural control is easy to recognize. For example, the topography of the Finger Lakes Region is dominated by a weathering scarp separating the fine-grained shale of the Ontario Lowlands from the coarser-grained sandstone and shale of the Allegheny Plateau. In other areas, however, minor differences in deformation may have influenced erosion more greatly than variations in ice thickness or basal sliding. This is not a great concern for distinctive landforms such as finger lakes. However, if bedrock geologic maps and structural data are available for a region, structural control should be evaluated through correlations between topography and structure.

Determining the controls on glacial bedrock-erosion rates is an active area of research. Boulton (1974) emphasized the role of ice thickness in determining erosion rates, while Hallet (1979, 1996) emphasized the importance of sliding velocity. In Boulton's model, the hydrostatic pressure induced by the ice overburden pressure increases the normal stress acting on subglacial debris, in turn increasing their ability to abrade the bed. In Hallet's model, ice thickness is not a factor because hydrostatic pressures are assumed to act equally on the tops and bottoms of rounded subglacial particles. In this case, the velocity of

the abrading particles, embedded in the ice, is the most important controlling factor for erosion. Hallet's model can be summarized as

$$\frac{\partial b}{\partial t} = -cV^n \qquad (6.50)$$

where $\partial b/\partial t$ is the erosion rate, V is the basal sliding velocity, and n and c are empirical coefficients. n may taken to be equal to 2 (e.g. MacGregor et al., 2000), although this value is not well-constrained empirically. The value of c may be constrained using sediment-flux estimates from modern glaciers (e.g. Hallet et al., 1996). In this chapter we use Hallet's model, reflecting the consensus that sliding velocity is the dominant control on bedrock erosion rates. Numerical modeling in glacial geomorphology is still in its infancy. Three of the most important contributions have been made by Harbor (1992), MacGregor et al. (2000), and Braun and his colleagues (Braun et al., 1999; Tomkin and Braun, 2002). All of these contributions focused on the alpine glacial environment, but they represent the previous work most similar to the work in this chapter. Harbor's model illustrated how U-shaped glacial valleys may form from pre-existing fluvial valleys. However, because he assumed a simple form for the glacier (i.e. a constant-area cross-sectional fill with a level surface), his model cannot readily be generalized to more complex 2D cases in which a level surface cannot be assumed. MacGregor et al. (2000) developed a 1D model to study the formation of overdeepenings using a model based on mass-balance, ice-creep, and stress-controlled basal sliding. They showed that several classic alpine-glacial landforms, including overdeepenings and hanging tributaries, could be reproduced by their model. Their results were limited to 1D, however.

Braun and colleagues (Braun et al., 1999; Tomkin and Braun, 2002) constructed the first 2D glacial-landform evolution model. These authors used a continuity equation to model local changes in ice thickness:

$$\frac{\partial h}{\partial t} = \nabla \cdot F + M \qquad (6.51)$$

where the flux F was dependent on creep and sliding velocity and M is the mass balance or rate of accumulation and ablation. Sliding velocities in their model were given by

$$u_s = \frac{2A_s\beta_c(\rho g h)^n}{N + P}|\nabla(h + b)|^{n-1}\nabla(h + b) \quad (6.52)$$

where h is the ice thickness, b is the bedrock elevation, and A_s, β_c, N, n, and P are empirical parameters. Importantly, basal sliding in this model is dependent only on the *local* ice depth, ice slope, and bed slope. In real glaciers, however, sliding velocity is controlled by the discharge of ice, increasing with distance from the divide and reaching a maximum beneath the ELA, even if the glacier and bed morphology remains constant along the glacier profile.

Here we illustrate the behavior of our glacial-landform evolution model with several numerical experiments. We distinguish three types of depressions: (1) lakes in ice sheet interiors that are flexurally compensated, (2) large lakes that are uncompensated ("great" lakes), and (3) elongate ice-marginal depressions ("finger" lakes). In each case, topographic analyses are introduced for comparison with the model predictions.

6.6.1 Length scales < flexural wavelength

First we consider bedrock erosion beneath a gently-sloping ice sheet at scales less than the flexural wavelength. In this example we assume that the lithosphere is perfectly rigid. The initial bed topography is flat with random microtopography characterized by a Gaussian white noise with a standard deviation of 10 m. This initial topography is certainly an idealization; no initial condition would look like this. However, all topography has some small-scale roughness, and it is essential to include these variations in the initial bed topography because the glacial-erosion instability amplifies initial topography over time. White noise variations in initial conditions are commonly used to initiate fluvial-geomorphic instabilities such as drainage-network evolution (e.g. Willgoose *et al.*, 1991) and hillslope rilling (Hairsine and Rose, 1992), for example. A key question in this section will be whether the glacial-erosion instability amplifies initial topography at all wavelengths or whether certain wavelengths are preferred. As an alternative to assuming variations in initial topography,

the basal shear stress can also be varied to simulate "sticky" spots on the bed. Both assumptions yield nearly identical results.

The ice-surface topography and basal-sliding velocities above a flat surface with white noise variations (shown in Figure 6.24c) are illustrated in Figures 6.24a and 6.24b for a basal shear stress of 1 bar. The grid for this example is 50 km × 50 km in size with a pixel resolution of 100 m. To construct this surface, we allowed all pixels in the grid to accumulate ice to simulate an ice sheet with no margin. The slopes in the y direction, S_y, were restricted to point *down* the grid, rather than in either the up or down direction, to simulate an ice-sheet flowing from top to bottom. The slope in the x direction, S_x, was unconstrained. The sandpile algorithm for this example was initiated with a uniform ice-surface topography of 2 km. From this initial state, the sandpile method was used to solve for the threshold-sliding ice-surface topography. The resulting ice-surface topography dips gently from the top of the grid to the bottom (Figure 6.24a). In addition, the ice surface has small-scale variations that reflect the white-noise variations in bed topography. Equation (6.46) requires that the ice-surface slope and ice thickness be inversely proportional. In the model reconstruction, therefore, pixels with slightly lower bed elevations have lower ice-surface slopes. In this way, variations in the bed topography are reflected in the ice-surface topography.

The pattern of basal-sliding velocities in Figure 6.24b was produced assuming uniform accumulation within the model domain. Although the ice-surface topography has only minor small-scale variations, the basal sliding is strongly localized into bedrock depressions in a braided pattern. This behavior results from an additive effect of the ice-surface topography on the basal sliding. The input of ice to the system as accumulation is uniform throughout the domain. At each pixel, however, slightly more flow is diverted to steeper downslope pixels than to pixels with a gentler slope. As ice flows from pixel to pixel downice, small variations in surface slope act cumulatively to divert ice into bedrock depressions. The resulting areas of intense sliding velocities result in enhanced scouring of the bed. This, in turn,

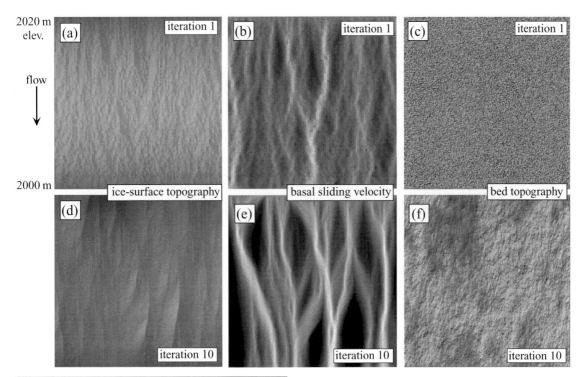

Fig 6.24 Erosion and basal sliding beneath ice-sheet interiors. (a) Shaded-relief image of ice-surface topography, (b) grayscale image of basal-sliding velocities, and (c) shaded relief image of bed topography in the first iteration of the model. (d) Ice-surface topography, (e) basal-sliding velocities, and (f) bed topography after ten model iterations. Bedrock depressions deepen and expand over time, focusing basal sliding and erosion in a positive feedback.

results in localized bedrock erosion in depressions. This is the fundamental glacial-erosion instability. In the case of a gently-sloping surface of an ice sheet with no margin, the instability results in lakes with little or no elongation and a wide range of sizes. Near the ice margin, in contrast, lakes are elongated and regularly spaced.

The ice-surface topography, basal sliding velocities, and bed topography of this model after 10 iterations are given in Figures 6.24d–6.24f. These figures illustrate that flow becomes more focused into bedrock depressions as the relief of the bed topography increases through time. Initially, all depressions are limited to a few pixels in width. Over time, large depressions deepen and expand more quickly than smaller depressions nearby. Large depressions expand faster because

they depress the ice-surface topography over a larger area, focusing more flow into the depression. The dynamics of the model, therefore, can be characterized as depressions of different size competing for ice drainage. The larger a depression is, the faster it expands to "capture" neighboring lakes. As a result, Figure 6.24f contains lakes or pits of many sizes, while Figure 6.24c contains only small pits. The braided patterns of Figures 6.24b and 6.24e meander and shift as the bed is eroded and depressions expand.

In Figure 6.25, the lakes of Figure 6.24f are compared with the lakes of glaciated topography southwest of Hudson Bay (Figures 6.25d–6.25e). In Figure 6.25a, a shaded-relief image of the model topography from Figure 6.24f is given, with all closed depressions filled. The binary image of Figure 6.25b includes all of the filled areas, or lakes, of Figure 6.25a.

The frequency-size distribution was introduced in Chapter 1 as a tool for characterizing the population of lakes in a given area. The frequency-size distribution is the number of lakes as a function of their size. This curve can be used to compare spatial domains (e.g. lakes, landslides, vegetation clusters) modeled on a computer with

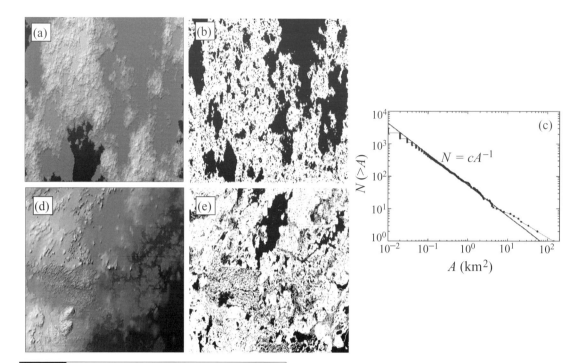

Fig 6.25 Comparison of modeled and observed topography resulting from glacial erosion beneath ice-sheet interiors. The pixel size in the model is assumed to be 100 m. (a) Shaded relief of bedrock topography with filled depressions. (b) A binary image of filled depressions (i.e. lakes) from (a). (c) The cumulative frequency-size distribution of lakes in (b), well fit by the same relationship observed for Canadian lakes less than 10^4 km^2 in area: $N(> A) \propto A^{-1}$. (d) Shaded relief of topography in central Canada for qualitative comparison with (a). (e) Binary image of lakes in (d).

those of populations observed in nature. The cumulative frequency-size distribution of lakes is the number of lakes larger than a given area. Figure 6.25c illustrates the cumulative frequency-size distribution for the lakes of Figure 6.25e. The data fit a power-law relationship $N(> A) \propto A^{-1}$. This compares well with the frequency-size distribution of small and medium-sized lakes in Canada shown in Figure 1.29.

6.6.2 Length scales > flexural wavelength

The largest lakes in Canada and the US are illustrated in Figure 6.26a, including (from northwest to southeast) Great Bear Lake, Great Slave Lake, Lake Athabaska, Lake Winnipeg, and the Great Lakes. In this section we refer to these lakes collectively as the "great" lakes of Canada. These lakes are roughly uniformly-spaced in a ring around Hudson Bay, with distances between lake centers ranging from 370 km to 1000 km. The map of Figure 6.26a was created by extracting all the lakes from a 1-km resolution DEM of Canada and the northern US. The shading of each lake is proportional to elevation, with black indicating lakes near sea level and light gray shades representing lakes as high as several hundred meters above sea level. The shading values indicate, for example, that the great lakes of western Canada are separated from Hudson Bay by a broad ridge of topography several hundred meters above sea level and parallel to the Hudson Bay shoreline.

To investigate the role that flexure may have in modifying the basic glacial-erosion instability, we performed a numerical experiment of glacial erosion beneath an initially semi-circular ice sheet. The 2D flexure of the lithosphere beneath the ice sheet was included in this model experiment. The grid for this model is 3000 km × 3000 km with a pixel resolution of 10 km. The

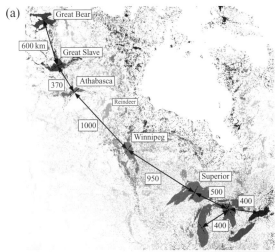

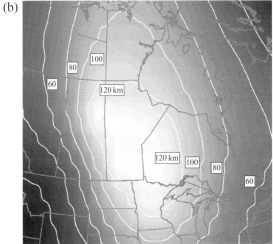

Fig 6.26 The "great" lakes of Canada and their relationship to the flexure of the lithosphere. (a) Topographic depressions of Canada (all non-white areas) with grayscale shading mapped to elevation (lakes in black are at sea level, lighter tones have water surfaces at higher elevations. The distances between the largest lakes are indicated. (b) Grayscale and contour map of the elastic thickness of the lithosphere from Tony Watts (personal communication, 2002), rectified to (a). The elastic thicknesses are greatest in the areas between Athabasca Lake, Lake Winnipeg, and Lake Superior, where the distances between lakes are also the greatest. The distances separating lakes are lower in the northwest and southeast portion of Canada where the elastic thickness values fall to approximately one half their values in central Canada. If Reindeer Lake is considered to be a great lake, however, the relationship between lake spacing and elastic thickness is less robust.

elastic thickness is 50 km. As in Section 6.6.1, we assumed a flat initial surface with white noise variations as the initial bedrock surface. The ice-surface topography, flow, and bedrock deflection in the first model iteration are given in Figures 6.27a–6.27c. The accumulation pattern was assumed to be uniform for elevations greater than 1 km.

The ice-surface topography, flow, bedrock deflection, and bedrock topography of the model after ten iterations are given in Figures 6.27d–6.27g. The bedrock topography does not include the deflection under the ice load. Therefore, it represents what the landscape would look like if the ice load was removed and only the effect of glacial erosion remained. In this experiment, the largest lakes are uniformly spaced in a ring around the center of the ice sheet, directly underneath the equilibrium line. In the center of the ice sheet the original rough topography remains because erosion is minimal there. Near the equilibrium line the glacial instability is amplified by the deflection of the lithosphere at wavelengths equal to ≈ 500 km. Bedrock depressions that form at wavelengths of 500 km are amplified faster than smaller wavelengths because their ice infill deflects the lithosphere, further depressing the ice surface and focusing ice drainage into the depressions.

A grayscale image and contour map of the elastic thickness of central Canada is given in Figure 6.26b. The elastic-thickness estimates were obtained by Tony Watts (personal communication, 2002) and are based on the coherence method between gravity and topography. The greatest elastic thickness in Canada exists just southwest of Hudson Bay with a maximum value of nearly 140 km. To the northwest of this maximum value, following the lines adjoining the great lakes from Winnipeg to Athabasca to Great Slave and Great Bear, elastic thicknesses decrease gradually to about 100 km at Athabasca, then more rapidly to about 60 km near Great Slave. Moving in the southeast direction from Winnipeg, elastic thicknesses also decrease, to 100 km at Superior and 60 km at Ontario. This pattern suggests that the distances between the great lakes may correlate with the local elastic thickness, with a

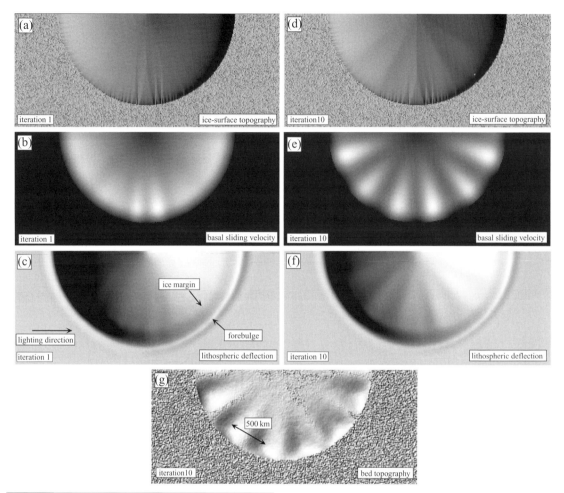

Fig 6.27 Numerical experiment illustrating the role of flexure in lake formation. (a) Shaded-relief image of ice-surface topography, (b) grayscale image of basal-sliding velocities, and (c) shaded-relief image of lithospheric deflection in the first iteration of the model. Basal sliding velocities and bedrock erosion are diffuse in this early stage of the model. (d) Ice-surface topography, (e) basal-sliding velocities, (f) bedrock deflection, and (g) bedrock topography (no deflection component) after 10 iterations of the model run. Domain is 3000 km in length and elastic thickness is 50 km.

proportionality factor of 10. For example, in areas where the elastic thickness is 100 km, the distance between lakes is $\approx$ 1000 km. It should be noted that if Reindeer Lake is considered as a great lake in this analysis, the relationship between great-lake spacing and elastic thickness is not as robust.

Is the relationship between great-lake spacing and elastic thickness consistent with flexure? The elastic thickness is related to the three-dimensional flexural parameter by the relationship:

$$\beta = \left(\frac{D}{\Delta \rho g} \right)^{\frac{1}{4}} \qquad (6.53)$$

where D is the flexural rigidity given by

$$D = \frac{E T_e^3}{12(1 - \nu^2)} \qquad (6.54)$$

Assuming $E = 70\,\text{GPa}$, $\nu = 0.25$, $\Delta \rho = 2300\,\text{kg/m}^3$, and $g = 10\,\text{m/s}^2$, and an elastic thickness $T_e = 50\,\text{km}$, the flexural parameter β is 77 km. The wavelength of deflection is related to the flexural parameter by $\lambda = 2\pi\beta = 480\,\text{km}$, or about ten times the elastic thickness. This analysis indicates that the observed relationship between great-lake spacing and elastic thickness is consistent with flexural control.

Although the relationship between lake spacing and elastic thickness suggests that flexure is the key process controlling great-lake formation, the ice margin may be home to many feedback processes that lead to instabilities. In addition to the focusing of flow by the bed topography (through its influence on the ice-surface topography), feedbacks between ice temperature, flow, meltwater content, and other variables may also be at work. For example, the EISMINT model comparison project noted that an ice-marginal instability driven by a feedback between ice temperature, flow, and ice-sheet geometry was reproduced in all of the models tested (Payne *et al.*, 2000). Since ice temperature is not a variable in our model, the instability found in the EISMINT experiments is independent from the flexural mechanism of our model. The wavelength of the thermomechanical instability observed in the EISMINT experiments varied between model experiments but was generally several hundred kilometers. As such, a thermomechanical instability could also generate "great" lakes.

6.6.3 Near ice margins

Finger lakes are elongate glacially carved lakes formed on the margins of ice sheets. The type example is the Finger Lakes Region of New York State. In this chapter we use the term more broadly to include elongate glacial troughs near the ice margin whether or not they impound water. For example, the Finger Lakes Region includes five major lakes, several smaller lakes, and dozens of glacial valleys of similar shape. By "finger lakes" we will be referring to these features collectively. In this section we present model results for the ice-sheet geometry, basal velocities, and bedrock erosion near linear and curved ice margins. We show that uniformly-spaced ice-flow channels result from flow over a rough surface near the ice margin.

The ice-surface topography and basal-sliding velocities for an ice sheet sliding over a rough surface near a linear margin are given in Figures

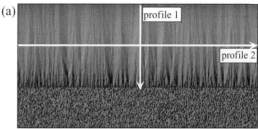

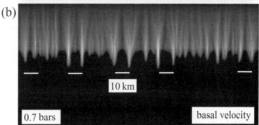

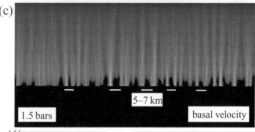

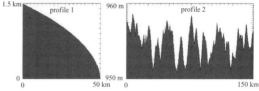

Fig 6.28 Ice-surface topography and flow above a rough surface with $\tau = 0.7$ bars. (a) Shaded-relief image of an ice sheet 150 km wide and 50 km from divide (top of image) to margin. Two profiles of the ice-surface topography are given: profile 1 is along-dip from divide to margin; profile 2 is along-strike from left to right. (b) Grayscale image of basal velocity corresponding to (a) with an ELA of 1 km. Ice-flow channels with a characteristic spacing of 10 km are present in the flow even though there is no characteristic scale to the bed topography. (c) A thicker ice sheet ($\tau = 1.5$ bars) results in narrower flow channels (assuming the ELA is the same as in (b)).

6.28a and 6.28b. The basal shear stress in this case is assumed to be 0.7 bars and the ELA is 1 km. Also given in Figure 6.28a are topographic profiles parallel and perpendicular to the margin. Profile 1 shows the parabolic profile characteristic of ice sheets on flat beds (e.g. Nye, 1951).

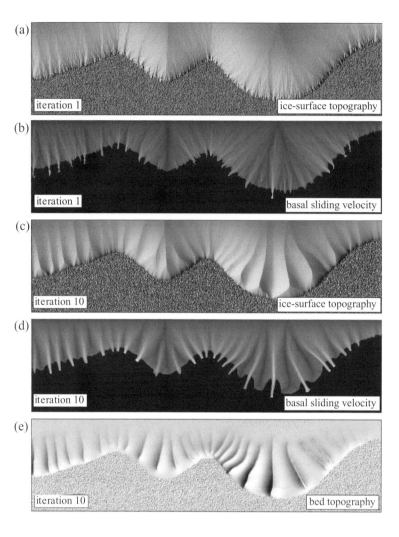

(a) iteration 1 ice-surface topography

(b) iteration 1 basal sliding velocity

(c) iteration 10 ice-surface topography

(d) iteration 10 basal sliding velocity

(e) iteration 10 bed topography

Fig 6.29 Numerical experiment illustrating the model behavior of focused erosion near a curved ice margin. (a) Shaded-relief image of ice-surface topography, (b) grayscale image of basal-sliding velocities at the beginning of the model run. (c) Ice-surface topography, (d) basal-sliding velocities, and (e) bed topography of the model after ten iterations. Finger lakes have formed with orientations parallel to the ice-flow directions and lake spacing is low in zones of converging ice and high in zones of diverging ice.

Profile 2 shows the microtopography of the ice surface along a direction parallel to the margin. One of the most important findings of the chapter is illustrated in Figure 6.28: even though the bed topography has no characteristic scale (i.e. it is white noise), the resulting ice-surface topography has a characteristic scale of about 10 km. This characteristic scale is even more apparent in the map of basal-sliding velocities (Figure 6.28b). Focusing of the ice flow occurs as a result of the additive deflection of ice flow along flow paths from the divide to the margin. This focusing involves several steps. First, the rough bed is reflected in the ice surface as microtopographic channels (Figure 6.28a). Second, these channels act to focus the flow a little bit as the ice is routed from pixel to pixel. The cumulative effect of this focusing

is to create strongly localized flow near the ice margin even though the ice-surface topography is nearly uniform along the margin. The width of ice-flow channels depends on the slope of the ice sheet at the ELA. A thicker ice sheet with the same ELA (1 km), for example, results in more closely-spaced channels, To illustrate this, Figure 6.28c is a basal-velocity map for an ice-sheet reconstruction with $\tau = 1.5$ bars. Ice-flow channels in this case are spaced at $\approx$ 5–7 km instead of 10 km.

How does the focusing effect differ for a curved ice margin? Figures 6.29a and 6.29b illustrate the ice-surface topography and basal-sliding velocities for a curved margin. The shape of this margin is identical to the 14 ka ice margin of the Finger Lakes Region. The bed topography in this

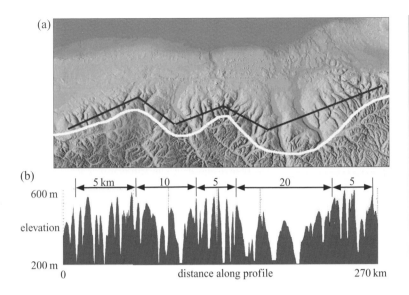

(a)

(b)

600 m

elevation

200 m

0   distance along profile   270 km

5 km   10   5   20   5

Fig 6.30 Spacing of finger lakes and troughs in upstate New York. (a) Shaded-relief DEM image of Finger Lakes area. White curve indicates approximate position of the Valley Heads Moraine (based on the Finger Lakes and Niagara sheets of the surficial geologic map of New York (New York Geological Survey, 1999)). Black lines indicate the location of the topographic transect given in (b). The transect shows alternating zones of narrow (≈ 5 km) and wide (up to 20 km) valley spacing. Zones of narrow spacing correspond to regions of converging flow in ice embayments. Conversely, wider spacings correspond with areas of diverging ice flow.

example was also chosen to be a flat surface with white noise variations. The curved margin results in ice-flow channels that are alternately closely spaced and widely spaced. Channels are more closely spaced in margin embayments where ice flow is convergent. Widely spaced channels occur beneath ice lobes where flow is divergent. Figures 6.29c–6.29e illustrate the ice-surface topography, basal velocities, and bedrock topography after 10 model iterations. The ice flow channels have carved distinct glacial troughs, depressing the ice-surface topography and focusing flow and erosion into the troughs.

Geomorphically, the Finger Lakes Region is dominated by subparallel, glacially scoured troughs with southernmost extents in Seneca and Cayuga Lakes, the two largest of the Finger Lakes. The five largest troughs of the region comprise the Finger Lakes proper, but there are numerous other troughs cut into the Allegheny Plateau of smaller size that are not deep enough to impound water. Figure 6.30a is a shaded-relief image of the topography of the region with the location of the 14 ka ice margin (the Valley Heads Moraine) shown in white. The troughs vary in spacing from 5 to 20 km along strike with the greatest spacing between Seneca and Cayuga Lakes. The southern tips of the Finger Lakes coincide with the Valley Heads Moraine (14 ka). As such, the scouring of the Finger Lakes Region most likely took place when the ice mar-

gin was coincident with this moraine, although several phases of glaciation may have contributed to their formation.

Figure 6.30b is the topographic profile of the region along the black line of Figure 6.30a. The profile of Figure 6.30b shows the same pattern as the model results of Figures 6.29c–6.29e: closely spaced (≈ 5 km) troughs in margin embayments and widely spaced (10–20 km), deeper troughs beneath ice-sheet lobes.

If finger lake formation is an intrinsic feature of ice margins, why don't finger lakes occur more commonly in formerly glaciated terrain? One possible reason is that intense glacial erosion and a stable ice margin are both required for finger-lake formation. If the equilibrium line is rapidly migrating during ice-sheet advance or retreat, for example, there may not be sufficient time for the instability between depressions, ice flow, and erosion to develop deep troughs. It may be that the Finger Lakes were formed because 14 ka was a period of sufficient margin stability for erosion to carve the topography we see today. Alternatively, the margin may not have been unusually stable, but erosion may have been particularly intense during this period. For example, meltwater pulses may have driven rapid erosion just prior to 14 ka followed by rapid filling of the Finger Lakes between 14 and 13 ka. This hypothesis is consistent with the lake stratigraphy of the region (Mullins and Hinchey, 1989).

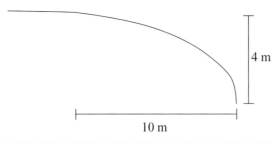

Fig 6.31 Schematic diagram of lava-flow margin in Exercise 6.1.

Exercises

6.1 Lava-flow margins in the vicinity of Lathrop Wells cone, Amargosa Valley, Nevada, are approximately 10 m wide and 4 m thick in areas of flat terrain (Figure 6.31). Estimate the yield stress for these flows based on a 1D perfectly plastic model.

6.2 Use a topographic map or DEM to extract a longitudinal profile of a steep fluvial valley. Import the profile in two-column format x, h into Excel. Using Excel, model the profile of a slow-moving debris flow with yield stress 0.1 bars flowing down the valley bottom using the perfectly plastic model. Choose several terminus positions and compute profiles for each position by numerically integrating from the terminus up the valley.

6.3 Using the code in Appendix 5 as a guide, model the evolution of a 1D gravity flow with temperature-dependent viscosity flowing down an inclined plane of slope S.

6.4 Download a DEM of an area formerly covered by alpine glaciers. Using the sandpile algorithm code given in Appendix 5 as a guide, map the thickness of glacier ice assuming a basal shear stress of 1 bar. Choose an ELA value and assume that ice covers topography only above the ELA.

6.5 The Boussinesq equation

$$\frac{\partial h}{\partial t} = \frac{k\rho g}{\mu \phi} \frac{\partial}{\partial x} \left(h \frac{\partial h}{\partial x} \right) \tag{E6.1}$$

is used to quantify the evolution of a water table of height h in an unconfined aquifer of permeability k and porosity ϕ, where ρ is the density of water, g is gravity, and μ is the dynamic viscosity of water (Turcotte and Schubert, 2002). Plot the shape of the *steady-state* water table in a 1D aquifer of length 10 km above an impervious stratum with a dip of 1%. Assume $k = 10^{-12}$ m^2 and $\mu = 10^{-3}$ Pa s. (Hint: map this problem onto the perfectly plastic model for alpine glaciers. See Section 9.5 of Turcotte and Schubert for more details on the Boussinesq equation.)

Chapter 7

Instabilities

7.1 | Introduction

Many geomorphic processes involve positive feedbacks. The incision of a river valley into an undissected landscape is an example of a positive feedback process in which an incipient channel erodes its bed and expands its drainage area in a positive feedback. In some cases, positive-feedback processes give rise to landforms with no characteristic scale. Many drainage networks, for example, are branching networks with no characteristic scale above the scale that defines the hillslope-channel transition. In other cases, periodic landforms are created (Hallet, 1990). Sand dunes, for example, form by a positive feedback between the height of the dune, the deflection of air flow above it, and the resulting pattern of erosion and deposition (i.e. enhanced windward erosion and leeward deposition). Unstable geomorphic systems all include some type of positive feedback. In many arid alluvial channels, for example, sediment becomes stored preferentially at certain points along the channel profile, thereby causing the channel to aggrade and the channel bed to widen. Channel widening further enhances sediment storage and bed aggradation in a positive feedback. This instability continues until a sufficiently steep slope develops downstream of the aggrading zone, triggering channel entrenchment and narrowing. The positive feedback between channel width and localized erosion/deposition leads to spatial and temporal oscillations in arid alluvial channels. The study of instabilities does not involve any fundamentally new types of equations. Rather, most equations involving geomorphic instabilities combine diffusion- and advection-type equations similar to those we have already considered. In most cases, however, two or more variables are coupled in such a way as to enhance each other in a positive feedback.

This chapter does include one fundamentally new tool: the linear-stability analysis. In this analysis, the fundamental equations that describe the evolution of a particular system are first linearized (i.e. nonlinear terms are neglected) and then solved for the case of an initially small perturbation of wavelength λ (where λ is a free parameter). Conceptually, a linear stability approach considers the evolution of the system with small, incipient landforms with a wide range of sizes (i.e. wavelengths) and determines whether and how fast each perturbation will grow to become full-sized landforms. For those wavelengths that do grow, the wavelength that grows the fastest provides an estimate of the characteristic scale of the landform as a function of the system parameters. This fastest-growing wavelength often, but not always, provides a good estimate of the spatial scale of full-sized landforms. As such, linear stability analyses often provide a useful starting point for understanding the emergence of characteristic scales in complex geomorphic systems.

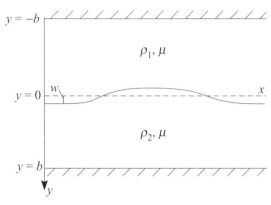

7.2 | An introductory example: the Rayleigh–Taylor instability

As an introduction to the study of instabilities in geomorphology, we consider the Rayleigh–Taylor instability of a dense fluid overlying a lighter fluid. The Rayleigh–Taylor instability describes the formation of salt domes in sedimentary basins (Figure 7.1) and it provides a simple model for the initiation of mantle plumes. When a dense layer overlies a lighter layer, the system is gravitationally unstable, i.e. the lighter fluid will rise and penetrate through the upper layer. Here we will quantify how this instability works following closely the approach of Turcotte and Schubert (2002).

The geometry of the two-layer system is shown in Figure 7.2. A fluid with a thickness b and density ρ_1 overlies a lighter fluid of thickness b and density ρ_2. Here we assume that both fluid layers have the same viscosity μ. We also assume that both the top and bottom of the system have rigid surfaces. The interface between the two fluid layers is initially centered on $y = 0$ but has a sinusoidal variation given by

$$w(x) = A_1 \cos \frac{2\pi x}{\lambda} \tag{7.1}$$

The goal of the linear stability analysis is to predict how w varies with time and to determine the wavelength λ associated with the fastest growth.

The first step of the Rayleigh–Taylor analysis is to solve for the flow field in the two-layer system given a growing interface given by Eq. (7.1). The flow fields are characterized by four velocity functions: u_1, v_1, u_2, v_2. These functions are the horizontal (u) and vertical (v) components of the velocity in the upper (1) and lower (2) layers. Turcotte and Schubert (2002) used the stream function method to obtain expressions for these flow velocities. In this method, the velocity functions u and v are obtained by differentiating a single "stream function" ψ:

$$u = -\frac{\partial \psi}{\partial y} \tag{7.2}$$

$$v = \frac{\partial \psi}{\partial x} \tag{7.3}$$

Turcotte and Schubert's (2002) analysis for this

problem gives

$$\psi_1 = A_1 \sin\frac{2\pi x}{\lambda}\cosh\frac{2\pi y}{\lambda} + A_1 \sin\frac{2\pi x}{\lambda}\left(\frac{y}{b}\left(\frac{\lambda}{2\pi b}\right)\tanh\frac{2\pi b}{\lambda}\sinh\frac{2\pi y}{\lambda} + \left(\frac{y}{b}\cosh\frac{2\pi y}{\lambda} - \frac{\lambda}{2\pi b}\sinh\frac{2\pi y}{\lambda}\right)\right.$$

$$\left. \times\left(\frac{\lambda}{2\pi b} + \frac{1}{\sinh\frac{2\pi b}{\lambda}\cosh\frac{2\pi b}{\lambda}}\right)\right)\left(\frac{1}{\sinh\frac{2\pi b}{\lambda}\cosh\frac{2\pi b}{\lambda}} - \left(\frac{\lambda}{2\pi b}\right)^2\tanh\frac{2\pi b}{\lambda}\right)^{-1} \tag{7.4}$$

The expression for ψ_2 is obtained from Eq. (7.4) by replacing y with $-y$.

Equation (7.4) can be used to solve for the evolution of the interface. The rate of change of the amplitude of the interface, $\partial w/\partial t$ must equal the vertical component of the fluid velocity at just above the interface. If not, a void would open up in the system. This condition can be written as

$$\frac{\partial w}{\partial t} = (v_1)_{y=0} \tag{7.5}$$

As an approximation, the velocity in Eq. (7.5)

$$(\rho_1 - \rho_2)gw = -\frac{4\mu A_1}{b}\left(\frac{2\pi}{\lambda}\right)\cos\frac{2\pi x}{\lambda}\left(\frac{\lambda}{2\pi b} + \frac{1}{\sinh\frac{2\pi b}{\lambda}\cosh\frac{2\pi b}{\lambda}}\right)\left(\frac{1}{\sinh\frac{2\pi b}{\lambda}\cosh\frac{2\pi b}{\lambda}} - \left(\frac{\lambda}{2\pi b}\right)^2\tanh\frac{2\pi b}{\lambda}\right) \tag{7.10}$$

is evaluated at the unperturbed position of the interface (i.e. $y = 0$). The value of v_1 at $y = 0$ is obtained by differentiating Eq. (7.4) with respect to x and evaluating the resulting expression at $y = 0$. The result is

$$\frac{\partial w}{\partial t} = \frac{2\pi A_1}{\lambda}\cos\frac{2\pi x}{\lambda} \tag{7.6}$$

In order to constrain the value of the coefficient A_1 in Eq. (7.6), we must incorporate the role of buoyancy on fluid motion. As the interface between the two fluids is disturbed downward (or upward) the less (more) dense fluid will displace the other. The pressure difference between the two sides of the interface resulting from this buoyancy force is given by

$$2(P_1)_{y=0} = -(\rho_1 - \rho_2)gw \tag{7.7}$$

The pressure on $y = 0$ in the upper layer can be found by using a force balance equation that equates the pressure gradient in the flow with the viscous stresses:

$$\frac{\partial P}{\partial x} - \mu\left(\frac{\partial^2 u}{\partial x^2} + \frac{\partial^2 u}{\partial y^2}\right) = 0 \tag{7.8}$$

Substituting Eq. (7.4) into Eqs. (7.3), (7.3), and (7.8) gives

$$(P_1)_{y=0} = \frac{2\mu A_1}{b}\cos\frac{2\pi x}{\lambda}\left(\frac{2\pi}{\lambda}\right)$$

$$\times\left(\frac{\lambda}{2\pi b} + \frac{1}{\sinh\frac{2\pi b}{\lambda}\cosh\frac{2\pi b}{\lambda}}\right)$$

$$\times\left(\frac{1}{\sinh\frac{2\pi b}{\lambda}\cosh\frac{2\pi b}{\lambda}} - \left(\frac{\lambda}{2\pi b}\right)^2\tanh\frac{2\pi b}{\lambda}\right) \tag{7.9}$$

Substituting Eq. (7.9) into Eq. (7.7) gives

Solving Eq. (7.10) for A_1 and substituting the result into Eq. (7.6) gives

$$\frac{\partial w}{\partial t} = w\frac{(\rho_1 - \rho_2)gb}{4\mu}$$

$$\times\frac{\left(\frac{\lambda}{2\pi b}\right)^2\tanh\frac{2\pi b}{\lambda} - \left(\sinh\frac{2\pi b}{\lambda}\cosh\frac{2\pi b}{\lambda}\right)^{-1}}{\frac{\lambda}{2\pi b} + \left(\sinh\frac{2\pi b}{\lambda}\cosh\frac{2\pi b}{\lambda}\right)^{-1}} \tag{7.11}$$

Equation (7.11) describes an exponential growth process with time scale τ given by

$$\tau = \frac{4\mu}{(\rho_1 - \rho_2)gb}\frac{\frac{\lambda}{2\pi b} + \left(\sinh\frac{2\pi b}{\lambda}\cosh\frac{2\pi b}{\lambda}\right)^{-1}}{\left(\frac{\lambda}{2\pi b}\right)^2\tanh\frac{2\pi b}{\lambda} - \left(\sinh\frac{2\pi b}{\lambda}\cosh\frac{2\pi b}{\lambda}\right)^{-1}} \tag{7.12}$$

Equation (7.12) provides the necessary relationship between the time scale of the instability and the wavelength λ.

Figure 7.3 plots the relationship between the dimensionless growth rate of a perturbation (i.e. $1/\tau$) and its dimensionless wavelength $2\pi b/\lambda$. The fastest growing wavelength can be determined by differentiating Eq. (7.12) as a

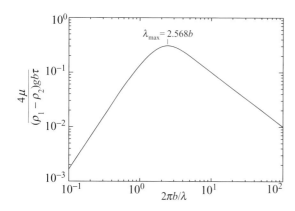

Fig 7.3 Plot of the inverse of the dimensionless time scale τ as a function of the dimensionless wavelength for the Rayleigh–Taylor instability (Eq. (7.12)).

function of τ and setting the result equal to zero. Turcotte and Schubert (2002) give the result of that equation as

$$\lambda_{max} = 2.568b \tag{7.13}$$

where λ_{max} is the fastest-growing wavelength. Perturbations of this wavelength have a growth time scale given by

$$\tau = \frac{13.04\mu}{(\rho_1 - \rho_2)gb} \tag{7.14}$$

Equation (7.12) predicts that the fastest growing wavelength is equal to about two and half times the layer thickness. The Rayleigh–Taylor instability provides a classic example of a linear stability analysis. We will use the linear stability analysis procedure again in Section 7.5, to explore the instability mechanism that creates oscillating arid alluvial channels.

7.3 | A simple model for river meandering

Alluvial rivers, supraglacial meltwater streams, Gulf Stream meanders, and lava channels on Earth, the Moon, and Venus (Komatsu and Baker, 1994) all exhibit meandering with similar proportionality between meander wavelength and channel width (Leopold *et al.*, 1964; Stommel, 1965; Parker, 1975; Komatsu and Bakerm, 1994). The ratio of meander wavelength and channel width is a constant equal to about ten for these diverse phenomena (Leopold *et al.*, 1964). This similarity suggests a common mechanism for meandering.

In this section we present a linear stability analysis which predicts channel migration rates and the proportionality between meander wavelength and channel width using a very simple geometric approach that applies to both river channels and lava channels. Most models of alluvial channel meandering stress the importance of the cross-sectional circulation or secondary flow in meander initiation (Callander, 1978; Rhoads and Welford, 1991). This model does not replace those more sophisticated models, but rather establishes sufficient conditions for meandering to occur in a wide variety of channel flows. Meanders of the correct scaling form when the bank shear stress (in the case of sediment transport) or temperature (in the case of thermal erosion) is proportional to the curvature of the bank and cross-sectional variations in shear stress or temperature have a linear form. For alluvial channels, this curvature dependence arises from the centripetal force of fluid rounding a bend, and in thermal erosion it arises through the excess latent heat of a curved bank. A linear stability analysis predicts that channels are unstable to growth at all wavelengths larger than a critical wavelength given by $\lambda = 2\pi w$, where w is channel width, and stable below it (provided that some channel flows exceed the critical condition required for erosion). The wavelength of fastest growth predicted by this analysis is equal to $2\sqrt{3}\pi w$ or $10.88w$, which is very similar to observations from alluvial and lava channels.

In alluvial rivers, the dependence of channel migration on the local curvature of the bed has been stressed in measurements of channel migration by Nanson and Hickin (1983) and in theoretical studies by Begin (1981) and Howard and Knutson (1984). Nanson and Hickin (1983) found that channel migration rates were a maximum within a narrow interval of R/w near 3, where R is the radius of curvature of the meander, and w is the channel width. We will compare our theoretical results to their data. Begin (1981) showed, by computing the centripetal force necessary to accelerate flow around a bend, that the

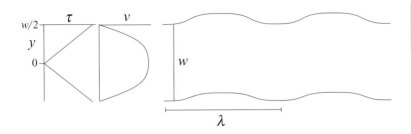

shear stress at the bed is proportional to the curvature. Howard and Knutson (1984) developed a simulation model of meandering where the migration rates were proportional to local curvature up to some maximum value. They obtained realistic meandering channels with their model.

In Figure 7.4 we present the geometry to be considered in our model. We assume symmetry within our 2D model, i.e. that the forces and erosion on one bank are equal in magnitude and opposite in direction to those on the other bank. First we consider the distribution of boundary shear stress in a straight channel of width w. The fundamental assumption of this analysis is that the cross-sectional shear stress profile is linear, decreasing from a maximum value at the bank ($y = w/2$) to a minimum value at the channel centerline:

$$\tau(y) = \tau_{b,s}\left(1 - \frac{2}{w}y\right) + \tau_c \quad (7.15)$$

where $\tau_{b,s}$ is the maximum shear stress at the bank for a straight channel and τ_c is the minimum shear stress at the centerline. In a curving channel, empirical data indicate that the bank shear stress is equal to (Richardson, 2002)

$$\tau_{b,c} = \tau_{b,s}\left(\frac{w}{2R} + 1\right)^2 \quad (7.16)$$

where $\tau_{b,c}$ refers to the bank shear stress for a curved channel, and R is the radius of curvature of the channel centerline. For incipient meanders, the radius of curvature of the centerline is much larger than the channel width. As such, we can use the approximation $(1 + \epsilon)^2 \approx 1 + 2\epsilon$ to rewrite Eq. (7.16) as

$$\tau_{b,c} \approx \tau_{b,s}\left(1 + \frac{w}{R}\right) = \tau_{b,s}(1 + wK) \quad (7.17)$$

where K is the planform curvature of the centerline (equal to the inverse of R). Combining

Eqs. (7.15) and (7.17) for the cross-sectional and along-channel variation in shear stress gives

$$\tau(x, y) = \tau_{b,s}(1 + wK)\left(1 - \frac{2}{w}y\right) + \tau_c \quad (7.18)$$

The shape of the curved-channel bank within this linear-stability approach is given by $y_b = \epsilon \sin(2\pi x/\lambda)$. The curvature K is given by $d^2 y_b/dx^2$. Substituting the sinusoidal bank shape into Eq. (7.18) gives

$$\tau(x, y) = \tau_{b,s}\left(1 - \frac{2}{w}y\right)\left(1 + w\left(\frac{2\pi}{\lambda}\right)^2\right)\epsilon$$
$$\times \sin\left(\frac{2\pi x}{\lambda}\right) + \tau_c \quad (7.19)$$

We further assume that spatial variations in bed shear stress produce erosion and accretion of the bank proportional to the variation in shear stress. In other words, the bank will erode in locations where the variation in shear stress downriver is positive, and a point bar will prograde in places where variations in shear stress are negative. The rate of migration, therefore, is proportional to the derivative of $\tau(x, y)$ with respect to x, evaluated at $y = w/2$:

$$\left.\frac{\partial \tau}{\partial x}\right|_{y=w/2} = \frac{d\epsilon}{dt}\sin\left(\frac{2\pi x}{\lambda}\right)$$
$$= \tau_{b,s}\left(\frac{w}{2}\left(\frac{2\pi}{\lambda}\right)^2 - \frac{2}{w}\right)\epsilon\sin\left(\frac{2\pi x}{\lambda}\right) \quad (7.20)$$

The normalized growth rate of perturbations of the bed is given by

$$\frac{\dot{\epsilon}}{\epsilon} = \frac{2\pi \tau_{b,s}}{\lambda}\left(\frac{2}{w} - \frac{w}{2}\left(\frac{2\pi}{\lambda}\right)^2\right) \quad (7.21)$$

which can be simplified to

$$\frac{\dot{\epsilon}}{\epsilon} \propto \frac{1}{\lambda}\left(1 - \left(\frac{2\pi w}{\lambda}\right)^2\right) \quad (7.22)$$

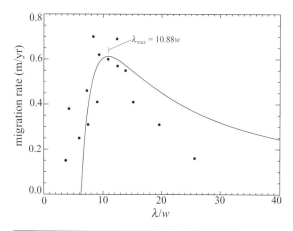

Fig 7.5 Plot of the normalized growth rate of perturbations as a function of wavelength λ (Eq. (7.22)). Also shown (dots) are the bank-migration data of Nanson and Hickin (1983) along the Beatton River for comparison.

Equation (7.22) is plotted in Figure 7.5 as the solid curve. The maximum growth rate occurs where the first derivative of $\dot{\epsilon}/\epsilon$ with respect to λ is zero. This occurs at $\lambda = 2\sqrt{3}\pi w = 10.88w$, which is close to the observed scaling relation (Leopold *et al.*, 1964).

For comparison, Figure 7.5 shows observed migration rate data for the Beatton River from Nanson and Hickin (1983). These authors employed dendrochronological techniques to estimate the migration rates for sixteen bends of that river. They expressed their data as a function of the average radius of curvature of the bend. To compare their results to our theoretical growth curve, we used the relation $\lambda = \pi R$ which is applicable for an ideal shape of fully developed meanders: the sine-generated curve (Langbein and Leopold, 1966). Their results show the same dependence with meander wavelength predicted by our linear stability analysis. The migration rate rises rapidly for small λ/w, reaches a maximum value, and then decreases. Migration rates decrease rapidly for tightly curved bends with $\lambda/w < 2\pi$, consistent with our model's prediction that channel bends become stable below that wavelength. It should be noted that some studies following Nanson and Hickin (1983) showed that migration rates are far less systematic that the results obtained for the Beatton River. Most of those studies have concluded that the Nanson and Hickin results are not generally applicable. It

is difficult, however, to compare migration rates from many different bends without first quantifying the effects of bank-material texture. Lumping data from meander bends with different erodibilities would increase the scatter compared to the Nanson and Hickin data, even if individual bends closely follow a universal growth curve. Migration rate data from within individual bends would help to resolve this question.

The analysis that led to Eq. (7.22) also applies to channels carved by thermal erosion. Supraglacial meltwater streams (Parker, 1975) and lava channels on the Earth, the Moon, and Venus (Hulme, 1973) form by thermal erosion. In the case of lava channels, melting of the channel bed by hot lava (and solidification on the opposite bank) causes channel migration to occur. Curved banks have a melting point that depends on the curvature of the bank. The melting temperature of the bank is given by

$$T = T_m + cK \tag{7.23}$$

where T_m is the melting point for a straight channel, K is the planform curvature of the bank, and c is a constant that depends on the latent heat of the bank material. The temperature required to melt a curved bank is elevated because eroding the bank by a given amount requires melting a larger volume of material per unit surface area than for a straight channel. As such, the temperature boundary condition in lava channels has the same curvature dependence as the shear stress boundary condition in alluvial channels. In addition, turbulent eddies will transport heat away from the hot center of the flow towards the cool sides of the flow in the same way that turbulent eddies transport shear stress through an alluvial channel, resulting in an approximately linear temperature profile. These underlying similarities between alluvial and lava channels suggests that the same geometrical instability mechanism is at work in both cases.

7.4 | Werner's model for eolian dunes

Eolian ripples, dunes, and megadunes obey a striking periodicity. Many attempts have been made to understanding the controls on eolian bedform spacing in terms of both microscale processes (e.g. the trajectories of individual grains in saltation and reptation) and macroscale processes (e.g. the interaction between a growing bedform, the wind flow above it, and the resulting pattern of erosion and deposition). Despite the efforts of many scientists, the mechanisms responsible for the formation of eolian ripples, dunes, and megadunes and the factors responsible for their size and spacing are not fully understood. Perhaps the most well studied type of bedform is the eolian ripple. Bagnold (1941) performed the pioneering work on the formation of eolian ripples. He proposed that ripples form by a geometrical instability in which a perturbation exposes the windward side of the perturbation to more impacts and faster surface creep than the leeward side, resulting in more grains being ejected into saltation on the windward slope. In his theory, the spacing of ripples is related to a characteristic saltation distance that is constant in time. Sharp (1963), however, observed that incipient ripples increase in spacing over time until a steady-state spacing is achieved. In addition, experimental studies of grain impacts suggest that a wide distribution of energies are imparted to grains on the bed during each impact, resulting in a wide distribution of saltation path lengths (Mitha *et al.*, 1986). As such, Bagnold's hypothesis is not well supported by available data, and the controls on ripple spacing are still not well understood. Anderson and Bunas (1992) developed a model that produced realistic ripples based on the microscale processes of grain movement and bed impact, but the spacing of ripples in their model was controlled by an ad hoc "ceiling" in the model domain that does not occur in nature.

The mechanisms controlling the formation and spacing of eolian dunes are also incompletely understood. Eolian dunes represent a distinct bedform type from eolian ripples since they occur at a larger scale with no transitional bedforms. It has been proposed that dunes result from the presence of pre-existing wave-like motions or secondary circulations in the atmospheric boundary layer. Periodic motions intrinsic to the flow or resulting from the response of the flow to a topographic disturbance upwind may produce periodic bedforms (Wilson, 1972;

Folk, 1976). Wilson (1972) has proposed the same mechanism for ripples and megadunes with the bedforms on these scales created by different scales of atmospheric circulation. However, careful field studies have not found evidence for the existence of these persistent atmospheric circulations (Livingstone, 1986).

In 1995, Werner made a significant advance in eolian geomorphology in a paper that described a very simple but powerful model of eolian dune formation (Werner, 1995). Werner's model begins with a set of discrete sand slabs on a rectangular grid. Slabs may represent individual sand grains or a collection of grains; the number of slabs on the grid is limited for computational reasons, but the model is not sensitive to the size of each slab. During each time step of the model, a sand slab is picked up from a randomly chosen pixel. If there is no sand at that pixel, nothing happens. If a slab is present, it is picked up and moved downwind a distance of l grid point and deposited back on the surface with a probability that depends on the presence or absence of sand at the new location. If sand is present at the new location, the slab is deposited with probability p_s. If no sand is present, it is deposited with a lower probability p_{ns}. In nature, sand is more likely to be deposited on a patch of sand than on a bare desert surface because the boundary layer above a sandy surface is usually rougher than above a bare, smooth surface, and because energy is absorbed from the saltating grain by the granular bed. If the slab is not deposited, it is transported another distance l and again deposited with a probability given by p_s or p_{ns}. One additional rule controls the probability of deposition: if a sand slab lands in a "shadow zone," it is deposited with 100% probability. A shadow zone is defined as the domain downwind of any topographic high that lies below a plane with an angle of 15° to the horizontal. To map the shadow zone, a path is traced from every local ridge down a 15° slope in the downwind direction until the plane intersects the surface. The 15° angle of the shadow zone can be varied, but its value should be significantly less than the angle of repose. Physically, the shadow zone represents the zone of recirculation on the lee side of a growing dune. In this zone, airflow velocities are greatly reduced and any sand in saltation will likely be deposited.

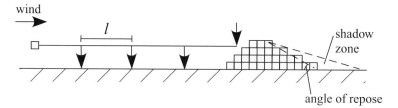

Fig 7.6 A schematic diagram of Werner's model for eolian dune formation and evolution. After Werner (1995).

Finally, when the sand slab is deposited it may create an oversteepened condition in which the local slope is higher than the angle of repose. If this occurs, the model moves the slab down the direction of steepest descent until a site is found that does not lead to an oversteepened condition. Figure 7.6 illustrates the model schematically. A code that implements Werner's model is provided in Appendix 6.

Initially, a rough surface develops in Werner's model from an initially flat surface due to the random entrainment of grains from the surface (i.e. sand from some areas will be entrained more than in others because of the stochastic nature of entrainment in the model). Incipient ridges then develop small shadow zones that encourage deposition on their lee sides. This deposition leads to the development of taller dunes with longer shadow zones in a positive feedback. Using different parameter values, Werner (1995) was able to reproduce a variety of dune types by varying the initial sand thickness (a relatively thin initial sand bed leads to barchan dunes, as in the example of Figure 7.7) and the direction of transport (e.g. star dunes form if the sand bed is initially thin and the wind direction is periodically reversed). Werner's model provides a nice illustration of the power of simplified models for understanding the behavior of complex systems with positive feedback. Werner's model can easily be modified to account for variable bed topography. Figure 7.8, for example, illustrates a dune field migrating through a crater. On Mars, the crestlines of dunes are deformed downwind of craters, similar to the effect shown in Figure 7.8.

The height and spacing of dunes in Werner's model increases continuously over time without ever reaching a steady-state value. Numerical models that incorporate the interaction between dunes and the airflow above them, in contrast, predict that dunes eventually achieve an optimum dune height and spacing. In nature, it is unclear which model is correct. In locations where dunes and megadunes are both present, these two bedform types appear to

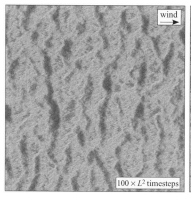 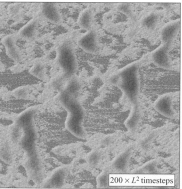 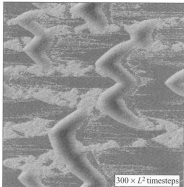

Fig 7.7 Results of Werner's model for a grid of size $L = 500$ and model parameters $l = 5$, $p_{ns} = 0.4$, $p_s = 0.6$, angle of repose = 30°, and angle of shadow zone = 15° for three model durations. The initial condition is uniform sand with a thickness of 10 sand parcels on each pixel.

occur at distinctly different spatial scales with no intermediate bedforms (Wilson, 1972). In other words, even in sand seas mature enough to have formed megadunes, dunes appear to

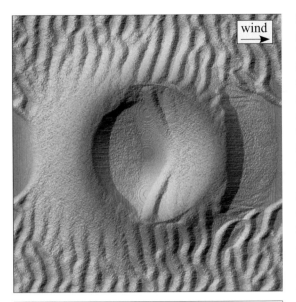

wind

Fig 7.8 Results of Werner's model modified to include variable bed topography. In this example, a dune field interacts with a crater. The bed topography causes deformation of dune crest lines within and downwind of the crater.

reach a steady-state spacing of approximately 10–200 m, with stronger winds and larger grain sizes associated with greater spacing. This observation suggests that the aerodynamic interaction between dunes and the airflow above them is essential for understanding the height and spacing of mature eolian dunes.

7.5 | Oscillations in arid alluvial channels

Most fluvial-geomorphic studies that have modeled the evolution of channel longitudinal profiles have assumed channels that are uniform in width or a prescribed function of drainage area (e.g. Begin *et al.*, 1981; Slingerland and Snow, 1987; Sinha and Parker, 1996; Dade and Friend, 1998). Actual channels often vary substantially in width, however, and these variations in width can strongly influence channel evolution. Schumm and Hadley (1957), for example, developed their conceptual "arid cycle of erosion" on the basis of the episodic cut-and-fill histories and oscillating geometries observed in the arroyos of

the American West. In their model, channels aggrade and widen until a threshold slope and/or width is reached, initiating incision in the oversteepened reach and aggradation in the downstream reach. In this section, we explore a mathematical model for Schumm and Hadley's arid cycle of erosion and compare the model predictions to the spatial pattern of oscillations commonly observed in alluvial channels of the southwestern US.

Oscillating alluvial channels in southern Arizona are classified as one of three basic types depending on their geomorphic position. Piedmont channels fed by montane drainage basins are continuous channels that alternate between braided and narrowly entrenched reaches with wavelengths on the order of 1 km. Wild Burro Wash on the Tortolita Piedmont is a classic example of this type of oscillating channel (Figure 7.9a). Channel width (measured from 1 m resolution US Geological Survey Digital Orthophotoquadrangles (DOQQs)) and bed elevation (measured from a high-resolution DEM with average channel slope subtracted) in Wild Burro Wash are plotted in Figure 7.9a. These data illustrate that oscillations in both bed elevation and channel width occur in Wild Burro and that the two oscillations are in phase.

The second type of oscillating channel is the piedmont discontinuous ephemeral stream. These channels are fed only by local piedmont runoff, and they alternate between incised channels and sheet-flow-dominated channel fans without well-defined banks (Bull, 1997). The transitions between channel fans and incised reaches occur abruptly at one or more steep "headcuts" (Figure 7.9b). Dead Mesquite Wash in southeast Tucson is the classic example given by Bull (1997) (Figure 7.9b). Packard (1974) documented in-phase oscillations in bed elevation and width in Dead Mesquite, similar to the behavior of Wild Burro.

The third type of oscillating channel is the classic southwestern valley-floor arroyo. These systems alternate between narrow, deeply incised channels and broad, shallow channel fans, but they typically drain one or more broad basins and adjacent ranges and have wavelengths on the order of 10 km. Abrupt headcuts also characterize distributary-to-tributary transitions in these

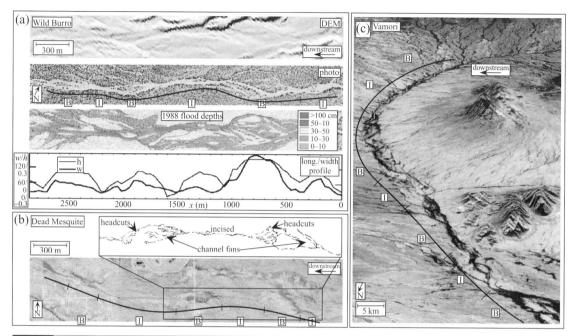

Fig 7.9 Examples of oscillating channels in southern Arizona. Alternating reaches: B–braided and I–incised. (a) Wild Burro Wash data, including (top to bottom) a high-resolution DEM (digital elevation model) in shaded relief (PAG, 2000), a color Digital Orthophotoquadrangle (DOQQ), a 1:200-scale flow map corresponding to an extreme flood on July 27, 1988 (House et al., 1991), and a plot of channel width w (thick line) and bed elevation h (thin line, extracted from DEM and with average slope removed). (b) Dead Mesquite Wash shown in a color DOQQ and a detailed map of channel planform geometry (after Packard, 1974). (c) Vamori Wash shown in an oblique perspective of a false-color Landsat image (vegetation [band 4] in red) draped over a DEM. Channel locations in Figure 7.11a. Modified from Pelletier and Delong (2005). For color version, see plate section.

systems. In contrast to the other oscillating-channel types, not all arroyos are periodically alternating; some are tectonically controlled by the elevations of basin divides. In southern Arizona, Cooke and Reeves (1976) identified Vamori Wash as perhaps the best example of an oscillating arroyo system (Figure 7.9c). Entrenched reaches are associated with low vegetation density, whereas channel fans support a broad, dense riparian zone in Vamori Wash.

The common oscillatory pattern in these channel types suggests a common underlying mechanism that acts across multiple spatial scales. This section explores this hypothesis in

two ways. First, a numerical model for the coupled evolution of the channel longitudinal profile and cross section is presented. This model predicts that arid alluvial channels are unstable over a range of wavelengths controlled by channel width, depth, and slope. Second, a database of channel geometries in southern Arizona is constructed that confirms the model predictions and places the three channel types within a single continuum from steep, small-wavelength-oscillating channels to gently sloping, long-wavelength-oscillating channels.

The equation describing conservation of mass for sediment states that erosion and deposition in an alluvial-channel bed are proportional to the gradient of total sediment discharge, or

$$\frac{\partial h}{\partial t} = -\frac{1}{C_0} \frac{\partial (w q_s)}{\partial x} \tag{7.24}$$

where h is the elevation of the channel bed, t is time, C_0 is the volumetric concentration of bed sediment, w is the channel width, q_s is the specific sediment discharge (i.e. the sediment discharge per unit channel width), and x is the distance downstream. In most bedload transport relations, specific sediment flux is proportional to the 3/2-power of the bed shear stress. This implies that specific sediment discharge is a linear function of channel gradient and a nonlinear function of the discharge per unit channel width, or

$$q_s = -B \left(\frac{Q}{w} \right)^b \frac{\partial h}{\partial x} \qquad (7.25)$$

where B is a mobility parameter related to grain size and Q is water discharge. The value of b is constrained by sediment rating curves and is between 2 and 3 for both suspended-load and bed-load transport. Here we consider the case $b = 2$. Equation (7.25) assumes that the bed shear stress is much larger than the threshold for particle entrainment.

If the channel width is assumed to be uniform along the longitudinal profile, the combination of Eqs. (7.24) and (7.25) gives the classic diffusion equation:

$$\frac{\partial h}{\partial t} = \kappa \frac{\partial^2 h}{\partial x^2} \qquad (7.26)$$

where the diffusivity is given by $\kappa = (B Q^2)/(C_0 w_0)$, and w_0 is the uniform channel width.

If the channel width is not uniform along the longitudinal profile, the chain rule must be used when differentiating q_s in Eq. (7.24). This approach introduces an additional nonlinear term into Eq. (7.26), to give

$$\frac{\partial h}{\partial t} = \kappa w_0 \left(\frac{1}{w} \frac{\partial^2 h}{\partial x^2} - \frac{1}{w^2} \frac{\partial w}{\partial x} \frac{\partial h}{\partial x} \right) \qquad (7.27)$$

It is mathematically easier to define h as the difference between the local bed elevation and that of a straight, equilibrium channel with uniform discharge. Equation (7.27) then becomes

$$\frac{\partial h}{\partial t} = \kappa w_0 \left(\frac{1}{w} \frac{\partial^2 h}{\partial x^2} - \frac{1}{w^2} \frac{\partial w}{\partial x} \left(S_0 - \frac{\partial h}{\partial x} \right) \right) \qquad (7.28)$$

where S_0 is the equilibrium channel slope. Equation (7.28) can be used to study the evolution of perturbations from the equilibrium geometry.

Channel widening and narrowing occurs by a complex set of processes including bank retreat and bed scouring. Scouring often leads to channel narrowing as flow is focused into the scour zone and parts of the former channel bed effectively become part of the bank. Bull's "threshold-of-critical-power" concept (Bull, 1979) can be used to relate the rate of channel widening and narrowing with the excess stream power if channels are assumed to widen as they aggrade and narrow as they incise. Bull's model states that channels incise if the stream power is greater than a threshold value and aggrade if the stream power is less than that value. Expressing this relationship in terms of channel widening (aggradation) and narrowing (incision) gives

$$\frac{\partial w}{\partial t} = -\frac{w q_s - w_0 q_{s0}}{h_0 - h} \qquad (7.29)$$

where q_{s0} is the equilibrium sediment flux or stream power, and h_0 is the equilibrium bank height. Equation (7.29) represents a cross-sectional mass balance: sediment removed from the bank contributes to the local sediment-flux deficit $w q_s - w_0 q_{s0}$ (i.e. the amount of sediment that cannot be transported out of the reach), promoting further aggradation in the reach.

The nonlinear term in Eq. (7.28) alters the dynamics of channels markedly compared to the diffusive behavior expressed in Eq. (7.26). Along a reach with uniform discharge and grain size, the diffusion equation smoothes out curvatures in the profile over time. The nonlinear term in Eq. (7.28), however, has the opposite effect through a positive feedback between channel width and slope. In this feedback, spatial variations in channel slope generate variations in width via Eq. (7.29). Large gradients in channel width, in turn, increase the nonlinear behavior in Eq. (7.28), further localizing erosion and deposition to complete the feedback cycle. This cycle is balanced by the diffusive term, and the balance between these two terms controls the oscillation wavelength.

Equations (7.28) and (7.29) may be solved by using linear stability analysis and direct numerical solution. Linear stability analysis works by solving the linear approximation to Eqs. (7.28) and (7.29) for the growth rate of a small-amplitude oscillation superimposed on the initial channel geometry (a channel with specified equilibrium width w_0, bank height h_0, and slope S_0).

Linear-stability analysis involves solving for the growth rate of spatially periodic, small-amplitude perturbations from an equilibrium geometry by retaining only linear terms in the model equations. In this linear stability analysis, we assume that the channel has an average width

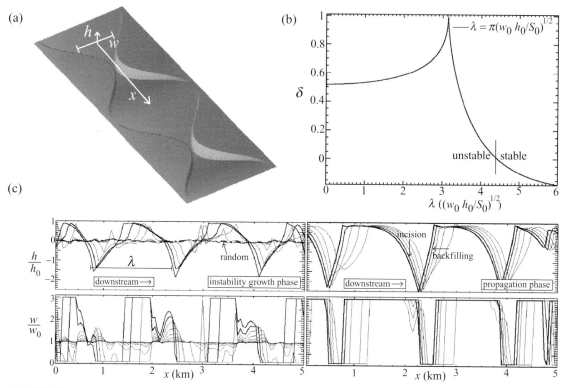

(a)

(b)

(c)

Fig 7.10 Model behavior. (a) Schematic diagram of model geometry. (b) Growth curve for linear-stability analysis, where δ is the nondimensional growth rate. This analysis predicts unstable behavior for small wavelengths, with a maximum value at $\lambda = \pi(w_0 h_0/S_0)^{1/2}$. (c) Numerical solution. Initial phase (left) is characterized by growth of small, random perturbations and an increase in oscillation wavelength with time (line thickness increases with time). Steady-state phase (right) is characterized by solitary-wave propagation of oscillations in upstream direction. Incision and channel narrowing on leading edge of each wave are balanced by backfilling and widening on trailing edge. Modified from Pelletier and Delong (2005).

w_0 and slope S_0. The variables h' and w' refer to the variations of h and w from their average values, and have solutions of the form

$$h' = h'_0 e^{i\left(\frac{2\pi x}{\lambda}\right)}e^{\delta t}, \quad w' = w'_0 e^{i\left(\frac{2\pi x}{\lambda}+\phi\right)}e^{\delta t} \quad (7.30)$$

where δ is the growth rate and λ is the wavelength. These solutions assume a spatially periodic form, and an exponential increase in amplitude with time as the instability grows. Substituting Eqs. (7.30) into Eqs. (7.28) and (7.29) and

retaining only linear terms gives

$$\delta h'_0 e^{i\left(\frac{2\pi x}{\lambda}\right)} = \frac{2\pi \kappa}{\lambda}\left(-\left(\frac{2\pi}{\lambda}\right)h'_0 e^{i\left(\frac{2\pi x}{\lambda}\right)} + \frac{S_0}{w_0}w'_0 e^{i\left(\frac{2\pi x}{\lambda}\right)}\right) \quad (7.31)$$

and

$$\delta w'_0 e^{i\left(\frac{2\pi x}{\lambda}+\phi+\frac{\pi}{2}\right)} = -\frac{\kappa}{w_0 h_0}\left(\frac{2\pi}{\lambda}\right)h'_0 e^{i\left(\frac{2\pi x}{\lambda}+\frac{\pi}{2}\right)} \quad (7.32)$$

Equation (7.31) implies $\phi = -\pi/2$ (i.e. h' and w' are 90° out of phase). Substituting Eq. (7.32) into Eq. (7.31) gives the following equation for growth rate:

$$\delta = \frac{\kappa}{w_0}\left(\frac{2\pi}{\lambda}\right)^2\left(1 - \frac{\kappa S_0}{h_0 w_0^2}\frac{1}{\delta}\right) \quad (7.33)$$

The solution to Eq. (7.33) is

$$\delta = \frac{2\pi\kappa}{w_0\lambda}\left(\frac{\pi}{\lambda} - \sqrt{\left|\left(\frac{\pi}{\lambda}\right)^2 - \frac{S_0}{w_0 h_0}\right|}\right) \quad (7.34)$$

Equation (7.34) is plotted in Figure 7.10b. This figure shows that the growth rate of perturbations has a maximum at

$$\lambda = \pi \sqrt{\frac{w_0 h_0}{S_0}} \tag{7.35}$$

This result can also be obtained by differentiating Eq. (7.34) with respect to λ and setting the result equal to zero.

This analysis predicts the growth curve shown in Figure 7.10b. The growth rate is positive (i.e., perturbations are unstable) for small wavelengths, rises to a steep maximum at

$$\lambda = \pi \sqrt{\frac{w_0 h_0}{S_0}} \tag{7.36}$$

and quickly becomes negative (i.e., perturbations decay to zero) for larger wavelengths. The growth rate is a function of κ (i.e., channels with larger values of κ develop oscillations more rapidly), but the wavelength corresponding to the maximum growth rate (Eq. (7.36)) is independent of κ. This independence is important for testing Eq. (7.36) against field measurements because κ is not well constrained, but the average channel width, bank height, and slope can be readily measured for any channel.

The two-step Lax–Wendroff scheme (e.g., Press et al., 1992) was used to integrate Eqs. (7.28) and (7.29) for the direct numerical solution. Code that implements this solution is presented in Appendix 6. The width was not allowed to go below 10 m or above 300 m in the model. Without bounds, the model develops infinitely small and large channel widths that are unrealistic. The model results are not sensitive to the specific values of the lower and upper bounds as long as they are small and large, respectively, compared to the equilibrium channel width.

Figure 7.10c shows plots of bed elevation h (top) and channel width w (bottom) for the early (left) and late (right) stages of the numerical model. We assumed an initial width, bank height, and slope of 100 m, 2 m, and 0.01, respectively, with small (1%) random variations superimposed on the initial channel width. The solutions are plotted for a temporal sequence, with thicker lines representing later times. The early stage of the model is characterized by

the amplification of spatial variations in width within a range of wavelengths close to 1 km. In the latter stage, oscillations achieve a steady-state amplitude and propagate upstream as a train of solitary (i.e., nondispersive) topographic waves. At any given instant, the channel geometry is characterized by alternating zones of narrow, deeply incised reaches and wide, shallow reaches. The most abrupt slope break is where the channel changes from distributary to incised. This break is similar to the headcuts often observed in arroyos and discontinuous ephemeral streams. The basic two-step model evolution is a robust feature of the model and was observed for a wide range of model parameters.

As the instability develops from random initial conditions, the dominant wavelength increases, and the channel geometry becomes more regularly periodic. The increase in oscillation wavelength indicates that nonlinear, finite-amplitude effects are an important part of the model behavior. This result is confirmed by Figure 7.11b, which compares the oscillation wavelengths predicted by the linear-stability analysis and the direct numerical solution for different values of $(w_0 h_0 / S_0)^{1/2}$. The lower solid line corresponds to the linear-stability prediction (Eq. (7.36)), and the filled squares and upper solid line are the fully nonlinear numerical results.

To test the model predictions for oscillation wavelength, we constructed a GIS database of DOQQs, rectified false-color Landsat imagery, and 30-m resolution DEMs for all of southern Arizona. This database was used to measure oscillation wavelengths and average channel widths, depths, and slopes for 15 oscillating channels. Our data set includes examples of each of the three channel types, including channels over a broad range of sizes. Figure 7.11b gives the average oscillation wavelength measured for each channel as a function of $\sqrt{w_0 h_0 / S_0}$. The data points corresponding to 10 yr flow depths are plotted with circles; 25 yr flow depths are plotted with triangles. The agreement between the observed and predicted trend in the data are quite good. This agreement provides validation for the quantitative prediction of the nonlinear model.

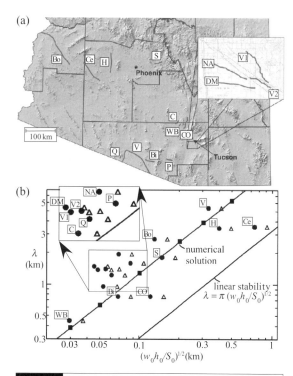

(a)

(b)

Fig 7.11 Database of oscillating channel geometries in southern Arizona. (a) Location map. Clockwise from upper left: (Bo) Bouse Wash; (Ce) Centennial Wash; (H) Hassayampa River; (S) Sycamore Creek; (NA) North Airport Wash; (VI) Vail Wash; (V2) Vail-Dead Mesquite Wash; (DM) Dead Mesquite Wash; (C) Cottonwood Wash; (WB) Wild Burro Wash; (CO) Canada del Oro Wash; (P) Penitas Wash; (Bi) Bobaquivari Wash; (V) Vamori Wash; (Q) East La Quituni Wash. (b) Plot of oscillation wavelength λ vs. $\sqrt{w_0 h_0/S_0}$. Lower line is linear-stability prediction. Upper line and filled squares are numerical results. Observed data for channels in (a) are plotted with filled circles (10 yr flood depth) and open triangles (25 yr flood depth). Modified from Pelletier and Delong (2005).

7.6 | How are drumlins formed?

Drumlins are subglacial bedforms elongated parallel to the ice-flow direction and composed primarily of subglacial till or sediment. Some drumlins have a bedrock core. These drumlins most likely formed by sediment accretion in the lee-side low-pressure zone caused by streaming ice flow around the core (Boulton, 1987). More enigmatic, however, are the drumlins composed entirely of sediment. Subsurface bedding in these

drumlins often parallels the topographic surface, suggesting that they formed by localized, upward flow of sediment. If so, what were the driving forces for this flow, and what factors controlled the morphology of the resulting drumlins? For example, why are drumlins so variable in size and shape, even over distances as short as a few kilometers (Figure 7.12)? Hindmarsh (1998) and Fowler (2001) proposed that shearing of an incompressible viscous till layer could be responsible for producing drumlins even in the absence of a bedrock core. Schoof (2002), however, has shown that bedforms produced by these models are not consistent with many aspects of natural drumlins.

Most conceptual drumlin models invoke either subglacial hydraulic processes or deforming bed processes. This section seeks to unite these two hypotheses by modeling sediment deformation caused by porewater migration. In this model, porewater migration generates buoyancy forces that drive converging, upward flow of the matrix to form hummocky moraine (under flat ice sheets) and drumlins (under sloping ice sheets) that have a characteristic scale that depends on the sediment texture and thickness of the till layer. This hypothesis builds on Menzies' (1979) conceptual model of drumlin formation by porewater expulsion. Menzies showed that sedimentary microstructures often indicate that drumlin sediments had a higher concentration of porewater than interdrumlin sediments. Menzies used this observation to propose that drumlin formation was driven by stresses associated with porewater migration and expulsion. To date, however, this process has not been quantified, nor is it known what drumlin morphologies result from this process.

The modeling of sediment deformation with porewater migration involves a complex set of diffusive and advective equations. The equations that describe deformable porous media such as subglacial sediment were first developed in the geophysics literature to describe the flow of magma through porous rock (Scott and Stevenson, 1986; Spiegelman and McKenzie, 1987) and, later, to fluid flow in sedimentary basins (e.g. Dewers and Ortoleva, 1994). Deformable porous media are comprised of two fluids with different

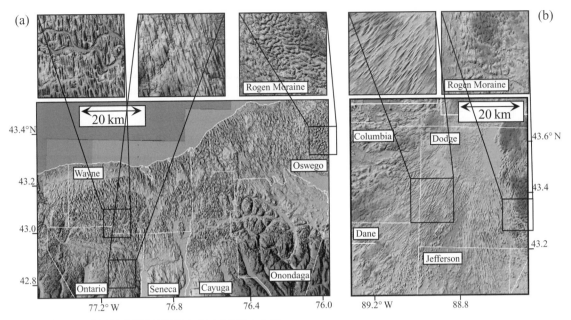

Fig 7.12 Shaded-relief maps of drumlin fields in (a) north-central New York and (b) east of Madison, Wisconsin. Inset regions illustrate the variety of drumlin shapes in each area. Rogen moraines (bedforms oriented perpendicular to the ice-flow direction) are present at the margins.

densities and viscosities. The low-density, low-viscosity "liquid" phase migrates according to Darcy's Law within a "matrix" of high-density, high-viscosity fluid governed by Stokes' Law. Deformable porous media exhibit two kinds of behavior not observed in incompressible fluids. First, the liquid can become segregated from the matrix by a positive feedback between porosity, permeability, and matrix expansion. In this feedback, preferential migration of the liquid into regions of the matrix with higher porosity (and hence permeability) expand the matrix to further increase the porosity. In a layer undergoing compaction, this feedback leads to alternating zones of high and low porosity with a characteristic spacing that depends on the layer thickness. Second, the expansion and contraction of the matrix generates buoyancy forces that drive convective flow of the matrix.

McKenzie (1984) first proposed a set of equations to describe partially molten rock as a deformable porous medium. His equations include: conservation of mass for the liquid and matrix phases, Darcy's Law for the migration of the liquid in the matrix, Stokes' Law for the viscous flow of the matrix, and an empirical power-law relationship between permeability and porosity:

$$\frac{\partial \phi}{\partial t} + \nabla \cdot \phi v = 0 \tag{7.37}$$

$$\frac{\partial \phi}{\partial t} - \nabla \cdot (1 - \phi)V = 0 \tag{7.38}$$

$$\phi(v - V) = -\frac{k_\phi}{\mu} \nabla P \tag{7.39}$$

$$\nabla P = -\eta \nabla \cdot \nabla \cdot V$$
$$+ \left(\zeta + \frac{4}{3}\eta\right)\nabla(\nabla \cdot V) - (1 - \phi)\Delta\rho gk \tag{7.40}$$

$$k_\phi = a\phi^n \tag{7.41}$$

where ϕ is the porosity, v is the liquid velocity, V is the matrix velocity, k_ϕ is the permeability, μ is the liquid kinematic viscosity, P is the excess pressure (above hydrostatic), ζ and η are the bulk and shear viscosities of the matrix, $\Delta\rho$ is the difference in density between the liquid and matrix, g is the gravitational acceleration, k is the unit vector in the vertical direction, and a and n are empirical parameters.

Equations (7.37)–(7.41) have been applied to 1D (vertical) compaction problems, but they are not easily solved in 2D or 3D. However, by introducing stream functions Φ and U for the incompressible and compressible components of the matrix velocity V (i.e. $V = \nabla \cdot \Phi + \nabla \cdot U$), Spiegelman (1993) developed a set of equations readily solved in 2D and 3D:

$$\nabla^4 \Psi = -\frac{\phi_0^2}{\eta} \nabla \cdot \phi k \tag{7.42}$$

$$-k_\phi \nabla^2 C - \nabla k_\phi \cdot \nabla C + C$$

$$= \nabla \cdot k_\phi \left(\frac{\eta}{\phi_0} (\nabla \cdot \nabla^2 \Psi) - (1 - \phi_0 \phi) k \right) \tag{7.43}$$

$$\nabla^2 U = \phi_0 C \tag{7.44}$$

$$\frac{\partial \phi}{\partial t} + (\nabla \cdot \Psi + \nabla U) \cdot \nabla \phi = (1 - \phi_0 \phi) C \tag{7.45}$$

$$k_\phi = a \phi^n \tag{7.46}$$

where ϕ is the initial porosity, ζ is the ratio of the shear to bulk viscosities, and C is the compaction rate (defined as $C = \nabla \cdot v$). Although originally developed for partially-molten rock, Eqs. (7.42)–(7.46) are applicable to any deformable porous medium governed by Darcian and Stokes' flow. Spiegelman's equations may appear formidable, but their solution is not fundamentally more difficult than solving the diffusion or wave equations. The strategy is to solve Eqs. (7.42)–(7.46) sequentially for the porosity at time t_1 given the porosity at an earlier time t_0, subject to the boundary conditions of the problem. In this study, Eqs. (7.42)–(7.44) were solved using the Alternating-Direction-Implicit (ADI) technique (Press et al., 1992), and Eq. (7.45) was solved using upwind differencing.

Equations (7.42)–(7.46) are scaled to the fundamental length scale of the problem: the compaction length, δ, given by

$$\delta = \sqrt{\frac{k_\phi \left(\eta + \frac{4}{3} v \right)}{\mu}} \tag{7.47}$$

The compaction length is the spatial scale at which compaction, requiring both viscous flow of the matrix and Darcian flow of the liquid, takes place most rapidly. At small spatial scales, liquid migrates through the matrix readily, but compaction occurs slowly because of viscous resistance of the matrix. At large spatial scales, the matrix can readily deform, but compaction is limited by the time required for the liquid to migrate long distances. The compaction

length is the length scale at which matrix deformation and liquid migration occur at similar rates and neither is a limiting factor. The compaction length is broadly analogous to the flexural parameter α, which characterizes lithospheric strength (Turcotte and Schubert, 2002). The flexural parameter represents a similar balance, in that case between flexural rigidity and buoyancy under crustal loading.

An order-of-magnitude estimate for δ in subglacial sediments can be obtained using a matrix shear viscosity of 10^{10} Pa s (Murray, 1997), a permeability of 10^{-10} m^2 (appropriate for sandrich deposits; Freeze and Cherry, 1979), and the kinematic viscosity of water: 10^{-2} Poise. The bulk viscosity of subglacial sediments (i.e. their viscous resistance to expansion and contraction) is not well constrained, but for simplicity we assume it to be equal to the shear viscosity. These values yield an order-of-magnitude estimate $\delta \approx 100$ m. Till thicknesses in New York and Wisconsin are between 0 and 50 m, or between $l = 0$ and $l = 0.5$ when scaled to this compaction length. The model results described below do not depend on the precise value of l or δ, as long as $l < \delta$. If this inequality holds, l is the length scale controlling the instability.

Spiegelman et al. (2001) solved Eqs. (7.42)–(7.46) in 2D and showed that partially molten rock undergoes an instability in which vertically-migrating melt focuses into regularly-spaced, high-porosity channels alternating with low-porosity channels. Channels develop in Eqs. (7.42)–(7.46) by a positive feedback in which ascending melt migrates preferentially into regions of higher initial porosity (and hence permeability), causing the matrix to expand locally to make room for the melt. Matrix expansion increases the local porosity, further enhancing the focusing of melt in a positive feedback. Although partially molten rock is governed by very different physics than subglacial sediments, both are composed of liquid embedded in a deformable matrix, and hence both may be expected to exhibit the same fundamental instability mechanism during compaction. Can this instability mechanism be responsible for drumlins?

Here we explore the results of three numerical experiments designed to illustrate how

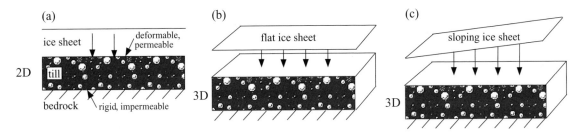

Fig 7.13 Model geometry and boundary conditions for the three model experiments. The upper boundary is deformable and permeable, and the lower boundary is rigid and impermeable.

subglacial sediment deforms under different sediment-thickness and ice-surface-slope conditions. In the first experiment (Figure 7.13a), 2D compaction was considered under uniform pressure. In the second experiment (Figure 7.13b), 3D compaction was considered under uniform pressure. In the third experiment (Figure 7.13c), a sloping ice sheet was considered in 3D.

In order to solve Eqs. (7.42)–(7.46), the values or derivatives of Ψ, U, C, and ϕ must be specified at the upper and lower boundaries of the grid. In all calculations, we assume an impermeable, rigid, no-slip lower boundary (the bedrock-sediment interface) and a permeable, deformable, free-slip upper boundary (the sediment–ice interface). The ice sheet plays only a passive role in this model, providing hydrostatic pressure but no shear. This contrasts with the models of Hindmarsh (1998) and Fowler (2001), in which drumlins were formed by the shearing of an incompressible viscous till layer by the overriding ice sheet. At the lower boundary, the condition $\Psi = 0$ specifies that both the horizontal and vertical components of the matrix velocity are zero (this combines both rigid and no-slip boundary conditions). The rigidity of the lower boundary also requires setting (i.e. the component of the matrix velocity perpendicular to a rigid boundary is zero). $C = 0$ and $\phi = 0$ define the impermeability of the lower boundary (Spiegelman, 1993). A deformable, permeable upper boundary requires that $U = 0$ and that the gradients of compaction and porosity normal to the upper boundary are zero: $\nabla C \cdot n = 0$ and $\nabla \phi \cdot n = 0$, where n is the unit normal vector (Spiegelman, 1993).

This boundary condition assumes the existence of a meltwater channel at the sediment–ice interface that conducts porewater out of the system as it leaves the matrix. The standard free-slip boundary condition for incompressible flows gives $\nabla \Psi \cdot n = 0$. The sediment–ice interface is a moving boundary in the model: during each time step, displacement occurs by an amount $V \Delta t$, which raises or lowers the interface locally, depending on the sign of V.

The model parameters for all the numerical experiments in this section are $\phi_0 = 0.5$, $n = 3$, and $\zeta = 1$. The sediment thickness was varied between $l = 0.2$ ("thin" sediment) and $l = 0.6$ ("thick" sediment) in the experiments. Porosities in subglacial sediments typically vary from 0.2 to 0.5. A relatively high porosity value was chosen to clearly illustrate the model drumlins in cross section (i.e. the initial porosity controls the relief of the bedforms relative to the sediment-layer thickness). The value $n = 3$ is a representative value for granular media. The value $\zeta = 1$ is constrained by the assumption of equal bulk and shear viscosities. Small (5%) random variations in porosity were superimposed on the uniform initial porosity to represent small-scale heterogeneity of the sediment matrix. These random variations play an important role in the model as "seeds" for the instability.

Figure 7.14 illustrates the evolution of the model using grayscale images of the compaction rate, C. Areas of white and yellow in these images are undergoing rapid expansion while red and black areas are undergoing contraction. Porosity and compaction rate are closely correlated in the model, so white and yellow regions also indicate high porosity. The 2D model evolution can be divided into three basic stages. In the early stage (Figure 7.14a), regions of slightly higher porosity (and hence permeability) near the top of the

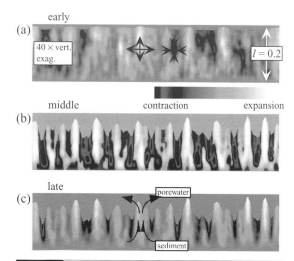

Fig 7.14 2D model evolution, with $40\times$ vertical exaggeration. Color plots of compaction rate C shown at (a) early, (b) middle, and (c) late times in the model. In the early phase, alternating zones of matrix expansion and contraction develop near the upper boundary as zones of initially higher porosity expand, capture upwardly migrating porewater, and further expand in a positive feedback. In the second stage, zones of expansion ascend and drive converging flow of the matrix into the high-porosity drumlin core. In the third stage, porewater is squeezed from the matrix until the sediment is fully compacted. For color version, see plate section.

layer expand as upwardly migrating porewater is focused into those areas. Viscous stresses suppress expansion and contraction near the rigid lower boundary of the model. The spatial distribution of compaction is largely random in this early stage, reflecting random variations in initial porosity. As the instability develops, however, regions of matrix expansion and contraction grow to a maximum wavelength equal to about ten times the layer thickness. As expanding regions increase in porosity, their buoyancy increases with respect to the surrounding sediment. This buoyancy imparts an upward velocity to both the liquid and matrix in these regions, deforming the ice–sediment interface above them. Conversely, compacting regions become heavier than surrounding regions and sink. This response can be thought of in terms of an isostatic balance. Mid-ocean ridges sit higher, topographically, than the surrounding ocean floor because of the lower density associated with focused melt beneath spreading centers. Subglacial bedforms

sit higher for the same reason. In the final stages of the model, high-porosity regions conduct the remaining porewater out of the upper boundary (Figure 7.14c). Deformation stops when all of the liquid has been squeezed from the matrix.

The time required for this instability to occur in natural drumlins can be estimated as $T \approx l/K$, where K is the sediment hydraulic conductivity and l is the layer thickness. Using a conductivity of $K = 10^{-9}\,\mathrm{m/s}$ (in the middle range of glacial till values, ranging from 10^{-12} to $10^{-6}\,\mathrm{m/s}$; Freeze and Cherry, 1979) and $l = 50\,\mathrm{m}$, $T \approx 1000\,\mathrm{yr}$. This estimate depends sensitively on the assumed value for conductivity; drumlins in coarse-grained sediments could develop in less than a year, while drumlins in fine-grained sediments could require tens of thousands of years to form. The overlying ice may also play a role in controlling the time scale for drumlinization. The Rayleigh–Taylor instability model of salt domes, considered in Section 7.2, provides a useful analogy here. It is clear that halite is less viscous and has a lower yield stress than typical clastic sedimentary rocks. Nevertheless, this does not prevent the salt from intruding into the overlying rock, nor does it prevent the instability wavelength from being controlled by the thickness of the salt. The thickness and viscosity of the salt control the wavelength because it is that layer that undergoes boundary-layer flow along the rigid lower boundary of the model. The overlying rock, in contrast, plays only a passive role, just as the ice plays a passive role in drumlin formation. Nevertheless, the rock must still flow in order for the instability to occur. Therefore, the viscosity of the rock (or ice, in the case of drumlins) can control the time for the instability to occur.

In the second model experiment, a 3D geometry was considered under uniform pressure (Figures 7.15a and 7.15b, final model topography shown in color shaded relief). This experiment results in "egg-carton" topography without a preferred orientation. This topography is not similar to drumlins, but it does closely resemble the hummocky moraine found in large parts of the Canadian prairie (Boone and Eyles, 2001). The formation of hummocky-moraine topography is not well understood, but this experiment suggests

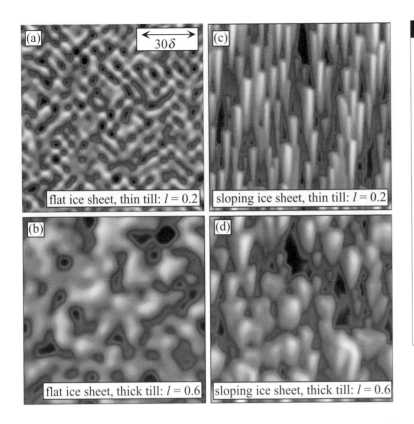

(a) flat ice sheet, thin till: $l = 0.2$

30δ

(b) flat ice sheet, thick till: $l = 0.6$

(c) sloping ice sheet, thin till: $l = 0.2$

(d) sloping ice sheet, thick till: $l = 0.6$

Fig 7.15 3D model evolution, map view. Shaded-relief and color maps of the model topography following compaction. (a) Flat ice sheet, thin sediment; (b) flat ice sheet, thick sediment; (c) sloping ice sheet, thin sediment; (d) sloping ice sheet, thick sediment. For a flat ice sheet, hummocky moraine is formed, with the size of the hummocks controlled by the sediment thickness. For sloping ice sheets, drumlins are formed, with the width of drumlins controlled by the initial sediment-layer thickness. A sloping ice sheet with thin sediment reproduces bedforms most similar to drumlins. With thicker sediment, wide bedforms are created, some of which join laterally to form barchanoid drumlins and Rogen moraine.

that it may be formed by porewater expulsion under the nearly uniform pressure conditions found beneath the interiors of great ice sheets.

The 3D model runs were varied by changing the initial till thickness to determine the effect on bedform morphology (Figures 7.15a and 7.15b). Varying the thickness results in a proportionate change in the width of hummocky moraine. In the third model experiment (Figures 7.15c and 7.15d), an ice sheet with a uniform slope of 0.01 was assumed. To model a sloping ice sheet, the vector k was replaced with $(k + Sj)$ in Eqs. (7.42) and (7.43), where S is the ice-surface slope and j is the unit vector in the down-ice direction. Sloping ice results in elongated, tapered bedforms very similar to drumlins (Figure 7.15b). In this model, the blunt up-ice end of the drumlin is associated with the early phase of sediment upwelling. As the upwelling plume migrates down ice and porewater is expelled from the matrix, the plume loses buoyancy and narrows. The streamlined drumlin "tail" represents the last vestige of porewater expulsion. The model result obtained with thick till (Figure 7.15d) also suggests that

drumlins can become wide enough to join laterally to form barchanoid drumlins and Rogen moraine (bedforms oriented perpendicular to the ice-flow direction). This suggests that Rogen moraine can be formed, in part, by the same instability mechanism as drumlins and hummocky moraine.

The instability of this model occurs when the effective pressure exceeds the yield stress, initiating viscous flow of the matrix. The effective pressure near the base of the sediment layer is given by

$$p_e = (1 - \phi)\Delta\rho gl \qquad (7.48)$$

or $\approx 10^5$ Pa for a till layer with a porosity between 0.2 and 0.5 that is at least 20 m thick. Typical yield stress values for subglacial sediments are on the order of 10^4–10^5 Pa, with higher values for well-drained sediments. These values indicate that saturated tills should commonly exceed the threshold stress required for viscous flow, as assumed in the model.

The fundamental model prediction is a linear relationship between drumlin width and

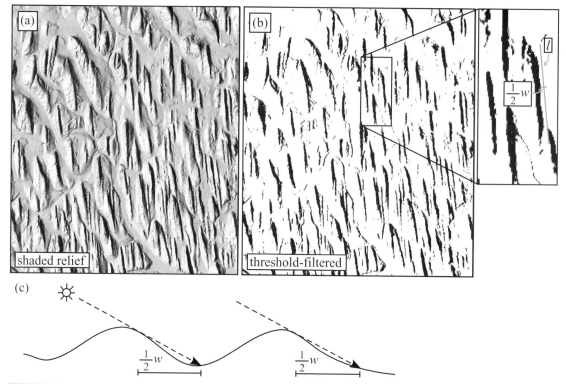

Fig 7.16 Drumlin mapping algorithm. (a) Step 1: Construct a grayscale shaded-relief image from a DEM with the illumination direction chosen to be perpendicular to the predominant drumlin axis. (b) Step 2: Apply a threshold filter to the shaded-relief image to make all shadows black and all other areas white. Step 3: Analyze the binary image for all connected clusters of black pixels.

initial sediment-layer thickness. This prediction is testable using morphological analyses of drumlins and groundwater well data that include depths to bedrock. To do this, drumlin widths were first mapped using a semi-automated algorithm (Figure 7.16). This algorithm uses the perpendicular shadow cast by each drumlin in a shaded-relief image as a proxy for its shape. First, a shaded-relief image was constructed using a high-resolution Digital Elevation Model (DEM). For the north-central New York area, 10-m resolution US Geological Survey DEMs were used. In Wisconsin, 30-m resolution DEMs were the best available. The illumination direction was chosen to be perpendicular to the predominant drumlin axis, and the azimuth was chosen to cast a

shadow that just barely covered the shaded side of drumlins completely. Several illumination directions were used in different portions of each drumlin field to properly illuminate each drumlin from a perpendicular direction. Second, the shaded-relief image was converted to a black and white image using a threshold filter. In this step, shadows remain black while shades of gray are converted to white. Third, the black-and-white image was input into a custom software program that identified all connected clusters of black pixels and computed their centroid positions, areas, lengths, and widths using standard definitions (Chorley, 1959; Smalley and Unwin, 1968). Drumlin widths were taken to be twice the cluster width because each cluster represents only one side of the drumlin in shadow. This algorithm leaves many small clusters remaining (e.g. Figure 7.16b), so the dataset was filtered to remove clusters that are too small (i.e. less than $0.01\,\text{km}^2$) to be drumlins. This procedure yielded a dataset of approximately 5800 drumlins in north-central New York, and 2900 drumlins in the Wisconsin study area.

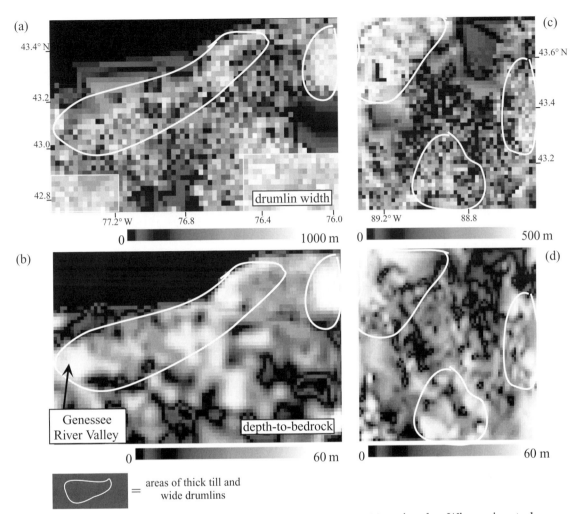

= areas of thick till and wide drumlins

Fig 7.17 Grayscale maps for average drumlin width and bedrock depth in (a) and (b) north-central New York and (c) and (d) Wisconsin, east of Madison. (a) and (c) Maps of drumlin width constructed by averaging widths within a 2-km square moving window. (b) and (d) Maps of bedrock depth also constructed with a 2-km square moving window and using USGS groundwater well data. Curves are drawn to highlight areas in which thick sediment and wide drumlins (including Rogen moraine) coincide. For color version, see plate section.

Finally, drumlin widths were spatially-averaged using a 2 km × 2 km moving window (Figures 7.17a and 7.17c). The resulting values for average drumlin width (including both drumlins and Rogen moraine) vary from 100 m to 1000 m in the New York study area (Figure 7.17a) and from

100 m to 500 m in the Wisconsin study area (Figure 7.17c).

The shadow-mapping algorithm does not produce an ideal representation of the drumlin form, but it is a robust method. The principal limitation of the shadow-based algorithm is that shadows created by other landforms may be inadvertently included in the analysis. This was not a significant problem for most of the study area, however. Two problematic areas were found in the southwest and southeast corners of the New York study area. In these areas near the Finger Lakes, bedrock slopes are comparable to drumlin slopes, resulting in some bedrock ridges and valleys being included as drumlins. These areas (shown as shaded in Figure 7.17a) were excluded from the analysis.

Till thicknesses were mapped using depth-to-bedrock (DTB) data from all publicly available groundwater well records. Data were obtained from the USGS New York office and the Wisconsin State Geological and Natural History Survey. In the New York study area, 2786 wells were available for six counties. In Wisconsin, 1349 wells were available for six counties. DTB data were averaged using a 2 km × 2 km moving window (Figures 7.17b and 7.17d). This averaging was done to minimize the small-scale spatial variability associated with well placement and to make the DTB maps directly analogous to the maps of drumlin width. DTB data in both study areas vary from 0 to 60 m. DTB data are a good proxy for till thickness except in areas with significant fluvial deposition. In the Genesee River Valley, for example, fluvial sediment thicknesses are comparable to or greater than glacial sediment thicknesses in the area, making any correlation with drumlin widths unreliable.

Maps of average drumlin width and sediment thickness show a strong correlation (except the Genesee River Valley where fluvial sediments are present). White curves in Figure 7.17 are used to highlight several regions in each study area where wide drumlins and thick till coincide. In the New York study area (Figures 7.17a and 7.17b) wide drumlins (and, in places, Rogen moraine) occur in the northeast corner of the study area and along a swath trending southwest-to-northeast near the Lake Ontario shoreline. In the Wisconsin study area (Figures 7.17c and 7.17d), the widest drumlins and thickest till occur in the northwest, south, and east portions of the study area. These results prove that the geometry of the till layer plays a dominant role in controlling the drumlin geometry and they lend string support to the proposed model.

Sedimentological and stratigraphic observations provide additional constraints on drumlin-forming mechanisms. Analyses of till fabric (the predominant orientation of elongated pebbles in the till) provide evidence for convergent flow of sediment upward and inward toward the drumlin axis (Evensen, 1971; Stanford and Mickelson, 1985). Based on observations in the Wisconsin study area of Figure 7.12b, for example, Evensen

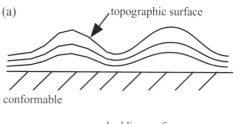

(a) conformable

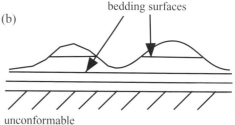

(b) unconformable

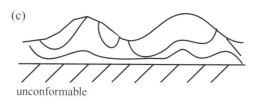

(c) unconformable

Fig 7.18 Drumlin stratigraphic end members (shown in cross section, perpendicular to the drumlin axis). Three types are possible: (a) subsurface bedding broadly parallels the topography (i.e. "concentric" drumlins), (b) subsurface bedding is undeformed, and (c) subsurface is deformed, but the deformation is poorly correlated with the topography. Only (a) is consistent with the proposed model.

(1971) proposed that till-fabric data are consistent with a low-pressure zone along the drumlin axis. Without a bedrock core to initiate the low-pressure zone, however, it is unclear how low pressure develops in Evensen's model. The buoyancy generated by concentrating porewater, however, provides a natural explanation for the alternating pressure zones and converging matrix flow inferred by Evensen and others using till-fabric analyses.

Figure 7.18 summarizes three possible end members for drumlin stratigraphy (shown in cross section, perpendicular to the drumlin axis). Figure 7.18a illustrates the case in which the subsurface bedding broadly parallels the topography. Drumlins with this stratigraphy are referred to as "concentric" (Hart, 1997). In Figure 7.18b, the subsurface bedding bears no indication of deformation. In Figure 7.18c, sediment deformation is pervasive, but the deformation is uncorrelated with topography. Of these three end members, only Figure 7.18a is consistent with the proposed model, although reworking of drumlins by subsequent ice-sheet advances often leads to erosional truncations and a more complex stratigraphy than the simplified case shown in Figure 7.18a. Nevertheless, concentric drumlins are commonly observed in many drumlin fields (e.g. Newman and Mickelson, 1994; Zelcs and Driemanis, 1997). The New York State drumlin field, in particular, is the classic location for concentric drumlins (Hart, 1997).

As a composite of porewater embedded within a matrix of sediment, subglacial sediments behave as deformable porous media. Porosity and buoyancy play a significant role in the evolution of these systems. First, spatial variations in porosity variations can be enhanced during compaction by a feedback between porosity, permeability, and matrix expansion. Bouyancy associated with the resulting regions of high porosity provide the driving force for converging flow of the sediment into regularly spaced bedforms. Near the margins of former ice sheets, where the ice surface is sloping, sediment and porewater are advected down the ice-surface-slope direction, creating bedforms very similar to the classic drumlins of New York and Wisconsin.

7.7 | Spiral troughs on the Martian polar ice caps

The spiral troughs on the Martian polar ice caps are one of the most fascinating landforms in the solar system. Spiral troughs form with an instability in which the ice-surface temperatures of steep, equator-facing slopes exceed $0\,^\circ$C during the summer, sublimating the ice locally to form steeper, lower-albedo scarps (through exposure of subsurface dust-rich layers) in a self-enhancing feedback (Howard, 1978). In this model, some of the water vapor released from the equator-facing scarp may accumulate on the pole-facing slope

to form a self-sustaining poleward-migrating topographic wave. The relationship between this model and the spiral morphology of troughs has not been fully established, but Fisher (1993) introduced an asymmetric ice-velocity field into Howard's model and obtained spiral forms. Recent observations from Mars Global Surveyor, however, suggest that ice flow near the troughs is not significant (Howard, 2000; Kolb and Tanaka, 2001) and it remains unclear precisely how the spirals form and what controls their spacing, orientation, and curvature.

In this section we explore a simple mathematical model designed to capture the essential feedback processes that couple the topography, the distribution of solar heating on the surface, and the resulting accumulation and ablation of ice. The model is based on the processes in Howard's migrating scarp model and includes the simplest mathematical descriptions of these processes in order to determine the necessary conditions for realistic spiral troughs. Lateral heat conduction is also included in the model, and this element is crucial for obtaining realistic troughs. The model does not include wind erosion or ice flow. The equations are similar to those from the field of excitable media, where solitary and spiral waves resulting from the dynamics of two interacting variables have been studied in detail.

The model equations consist of two coupled, scaled equations for the deviation of local ice-surface temperature, T, from its equilibrium value (the temperature at which no accumulation or ablation takes place) and the deviation of local ice-surface topography, h, below its equilibrium value (i.e. troughs have positive h):

$$\frac{\partial T}{\partial t} = \kappa \nabla^2 T + f(T, h) \tag{7.49}$$

$$\frac{\partial h}{\partial t} = \frac{1}{\tau_f} T \tag{7.50}$$

$$f(T, h) = \frac{1}{\tau_i}(T(T - T_0)(1 - T) - h) \text{ if } \nabla h \cdot \vec{r} > 0 \tag{7.51}$$

$$= \frac{1}{\tau_i} h \text{ if } \nabla h \cdot \hat{r} < 0 \tag{7.52}$$

The deviations from equilibrium temperature do not include diurnal or annual cycles; these

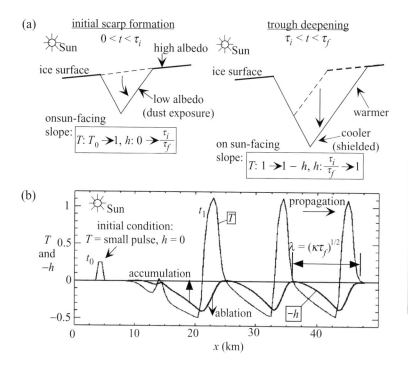

Fig 7.19 Diagram of trough evolution and solution to Eq. (7.52) in 2D. (a) Trough initiation involves a rapid increase in ice-surface temperature to its maximum value as the scarp is steepened and dust layers are exposed. During trough deepening, the value of h steadily increases to its maximum value over a time scale $\tau_f = 5$ Myr, while the temperature decreases in the deeper portions of the trough due to shielding. (b) Solution to (1) with Runga–Kutta integration and $\Delta x = 0.2$ km, $\kappa = 175$ km^2, $\tau_i = 0.05$, $\tau_f = 1$, and $T_0 = 0.2$ at $t_1 = 8$. The initial condition is a small pulse with $T > T_0$ at the left side of the domain. Modified from Pelletier (2004b).

are assumed to be included in the equilibrium temperature. Deviations from the equilibrium temperature initiate ablation and accumulation at any time, t, in the model, although accumulation and ablation of water ice is understood to be restricted to the summer months on Mars.

Equation (7.52) includes three basic processes: lateral heat conduction (the diffusion equation), accumulation and ablation ($\partial h / \partial t = T / \tau_f$), and enhanced heat absorption on equator-facing scarps ($\partial T / \partial t = f(T, h)$). The initiation and deepening of troughs is assumed to take place in two steps illustrated in Figure 7.19a. In the first step, slope steepening and dust exposure are initiated from an undissected ice surface through a positive feedback between ice-surface temperature, slope gradient, and albedo. In this step, the ice-surface temperature is assumed to increase from its threshold value for sublimation, T_0, to a maximum value of 1 over a time scale τ_i. This instability is represented mathematically in $f(T, h)$ by a cubic polynomial that generates a negative feedback if T is below the sublimation temperature, T_0, and a positive feedback for larger

temperatures, bounded by a maximum value of 1. In the second step, the temperature remains near its maximum value of 1 as the trough deepens over a longer time scale, τ_f. Trough deepening does not continue indefinitely, however, because deep portions of equator-facing scarps cool due to partial shielding from the pole-facing slope. The heat-absorption function $f(T, h)$ includes an h term to represent this negative feedback. As h increases in value, the temperature eventually falls below the sublimation temperature and accumulation begins. The heat-absorption term only applies to equator-facing slopes (i.e. $\nabla h \cdot \hat{r}$, where $\hat{r}$ is the radial unit vector from the pole).

Equation (7.52) includes four parameters: κ, τ_i, τ_f, and T_0. The thermal diffusivity of ice is $\kappa = 35$ m^2/yr. To estimate the vertical ablation rate on sun-facing slopes, Howard (1978) used Viking observations of summertime atmospheric water-vapor concentration to obtain 10^{-4} m/yr, implying $\tau_f = 5$ Myr for an average trough of 500 m depth. The time of scarp initiation is uncertain, but may be small compared with trough deepening. Here we assume $\tau_i = 0.05$, $\tau_f = 0.25$ Myr, which implies that ≈ 25 m of ablation must occur before an incipient scarp is sufficiently steepened for subsurface dust to be exposed and the maximum ice-surface temperature to be reached. The model behavior is not sensitive to the precise value of τ_i as long as it is small compared with τ_f. The threshold temperature for sublimation, relative to the equilibrium and maximum temperatures, is between 0 and 1 and varies as a function of latitude. At 75° latitude, the maximum summer temperature of a nearly flat, high-albedo slope is about $-10\,°C$ (Howard, 1978). The temperature difference between equator-facing troughs and nearby flats is about $40\,°C$ (Howard, 1978). Using these values, T_0, or $0\,°C$, is equal to 0.2. We assume this value applies to the entire ice cap for simplicity. Equation (7.52) can be further simplified by scaling time to the value of τ_f. With this rescaling, the model is reduced to three independent parameters: $\kappa = 175\,km^2$, $\tau_i = 0.05$, and $T_0 = 0.2$.

Equation (7.52), without the directional dependence in $f(T, h)$, is one example of the FitzHugh–Nagumo (FHN) equations, a well-studied class of equations that describe excitable media. The FHN equations have been applied to predator-prey systems (Murray, 1990), electrical conduction in nerve fibers (Winfree, 1987), chemical oscillators (Fife, 1976), and electrical transmission lines (Nagumo et al., 1965). The FHN equations come in several forms that all reproduce solitary waves in 2D (i.e. $h(x)$) and spiral waves in 3D ($h(x, y)$) for a range of parameter values. Excitable media are composed of two variables: an "activator" (T in Eq. (7.52)) and an "inhibitor" (h in Eq. (7.52)). Triggering or excitation of the medium occurs when a positive feedback is initiated above a threshold value of the activator. Local triggering may initiate the triggering of adjacent zones through diffusive spreading of the activator. The value of the inhibitor increases during excitation and eventually returns the system to the unexcited state through the negative coupling between h and T in Eq. (7.52). The FHN equations exhibit both excitable behavior, characterized by wide swings in h and T when $\tau_i \gg \tau_f$, and oscillatory behavior, characterized by smaller fluctuations when $\tau_i \approx \tau f$ (Gong and Christini, 2003). Although both regimes may have appli-

cation to Mars, here we focus on the excitable regime.

The solution of Eq. (7.52) in 1D starting with a narrow pulse of localized sublimation ($T = 0.25$ in the pulse, $T = 0$ elsewhere) is given in Figure 7.19b. The pulse initiates a self-sustaining train of waves that propagate with the trough trailing the temperature wave. The width of the troughs is given by $\lambda \approx \sqrt{\kappa \tau_f}$, or 13 km for the parameters characterizing the Martian polar ice caps. This value is in good agreement with observed trough widths on Mars. This behavior contrasts with that of previous models, in which troughs widened indefinitely without reaching a steady state. Another similarity with the observed topography of spiral troughs is trough asymmetry: both in the model and on Mars, equator-facing scarps are about twice as steep as pole-facing scarps. Code for implementing the 1D spiral trough model is given in Appendix 6.

Spiral formation in Eq. (7.52) is illustrated in Figures 7.20b–7.20e for a circular region with a permanently-frozen region near the pole and starting from random initial conditions. A shaded-relief image of a polar-stereographic DEM of MOLA topography is given in Figure 7.20a for comparison. In these calculations we have used the Barkley (1991) approximations to Eq. (7.52), which use a semi-implicit algorithm for the cubic function in $f(T, h)$ in order to gain computational efficiency and investigate spiral evolution over long time scales. The early-time evolution of the model is characterized by the initiation, growth and merging of spirals. It should be noted that heat conduction in the neighborhood of an equator-facing scarp may facilitate the sublimation of nearby regions, even if those regions face toward the pole. The late-time evolution of the model is characterized by continued spiral merging, alignment of troughs to face the equator, and a gradual increase in trough spacing as "defects" are removed from the system. Mature spirals face the equator at low latitudes but curve towards the pole at higher latitudes, just as those on Mars do. In the model, this poleward orientation is required for the maintenance of steady-state, rigidly rotating spiral forms. By developing poleward orientations, high-latitude scarps face

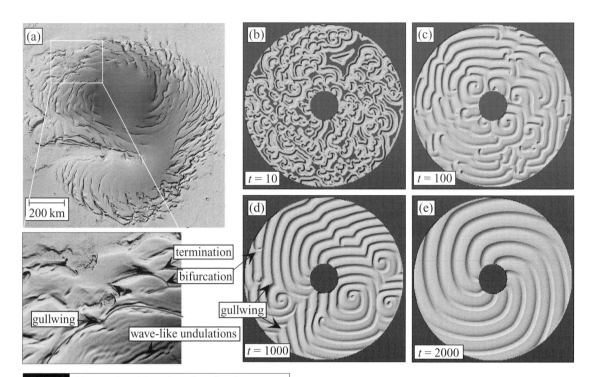

Fig 7.20 Numerical model results and north-polar topography. (a) Shaded-relief image of Martian north-polar ice cap DEM constructed using MOLA topography. The large-scale closeup indicates examples of gullwing-shaped troughs, bifurcations, and terminations. Highest elevations are red and lowest elevations are green. (b)–(e) Shaded-relief images of the model topography, $-h$, for (b) $t = 10$, (c) $t = 100$, (d) $t = 1000$, and (e) $t = 2000$ starting from random initial conditions. The model parameters are $L = 250$ (number of pixels in each direction), $\Delta x = 0.4$, $T_0 = 0.3$, $\tau_i = 0.05$, $\tau_f = 1$. In the Barkley approximations to Eq. (7.52), T_0 is combined into two parameters, $a = 0.75$ and $b = 0.01$. The model evolution is characterized by spiral merging and alignment in the equator-facing direction. A steady state is eventually reached with uniformly rotating spirals oriented clockwise or counter-clockwise depending on the initial conditions. For color version, see plate section. Modified from Pelletier (2004b).

troughs, bifurcations, and terminations (Figures 7.20a, 7.20c, and 7.20d). The steady state of the model after $t = 2000$ is a set of seven rigidly-rotating spirals. The sense of spiral orientation has an equal probability of being clockwise or counter-clockwise in the model depending on the random number seed used to generate the initial conditions. The spiral orientation of Figure 7.20e is opposite to those on the north pole of Mars, but 50% of random initial conditions

perpendicular to the equator, thereby reducing their ablation rates and decreasing their migration speeds. The decreased migration speed offsets the smaller distance required to complete a revolution at high latitudes, enabling spirals to rotate at a uniform angular velocity and preserve their steady-state form. The model also reproduces a number of secondary features of polar troughs on Mars, including gullwing-shaped

Fig 7.21 Vegetation bands of Ralston Playa, Nevada. Dirt road provides scale. Individual bands are 1–3 m wide.

lead to spirals that rotate in the same sense as those on Mars (i.e. clockwise away from the pole). This model behavior suggests that a uniform sense of spiraling need not represent an asymmetrical process of trough evolution. Instead, it could be the result of a minor asymmetry in the initial pattern of sublimation, as in the model.

Exercise

7.1 Figure 7.21 illustrates the curious phenomenon of vegetation bands. These bands form on some playas of the western US and are comprised of brushy, salt-tolerant vegetation. These vegetation bands most likely form by a positive feedback in which vegetation growth promotes mounding of the surface beneath the vegetation. Mounds trap occasional runoff into the playa, which promotes more vegetation growth. Bands evolve because a line of shrubs oriented along a contour pond the most water for the least biomass. Develop a model for the formation of these bands based on a simplified model for the positive feedback between vegetation, mounding, and ponding. Actual vegetation bands follow contours closely and are more closely spaced on more-steeply dipping portions of the playa. Is your model consistent with these observations?

Chapter 8

Stochastic processes

8.1 | Introduction

Thus far we have considered models in which the future behavior of the system can be determined using equations and boundary conditions known at some initial time. In many Earth surface systems, however, the future behavior of the system cannot be predicted with certainty, either because the system behavior is sensitive to small-scale processes that cannot be fully resolved and/or because there is significant uncertainty in model input parameters. The climate system is a good example of a system that depends on small-scale processes (i.e. turbulence) that cannot be fully resolved in any numerical model. As a result, climate and weather models are inherently limited in their ability to predict the details of future climate and weather patterns. Soil permeability is a good example of a model input parameter with significant uncertainty. Soil permeability depends on the detailed structure and composition of the soil at a range of spatial scales, making an exact determination of permeability very difficult over large areas. As such, numerical models that require soil permeability as an input (e.g. models for runoff, infiltration, aquifer recharge, etc.) are limited in their precision no matter how finely they resolve the underlying physics of the problem.

In some applications where deterministic models cannot predict the future behavior of a system precisely, stochastic models can be useful for understanding the range of possible system behaviors. Stochastic models based on observed statistical distributions of rainfall events, for example, can be used to generate synthetic storms that match the statistical behavior of actual storms. In cases where there is significant uncertainty in model input parameters, Monte Carlo methods can be used to determine a range of model outputs corresponding to a range of input parameters. Monte Carlo methods are based on deterministic models; they work by running many successive examples of a deterministic model using different input parameters sampled from prescribed probability distributions.

8.2 | Time series analysis and fractional Gaussian noises

Climatological and hydrological time series provide many good examples of phenomena that require a stochastic modeling approach. Time series data for rainfall and discharge at a point in space are the result of complex processes, including convective instabilities in the atmosphere. Although some elements of the climate system can be quantified deterministically, convective instabilities are very difficult to model more than a few days into the future. The occurrence of convective instabilities is one of the fundamental limitations on accuracy of weather forecasts.

Figure 8.1 plots the daily mean discharge in the Gila River near the town of Winkelman, Arizona from 1942 to 1980. Discharge in this case is plotted on a logarithmic scale because of its highly skewed distribution, with peak

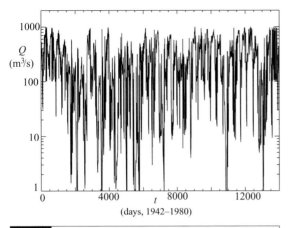

Fig 8.1 Streamflow of the Gila River near Winkelman, Arizona, from 1942–1980.

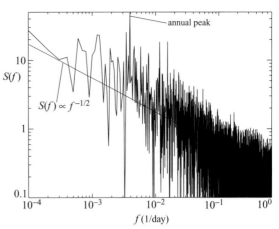

Fig 8.2 Power spectrum of the Gila River discharge plotted in Figure 8.1.

values many times larger than average values. Many kinds of datasets that arise in climatological and Earth surface studies involve spatial or temporal series data. Figure 8.1 is a time series because it is a quantity that varies as a function of time. Spatial series data, such as topographic profiles, can be analyzed using many of the same tools used to analyze time series data. Time series data can be characterized by their probability distributions and autocorrelation functions. The probability distribution of a dataset quantifies its mean value and its variability from the mean. The Gila River near Winkelman is characterized by a mean streamflow value of approximately $100 \, \text{m}^3/\text{s}$. The mean value is not very representative of the typical streamflow in this case, however, because the river also has extended dry periods with discharge values less than $10 \, \text{m}^3/\text{s}$ and occasional large storm events that trigger discharges greater than $1000 \, \text{m}^3/\text{s}$. This river, like most in the southwestern US, has a large variance in streamflow values due to the episodic nature of precipitation in semi-arid environments. The autocorrelation function of a time series quantifies how strongly points in the series are correlated over different time scales. In the case of river discharge, significant correlations can be expected over many time scales due to the nature of rainfall and hydrologic response in a large river basin. During an individual flood event that lasts several days or more streamflow values can be expected to have significant correlation

because they are controlled by a single meteorological event. At longer time scales, a seasonal cycle can be expected and is apparent in Figure 8.1. At still longer time scales, interannual and decadal variations in ocean temperatures (resulting from known periodic processes such as ENSO and also aperiodic variations) lead to clusters of wet years and clusters of dry years. Hurst *et al.* (1965) quantified this multi-scale autocorrelation pattern in hydrological data sets, now called the Hurst phenomenon. In this section we explore a particular type of stochastic model, i.e. the fractional Gaussian noise, that reproduces the autocorrelation structure common to many hydrologic and climatic time series.

The autocorrelation structure of a time series can be quantified in a number of ways, but the power spectrum is one of the simplest. The power spectrum is defined as the square of the coefficients in a Fourier series representation of the time series. The power spectrum quantifies the variance of a time series at different frequencies. Figure 8.2 plots the power spectrum of the Gila River streamflow dataset on logarithmic scales. This spectrum shows a prominent seasonal cycle (i.e. a large peak at a period of 1 year) superimposed on a broad spectrum that decreases with larger frequencies. Aside from the seasonal cycle, the background spectrum follows a power-law relationship with frequency given by $S(f) \propto f^{-1/2}$. This relationship turns out to be nearly universal for streamflow datasets worldwide. In a

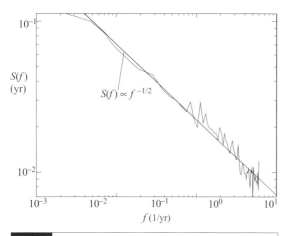

Fig 8.3 Average of power spectra of monthly average discharge data from 636 gage stations with the annual periodicity first removed.

study of 636 gage stations, Pelletier and Turcotte (1997) found that, when the annual peak was removed, the average power spectrum followed a nearly straight line on log–log scales with a slope of −0.50 (Figure 8.3). The data for this analysis were chosen from all complete records with durations greater than or equal to 512 months, including 636 records in the analysis. Pelletier and Turcotte (1997) took advantage of the large number of available stations to investigate the possible regional variability of the power spectra. All of the regions exhibited the same power-law power-spectral form with an average exponent of −0.52 and a standard deviation of 0.03, indicating little variation. Proxy data for hydrologic datasets (e.g. tree ring data in semi-arid environments) also exhibit that same spectrum out to time scales greater than 1000 years. Pelletier and Turcotte (1997) also showed similar power-spectral behavior in precipitation time series at time scales greater than 10 years, indicating that the correlation structure in streamflow data are primarily a result of correlations in the climate system on decadal time scales and greater, while shorter-term correlations are at least partially the result of surface and subsurface flow pathways.

Thus far we have shown that hydrological data sets have a common underlying autocorrelation structure. It is useful for many applications to generate synthetic data sets that match

this observed underlying structure. Figure 8.4 plots examples of three synthetic time series with power spectra given by $S(f) \propto f^{-\beta}$, with $\beta = 0$, $1/2$, and 1. Fractional Gaussian noises are simply time series that have Gaussian probability distributions and power spectra have a power-law dependence on frequency f. As the magnitude of the power-spectral exponent increases, the correlations between adjacent points in the series become stronger. Fractional Gaussian noises can be generated with the Fourier-filtering technique. This technique proceeds in several steps. First, a Gaussian white noise sequence is generated. Second, a discrete Fourier transform of the Gaussian white noise sequence is computed. The resulting Fourier spectrum will be flat, i.e. $\beta = 0$ in the expression $S(f) \propto f^{-\beta}$. Except for the statistical scatter, the amplitudes of the Fourier coefficients, $|Y_m|$ will be equal. Third, the Fourier coefficients are multiplied by a power-law function of frequency:

$$Y'_m = \left(\frac{m}{N}\right)^{-\frac{\beta}{2}} Y_m \qquad (8.1)$$

The exponent $\beta/2$ is used because the power spectrum is proportional to the square of the Fourier coefficients. Fourth, an inverse discrete Fourier transform is computed using the new coefficients Y'_m. This technique is called Fourier filtering because Eq. (8.1) has the effect of removing, or filtering out, some of the variability at high frequencies (if $\beta > 0$) or low frequencies (if $\beta < 0$). Two-dimensional fractional noises can also be constructed (Figure 8.5). These noises have applications to stochastic models of topography, soil moisture, and the porosity structure of sedimentary basins. The details of this technique, together with codes for its implementation, are given in Appendix 7.

Fractional noises have many applications in geosciences (Pelletier and Turcote, 1999). For example, generating synthetic fractional noises can help to assess the likelihood of future hydrological events that depend on the cumulative discharge over a series of years. A drought of a given duration, for example, is the result of several consecutive years of below-average flow. The correlation structure of hydrological time series plays a very significant role in the frequency of drought

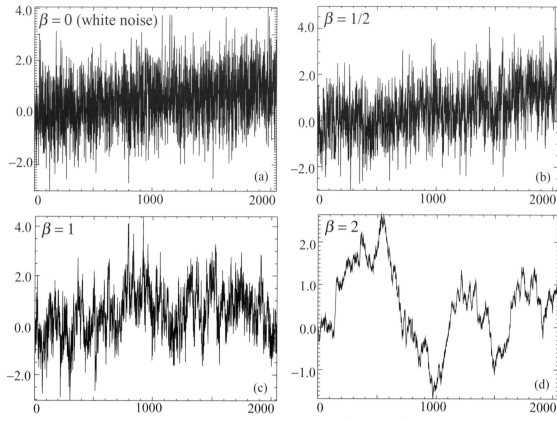

Fig 8.4 Examples of fractional Gaussian noises with $\beta = 0$, 1/2, 1, and 2.

occurrence. If hydrological time series were completely uncorrelated, then the probability of 10 years of below-average streamflow would be $1/2^{10}$ or 1/1024. However, the fact that hydrologic data sets have positive correlation over a range of time scales makes the probability of droughts, especially long droughts, far more frequent (Pelletier and Turcotte, 1997).

8.3 | Langevin equations

If a broad range of climatological and hydrological time series data have power spectra of the form $S(f) \propto f^{-1/2}$ from time scales of decades to centuries, we must look for some general underlying climatological mechanism. In this section we present one possible model of fluctuations in local temperature and water vapor based on a stochastic diffusion model of convective transport. Fluctuations in precipitation, river discharge, and tree ring growth are related in complex but direct ways to variations in temperature and/or humidity.

The simplest approach to the turbulent transport of heat and water vapor vertically in the atmosphere is to assume that the flux of heat or water vapor concentration is proportional to its gradient. The mixing of heat energy and humidity will then be governed by the diffusion equation. Experimental and theoretical observations support the hypothesis that diffusive transfer accurately models vertical transport in the atmosphere. Kida (1983) has traced the dispersion of air parcels in a hemispheric GCM and found the dispersion of those air parcels to obey the diffusion equation in the troposphere. Also, Hofmann and Rosen (1987) argued that turbulent diffusion of aerosols from the El Chichon volcanic eruption was consistent with the exponential decay of aerosol concentrations with time (i.e. turbulent diffusion and gravitational settling combine

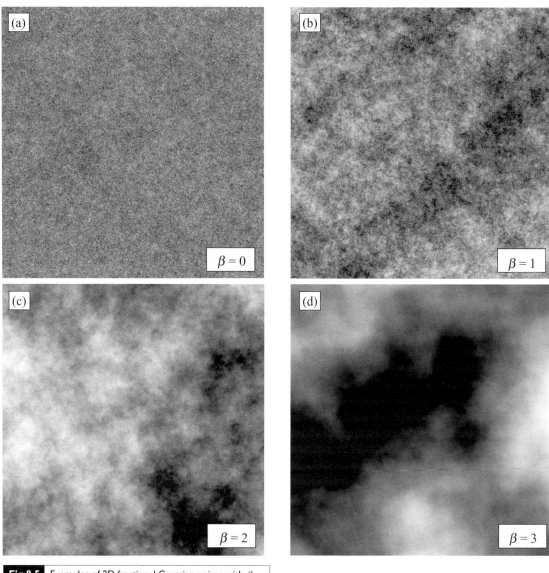

Fig 8.5 Examples of 2D fractional Gaussian noises with $\beta =$ (a) 0, (b) 1, (c) 2, and (d) 3.

to predict exponential concentrations in time). Vertical atmospheric turbulent diffusion is superimposed, as it is in oceanic diffusion, on large-scale Hadley and Walker circulations (Peixoto and Oort, 1992).

To see how time series with power-law power spectra arise from a stochastic diffusion process, we explore the behavior of a discrete, one-dimensional stochastic diffusion process. A discrete version of the diffusion equation for the density of particles, c, on a one-dimensional grid is given by

$$c_i^{n+1} - c_i^n \propto c_{i+1}^n - 2c_i^n + c_{i-1}^n \qquad (8.2)$$

where i is an index for space and n is an index for time. In our model we establish a one-dimensional lattice of 32 sites with periodic boundary conditions at the ends of the lattice. At the beginning of the simulation, we place ten particles on each site of the lattice. At each timestep, a particle is chosen at random and moved to the left with probability 1/2 and to the right if it does not move to the left. In this way, the average rate at which particles leave a grid point is proportional to the number of particles at the grid point. The average rate at which particles enter a grid point is proportional to the number of particles on each side multiplied by one half (since the particles to the left and right of site i move into site i only half of the time). This is a stochastic model satisfying Eq. (8.2). The probabilistic nature of this model causes fluctuations to occur in the local density of particles that do not occur in a deterministic model of diffusion. This simple model produces fluctuations in the number of particles in the central site of the 32-site lattice that have a power-law power spectra with $\beta = 1/2$ (Pelletier and Turcotte, 1997).

A stochastic diffusion process can be studied analytically by adding a noise term to the flux of a deterministic diffusion equation (van Kampen, 1981):

$$\rho c \frac{\partial \Delta T}{\partial t} = -\frac{\partial J}{\partial x} \tag{8.3}$$

where the flux, J, is given by

$$J = -\sigma \frac{\partial \Delta T}{\partial x} + \eta(x, t) \tag{8.4}$$

The variable ΔT represents the fluctuations in temperature from equilibrium and $\eta(x, t)$ is a Gaussian, white noise. Equations (8.3)–(8.4) are one example of a Langevin equation (i.e. a partial differential equation with stochastic forcing).

We will use Eqs. (8.3)–(8.4) to calculate the power spectrum of temperature fluctuations in a layer of width $2L$ exchanging heat with an infinite, one-dimensional, homogeneous space (Voss and Clarke, 1976). The Fourier transform of the heat flux of the stochastic diffusion equation is

$$J(k, \omega) = \frac{i\omega \eta(k, \omega)}{Dk^2 + i\omega} \tag{8.5}$$

The rate of change of heat energy in the layer will be given by the difference in heat flux out of the boundaries, located at $\pm L$: $dE(t)/dt = J(L, t) - J(-L, t)$. The Fourier transform of $E(t)$ is then

$$E(\omega) = \frac{1}{(2\pi)^{\frac{1}{2}}\omega} \int_{-\infty}^{\infty} sin(kL) J(k, \omega) \, dk \tag{8.6}$$

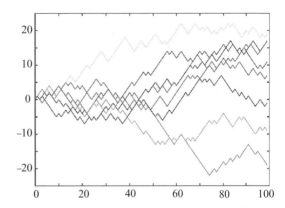

Fig 8.6 Five examples of a simple random walk process. Plotted in Gnuplot by mdoege@compuserve.com. Reproduced from http://en.Wikipedia.org/wiki/Image:Random_Walk_example.png.

The power spectrum of variations in $E(t)$, $S_E(\omega) = <|E(\omega)|^2>$ is

$$S_E(\omega) \propto \int_{-\infty}^{\infty} \frac{sin^2(kL)}{D^2 k^4 + \omega^2} \, dk \propto \omega^{-\frac{1}{2}} \tag{8.7}$$

for low frequencies. Since $\Delta T \propto \Delta E$, $S_{\Delta T}(\omega) \propto \omega^{-1/2}$ also. This is the same result as the discrete model.

8.4 | Random walks

The simplest type of random walk is a stochastic process consisting of a sequence of discrete steps of fixed length. Figure 8.6 illustrates five examples of a random walk which begins at zero and involves unit jumps along the y axis up or down with equal probability for each step along the x axis. Each realization of the random walk is different, but the fact that the distance from the origin tends to increase with each step in all of the examples gives the random walk a certain element of predictability. What is remarkable about random walks and other stochastic processes is that the average behavior of a large collection of random walks is entirely predictable. In a simple random walk, the expected distance from the origin, $|\Delta y|$ is proportional to the square root of the number of steps along the x axis, Δx:

$$< |\Delta y| > \propto \sqrt{\Delta x} \tag{8.8}$$

By *expected*, we mean that this is the average behavior obtained by many realizations of the random walk process.

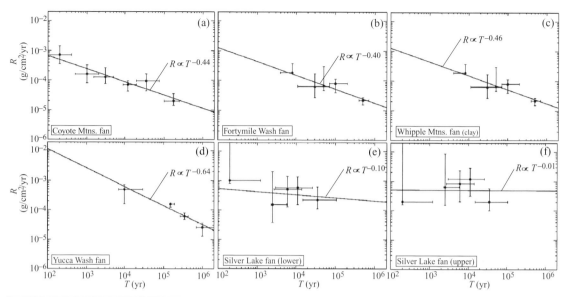

Fig 8.7 Plots of silt accumulation rates versus time interval on alluvial fan terrace surfaces from data in Reheis *et al.* (1995): (a) Fortymile Wash fan, (b) Coyote Mtns. fan, (c) Whipple Mtns. fan (clay fraction shown), (d) Yucca Wash fan, (e) lower fan near Silver Lake Playa, and (f) upper fan near Silver Lake Playa. Best-fit lines are also shown indicating power-law scaling with exponents close to −0.5 for (a)–(d). Trends in (e) and (f) are more consistent with linear accumulation or cyclical-climate models. Modified from Pelletier (2007a).

8.5 | Unsteady erosion and deposition in eolian environments

The accumulation of dust on alluvial fan terraces in desert environments provides a good example of the application of a simple random walk in geomorphology. In this example, the *y* axis in Figure 8.7 represents the accumulation of wind-blown dust (as a thickness or mass per unit area) underneath the desert pavement, and the *x* axis represents time.

Dust accumulation on desert alluvial fan terraces can be affected by changes in either deposition or erosion. Previous studies on dust accumulation of alluvial fans terraces have generally argued that climatically controlled dust deposition plays the predominant role in controlling

dust accumulation on alluvial fan terraces. Low rates of long-term (10^5–10^6 yr) dust accumulation in the southwestern US, for example, have been interpreted as a direct result of low dust deposition rates during the predominantly cool, wet Pleistocene, whereas high rates of shorter-term (10^3–10^4 yr) dust accumulation have been interpreted as an early Holocene pulse of dust deposition related to pluvial-lake dessication (McFadden *et al.*, 1986; Chadwick and Davis, 1990; Reheis *et al.*, 1995). In this section we compare the observed eolian dust accumulation rates at six alluvial fan study sites with the predictions of a climatically controlled deterministic model and a simple random-walk model of eolian dust accumulation. The results suggest that eolian erosion and deposition on alluvial fans is a highly episodic process that closely resembles a random walk.

Reheis *et al.* (1995) synthesized data on eolian dust accumulation from alluvial fan terrace study sites in the southwestern US with optimal age control. These sites are located on gently sloping, planar alluvial fan terraces that have not been subject to flooding since abandonment by fan-head entrenchment. The time since fan-head entrenchment is the surface "age" and it corresponds to the time interval of eolian dust accumulation. Figure 8.7 plots silt accumulation rates versus time interval on logarithmic scales for these locations. Error bars represent time

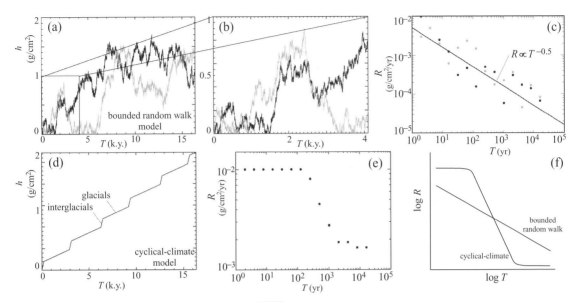

Fig 8.8 (a) Plot of dust accumulation versus time for two runs of the bounded random walk model. (b) Inset of (a) illustrating the scale-invariance of the bounded random walk model by rescaling time by a factor of 4 and accumulation by a factor of 2. (c) Plot of accumulation rate versus time interval illustrating power-law scaling behavior. (d) Plot of dust accumulation versus time in the cyclical-climate model. (e) Plot of accumulation rate versus time interval in the cyclical-climate model, illustrating an inverted S-shaped curve. (f) Schematic diagram summarizing the results of the two models for accumulation rate versus time interval. Modified from Pelletier (2007a).

interval (i.e., surface age) uncertainties. The first four plots in Figure 8.7 exhibit a power-law trend (i.e., a straight line on log–log scales) given by

$$R \propto T^{-\alpha} \tag{8.9}$$

where R is the accumulation rate, T is the time interval of accumulation, and α is an exponent close to 0.5. Values of α for each data set were obtained by a linear fit of the logarithm of accumulation rate to the logarithm of time interval, weighted by the age uncertainty for each point. Accumulation rates plotted in Figures 8.7e and 8.7f (lower and upper fans near Silver Lake Playa) show only a weak dependence on time interval, however. Figure 8.7e, for example, has a nearly constant rate across time intervals, while Figure 8.7f shows a peak in accumulation rates at time

intervals of 10^3–10^4 yr, consistent with an early Holocene dust pulse.

In order to interpret the temporal scaling of dust accumulation rates with time interval, consider two end-member models. The first model assumes that the magnitude of dust deposition and erosion is constant through time (with a rate equal to $0.01 \, \text{g/cm}^2/\text{yr}$ in this example) and that erosion or deposition takes place with equal probability during each time step (1 yr). Figure 8.8a illustrates two runs of this "bounded random walk" model. The walk is bounded because the net accumulation is always positive (in an unbounded random walk, negative values are also allowed). At the longest time interval (10^4 yr), both model runs have accumulated approximately $1.5 \, \text{g/cm}^2$ of dust (equal to a thickness of 1.5 cm if a density of $1 \, \text{g/cm}^3$ is assumed). On shorter time intervals (10^3 and 10^2 yr), net accumulations are smaller than the 1.5 value, but they are not linearly proportional to the time interval. For example, net accumulation is 0.3–$0.5 \, \text{g/cm}^2$ over 10^3 yr for the two model runs shown, a much larger change *per unit time* than the $1.5 \, \text{g/cm}^2$ observed over 10^4 yr. Figure 8.8b illustrates the temporal scaling of accumulation rates versus time interval in this model. The results in Figure 8.8b illustrate that accumulation rates decrease, on average, at a rate proportional to one over the square root of the time interval, similar to that in Figures 8.7a–8.7d. This power-law scaling behavior can also be derived theoretically. The average distance from zero in a bounded random walk, $< h >$, increases with the square root of the time interval, T (van Kampen, 2001):

$$< h >= aT^{1/2} \qquad (8.10)$$

where the brackets denote the value that would be obtained by averaging the result of many different model runs and a is the magnitude of erosion or deposition during each time step. As applied to eolian dust accumulation, the average distance $< h >$ represents the total accumulation of eolian dust (in g/cm^2 or cm) on the surface, including the effects of episodic erosion. The average accumulation rate is obtained by dividing Eq. (8.10) by T to give

$$< R >= aT^{-1/2} \qquad (8.11)$$

In the second end-member model, dust deposition is assumed to have a high value during interglacial periods ($0.01\,g/cm^2/yr$), a low value during glacial periods ($0.002\,g/cm^2/yr$) and no erosion. Figure 8.8d illustrates the accumulation rate versus time interval for this model. The plot follows an inversed S-shaped curve shown in Figure 8.8e. This relationship is characteristic of accumulation rate curves for all periodic functions (Sadler and Strauss, 1990) and it is markedly different from the observed scaling in Figures 8.7a–8.7d.

The results of Figure 8.7 suggest that eolian dust accumulation on alluvial fans located close to playa sources (Figures 8.7e–8.7f) is fundamentally different from that on fans located far from playa sources (Figures 8.7a–8.7d). The fan study sites close to Silver Lake Playa exhibit a linear accumulation trend on the lower fan site and a cyclical-climate trend on the upper fan site while the remaining fan sites exhibit accumulation trends consistent with the bounded random walk model. These results suggest that the bounded random walk model is representative of fans located far (i.e., greater than a few km) from playas, and that the classic linear-accumulation or cyclical-climate models are representative of accumulation on fans proximal to playa sources where dust deposition is rapid enough to overwhelm the effects of episodic disturbances.

The scaling behavior of accumulation rates shown in Figures 8.7a–8.7d has important implications for the completeness of eolian deposits. Sedimentary deposits are said to be "incomplete" if unconformities or hiatuses with a broad distribution of time intervals are present (Sadler, 1981). The accumulation rate curve can be used to quantify the completeness of a given deposit using the ratio of the overall accumulation rate to the average rate at a particular time interval T (Sadler and Strauss, 1990). In the bounded random walk model, for example, a complete stratigraphic section (i.e., one with no hiatuses) of 10 000 yr duration would have a total accumulation of $100\,g/cm^2$ based on an annual accumulation rate of $0.01\,g/cm^2$. The bounded random walk model, in contrast, has an average accumulation of $1\,g/cm^2$ over 10 000 yr based on Eq. (8.11). Therefore, a 10 000 yr sequence is only 1% complete (i.e., 99% of the mass that was once deposited has been eroded). A power-law accumulation-rate curve (e.g., Eq. (8.10)) specifically implies a power-law or fractal distribution of hiatuses, with few hiatuses of very long duration and many of short duration.

8.6 | Stochastic trees and diffusion-limited aggregation

Stochastic models have featured prominently in our understanding of drainage networks. Horton (1945) was among the first to quantitatively study the geometry of drainage networks using Strahler's classification scheme. In this scheme, channels with no upstream tributaries are classified as order 1. When two streams of like order join together they create a stream of the next highest order. Horton showed, remarkably, that the ratios of the number, length, and area of streams within a drainage basin nearly always follow common values independent of Strahler order, i.e.

$$\frac{N_i}{N_{i+1}} \approx 4, \quad \frac{A_{i+1}}{A_i} \approx 4, \quad \frac{L_{i+1}}{L_i} \approx 2 \qquad (8.12)$$

where i is the Strahler order. In other words the ratio of the number of order 1 streams to order 2 streams is nearly the same as the number of

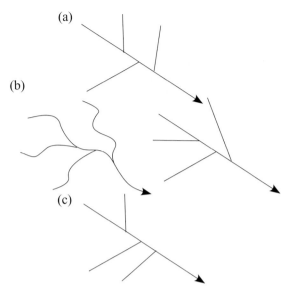

Fig 8.9 Illustration of the concept of Topologically Distinct Channel Networks (TDCNs). (a) A hypothetical channel network of magnitude 4. (b) Two channel networks that are topologically identical to that in (a) because they can be transformed into (a) by moving stream segments around within the plane. (c) A network that is topologically distinct from (a).

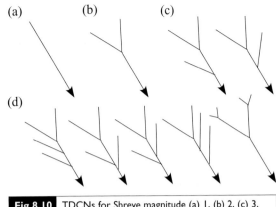

Fig 8.10 TDCNs for Shreve magnitude (a) 1, (b) 2, (c) 3, and (d) 4.

order 2 streams to order 3 streams (and so on). The fact that drainage basins have common ratios of number, length, and area across Strahler orders (i.e. spatial scales) is the basis for fractal models of drainage network geometry.

Following the establishment of Horton's Laws, stochastic models of drainage network geometry were proposed. One such framework is Topologically Distinct Channel Networks (TDCNs). TDCNs are tree structures that cannot be topologically transformed into one another within their plane. This concept is best illustrated with an example. In Figure 8.9a, a possible tree structure of order 2 is shown. Now, imagine that this network is made of a series of strings knotted together at junctions, and that the strings are laying on a table. A structure is topologically indistinct from another tree structure if the strings can be rearranged without lifting any of the strings off of the table. Figure 8.9b illustrates two possible tree structures that are topologically identical to the structure in Figure 8.9a. Figure 8.9c, in contrast, is topologically distinct because there is no way

to transform this structure into that of Figure 8.9a without lifting one of the strings off of the table. What is nice about the TDCN framework is that the entire set of all possible TDCNs can be identified for a given Strahler order or Shreve magnitude. In the Shreve classification scheme, channels without tributaries are defined to be magnitude 1 and when two streams join they form a stream with magnitude equal to the sum of the two tributaries. Figure 8.10, for example, shows all of the TDCNs up to Shreve magnitude 4. The number of TCDNs of Shreve magnitude M is given by

$$N = \frac{(2N - 2)!}{M!(M - 1)!} \tag{8.13}$$

where $M!$ is equal to $M*(M-1)*(M-2)\ldots2*1$.

An analysis of tree structures randomly selected from all possible TDCNs indicates that these structures obey Horton's Laws. Since TDCNs are a general mathematical model for tree structures and do not involve any geomorphic processes, the natural conclusion is that Horton's Laws are not really a consequence of drainage network evolution, but instead are a general feature of tree structures (Kirchner, 1993). Further studies of drainage network geometry have discovered other kinds of statistical relationships commonly found in real drainage networks. Tokunaga (1984) and Peckham (1996) expanded the Horton analysis to quantify not just the number of streams of a given order, but the number of "side branches" of a given order i that feed into streams of order j. This approach recognizes the fact that small streams that join with small streams in the headwaters of the Mississippi River basin in Minnesota are fundamentally different from small streams that join with the mainstem Mississippi River in Louisiana.

Diffusion-Limited Aggregation (DLA) provides one simplified stochastic model of drainage network evolution that matches the observed Tokunaga statistics of real drainage networks. In the DLA model for drainage networks (Masek and Turcotte, 1993), incipient channels are defined on a rectangular grid using "seed" pixels along one boundary (Figure 8.11). Then, a particle is introduced into an unoccupied cell chosen randomly from all unoccupied cells on the grid. The particle then undergoes a random walk until it encounters a cell with one of the four nearest neighbors occupied with a channel pixel. The new cell is then accreted to the network at this position (Figure 8.11).

Clearly, the DLA model is a very abstract model of the processes involved in the evolution of real drainage networks. So, how are we to interpret the fact that the model gives rise to drainage networks that look realistic and also match some of the statistical features of real drainage networks (Figure 8.12)? One explanation is that the random walk process mimics the particle pathways taken by groundwater flows that drive headward growth of channels by spring sapping.

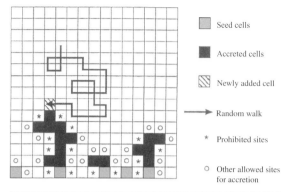

Seed cells

Accreted cells

Newly added cell

→ Random walk

* Prohibited sites

○ Other allowed sites for accretion

Fig 8.11 Illustration of the mechanism for network growth in the DLA model. A particle is randomly introduced to an unoccupied cell. The particle undergoes a random walk until a cell is encountered with one (and only one) of the four nearest neighbors occupied (hatched cell). The new cell is accreted to the drainage network. From Masek and Turcotte (1993). Reproduced with permission of Elsevier Limited.

In Dunne's model of drainage network evolution, headward-growing channels cause groundwater to be deflected towards the channel head, thereby increasing the headward growth rate of "master" streams at the expense of smaller streams nearby. The DLA model works in a broadly similar fashion. Random walkers dropped randomly on the surface will tend to accrete new channels close to the tips of existing channels because these channels extend out farther into the grid. The behavior of random walkers on the grid acts as a

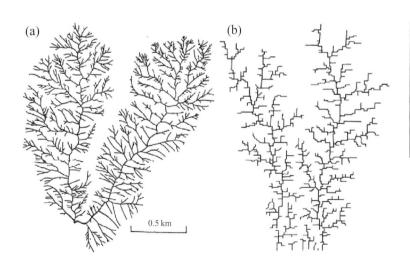

(a)

(b)

0.5 km

Fig 8.12 (a) Drainage network of the Volfe and Bell Canyons, San Gabriel Mountains, near Glendora, California, obtained from field mapping. (b) Illustration of one realization of the DLA model. From Masek and Turcotte (1993). Reproduced with permission of Elsevier Limited.

simplified model for the groundwater flow pathways in Dunne's model.

8.7 | Estimating total flux based on a statistical distribution of events: dust emission from playas

The relationship between the hydrological state of a playa and its dust-emitting potential is a nice example of the coupling of hydrology and geomorphology. In this section we consider a model for the long-term average dust emission from playas. Specifically, the goal of the model is to calculate the long-term average dust emission rate and the role of the water table depth in controlling dust emission rate. This example illustrates the use of probability distributions to calculate long-term geomorphic rates.

In nature, dust storms often originate when coarse sand from the playa margin enters into saltation. As sand blows across the playa surface, the crust is disturbed, releasing both sand and fine-grained sediments (silt and clay). Sand from the playa surface becomes part of a self-sustaining saltation cloud that drives dust production from the surface. The model we consider in this section does not resolve different components of that playa. Rather, it assumes that the flux of blowing sand along the playa margin can be used as a proxy for dust emission from the playa. This assumption is supported by data showing a strong correlation between horizontal saltation flux and the vertical dust flux (Gillette *et al.*, 2003). However, the model is not applicable to playas with no upwind sand available for transport.

Soil moisture affects dust emission from playas through its effects on soil cohesion. In order to model soil moisture at the surface of a playa, we consider the physics of steady-state capillary rise from a water table. For a shallow water table, capillary rise leads to a moist surface that suppresses dust emissions under all but the most extreme wind conditions. For a deep water table, the dry surface is at or near its maximum dust-producing potential and changes in subsurface moisture have little or no effect on dust emissions. Our model aims to quantify the range of water-table depths over which this transition from a wet to dry playa takes place. An important assumption in this analysis is that precipitation events are rare enough that the surface is near its steady-state moisture value most of the time.

The vertical transport of moisture in an unsaturated soil is governed by Richards' equation:

$$\frac{\partial \psi}{\partial t} = \frac{\partial}{\partial z}\left(K_\psi \frac{\partial \psi}{\partial z} - 1\right) \tag{8.14}$$

where ψ is the suction, t is time, z is height above the water table, and K_ψ is the hydraulic conductivity. For steady upward flow, Eq. (8.14) can be written as

$$E = \frac{\partial}{\partial z}\left(K_\psi \frac{\partial \psi}{\partial z} - 1\right) \tag{8.15}$$

where E is the steady-state evaporation rate. Gardner (1958) proposed the following relationship between K_ψ and ψ to solve Eq. (8.15):

$$K_\psi = \frac{a}{\psi^n + b} \tag{8.16}$$

where a and b are empirical constants for each soil. Gardner (1958) provided analytical solutions for suction profiles at several values of n. Most soils have n values between 2 and 3 (Gardner and Fireman, 1958). Here we assume $n = 2$, the most appropriate value for fine-grained playa sediments according to Gardner (1958). The solution to Eq. (8.15) for $n = 2$ and with boundary condition $\psi = 0$ at $z = 0$ is (Gardner, 1958):

$$z = \frac{1}{\sqrt{\frac{E}{a}\left(\frac{E}{a}b + 1\right)}}\tan^{-1}\left(\sqrt{\frac{E/a}{(E/a)b + 1}}\psi\right) \tag{8.17}$$

The value of E is determined by the maximum soil moisture flux within the profile, given by Gardner (1958) and Warrick (1988) as:

$$\frac{E}{a} = \left(\frac{\pi}{2d}\right)^2 \tag{8.18}$$

where d is the water table depth. Substituting Eq. (8.18) into Eq. (8.17) and solving for the suction at $z = d$ gives

$$\psi_{z=d} = \sqrt{b + \left(\frac{2d}{\pi}\right)^2}\tan\left(\sqrt{\left(\frac{\pi}{2}\right)^2\left(\left(\frac{\pi}{2d}\right)^2 b + 1\right)}\right) \tag{8.19}$$

Van Genuchten (1980) predicted the surface soil moisture from the suction to be

$$\frac{\theta_{z=d} - \theta_r}{\theta_s - \theta_r} = \left(1 + (\alpha \psi_{z=d})^{1+\lambda}\right)^{-\frac{\lambda}{\lambda+1}} \quad (8.20)$$

where θ_r is the residual soil moisture, θ_s is the saturated soil moisture, α is the inverse of the bubbling pressure, and λ is the pore-size distribution parameter. Equations (8.19) and (8.20) provide an analytical solution for the surface soil moisture as a function of water-table depth and soil-hydrologic parameters.

The second model component relates the surface soil moisture to the threshold friction velocity. Chepil (1956) developed the first empirical relationship between these variables. He obtained:

$$u_{*t} = \left(u_{*td}^2 + \frac{0.6}{\rho}\left(\frac{\theta_{z=d}}{\theta_w}\right)^2\right)^{\frac{1}{2}} \quad (8.21)$$

where u_{*td} is the dry threshold friction velocity,

$$u_{*t} = \left(u_{*td}^2 + \frac{0.6}{\rho}\frac{1}{\theta_w^2}\left((\theta_s - \theta_r)\left(\left(1 + \left(-\alpha\sqrt{b + \left(\frac{2d}{\pi}\right)^2}\tan\left(\sqrt{\left(\frac{\pi}{2}\right)^2\left(\left(\frac{\pi}{2d}\right)^2 b + 1\right)}\right)\right)^{1+\lambda}\right)^{-\frac{\lambda}{\lambda+1}} + \theta_r\right)^2\right)^{\frac{1}{2}}$$

$$(8.27)$$

ρ is the density of air (1.1 kg/m^3), and θ_w is the wilting-point moisture content (typically 0.2–0.3 for fine-grained soils (Rawls *et al.*, 1992)).

The final model component is the saltation equation for transport-limited conditions, given by Shao and Raupach (1993) as:

$$q = \frac{\rho}{g}u_*(u_*^2 - u_{*t}^2) \quad (8.22)$$

where q is the flux in kg/m/s, g is the gravitational acceleration (9.8 m/s^2), and u_* is the friction velocity. The friction velocity can be obtained from measured velocities at a height z_m above the ground using a modified law of the wall:

$$u_* = \frac{\kappa}{\ln\left(\frac{z_m}{z_0}\right)}u \quad (8.23)$$

where u is the measured velocity, κ is von Karman's constant (0.4), and z_0 is the aerodynamic surface roughness.

A two-parameter Weibull distribution, commonly used to quantify the distribution of wind velocities (Takle and Brown, 1978; Bowden *et al.*, 1983), was used for the probability distribution of friction velocities:

$$f_w(u_*) = \gamma \beta^{-\gamma} u_*^{\gamma-1} e^{-(u_*/\beta)^\gamma} \quad (8.24)$$

where γ and β are parameters fit to measured wind data. The long-term average saltation flux is obtained by integrating Eq. (8.22):

$$<q> = \frac{\rho}{g}\int_{u_{*t}}^{\infty} du_* f_w(u_*)u_*(u_*^2 - u_{*t}^2) \quad (8.25)$$

Substituting Eq. (8.24) into Eq. (8.25) yields a closed-form solution for the long-term average saltation flux:

$$<q> = \frac{\rho}{g}\frac{\beta}{\gamma^2}\left[3\beta^2\left(1 + \Gamma\left(\frac{3}{\gamma}, \left(\frac{u_{*t}}{\beta}\right)^\gamma\right)\right)\right.$$
$$\left. - u_{*t}^2\Gamma\left(\frac{1}{\gamma}, \left(\frac{u_{*t}}{\beta}\right)^\gamma\right)\right] \quad (8.26)$$

where $\Gamma(x, y)$ is the incomplete gamma function, and

In many cases of interest, including the Soda Lake example described below, the values of the incomplete gamma function in Eq. (8.26) are nearly equal to 1. In such cases, Eq. (8.26) can be approximated as

$$<q> \approx \frac{\rho}{g}\frac{\beta}{\gamma^2}(6\beta^2 - u_{*t}^2) \quad (8.28)$$

Relative changes in saltation flux, given by Eq. (8.26), are also expected to apply to dust emissions because of the proportionality between saltation and dust emissions observed in many areas (Gillette *et al.*, 2003). Dust emissions can also be explicitly calculated using Eq. (8.26) if an estimate is available for the K factor, or the ratio of the vertical dust flux to the horizontal saltation flux. The value of K depends primarily on the surface texture and must be determined empirically.

Three sites at the margins of Soda (dry) Lake, California (Figure 8.13) were selected as an illustration of the model calibration procedure. Soda

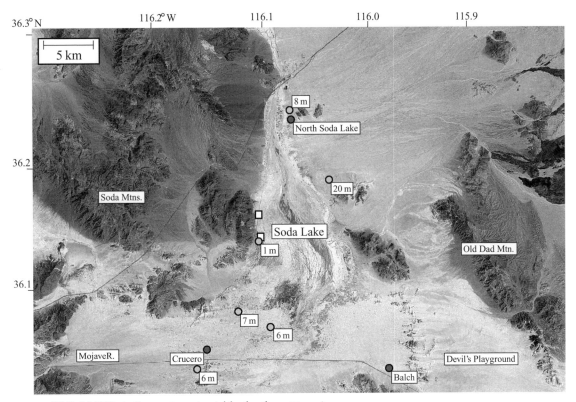

● CLIM-MET stations   ◐ water table depths   □ springs

Fig 8.13 LANDSAT image of Soda (dry) Lake study site, with locations of CLIM-MET stations and water table depths indicated. Modified from Pelletier (2006). Reproduced with permission of Elsevier Limited.

Lake is located at the western edge of the Mojave National Preserve just south of Baker, California. CLIM-MET data collected from 1999 to present by a team from the US Geological Survey are available for these areas. The water table beneath Soda Lake is at or near the ground surface along the western edge of the playa, deepening to the east to 20 m and lower beneath the alluvial fans on the eastern side of the basin. Water-table depths indicated in Figure 8.13 were obtained from the US Geological Survey National Groundwater Database.

Figure 8.14 illustrates the CLIM-MET station data for hourly rainfall, soil moisture, peak wind speed, and the number of particles in saltation measured at the Balch and Crucero stations near the southern playa margin. Figures 8.15b and 8.15e illustrate that the soil moisture at both

sites has a residual value of $\theta_r = 0.05$, and that the soil moisture is within a few percent of that value most of the time. For this reason, the steady-state approximation is an appropriate estimate for long-term-average moisture conditions, although it does not represent transient effects of moisture in the days to weeks following rare precipitation events.

Wind speeds measured 2 m above the ground were converted to friction velocities using Eq. (8.23) with $z_0 = 0.005$ m. This value was obtained from Marticorena et al.'s (1997) relationship $z_0 = 0.005 h_c$, together with a canopy height of $h_c = 1$ m. Friction velocities inferred from the measured wind speed data were fit to the two-parameter Weibull distribution. The best-fit parameters obtained for the Balch and Crucero stations are $\beta = 0.398$ and 0.405, and $\gamma = 2.17$ and 2.15, respectively. Based on these numbers, we chose $\beta = 0.40$ and $\gamma = 2.15$ as representative values for the meteorological conditions at Soda Lake. The dry threshold friction velocities varied from 0.5 to 0.8 m/s at the three CLIM-MET

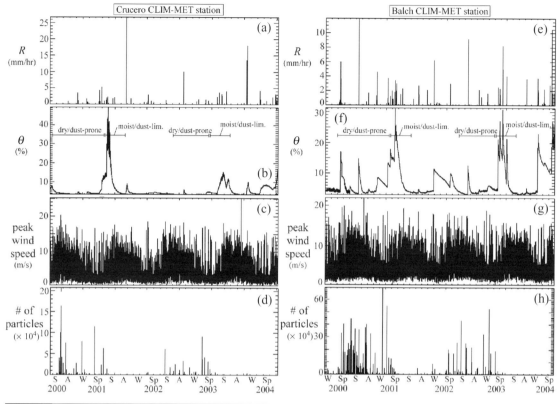

Fig 8.14 Hourly CLIM-MET data from (a)–(d) Crucero and (e)–(h) Balch stations. (a) and (e) Rainfall, (b) and (f) volumetric soil moisture, (c) and (g) peak wind speed at 2 m above ground, (d) and (h) number of particles in saltation, as measured by SENSIT instruments. Modified from Pelletier (2006). Reproduced with permission of Elsevier Limited.

stations. We chose a value of $u_{*td} = 0.5$ for our calculations, but clearly a spatially distributed model that resolves variations in threshold friction velocity would be the most accurate approach for characterizing the playa basin as a whole.

Soil-hydrologic parameters were estimated from values given in Gardner and Fireman (1958) and Rawls *et al.* (1992). Gardner and Fireman (1958) studied soils with fine sandy loam and clay textures, and obtained b values of 0.04 and $0.055\,m^2$, respectively. Rawls *et al.* (1992) provided estimates of α and λ appropriate for fine-grained soils. Representative values of $\alpha = 0.3\,m^{-1}$, $\lambda = 0.2$, and $\theta_w = 0.2$ were chosen.

Figures 8.15a, 8.15b, and 8.15c illustrate the solution to Eqs. (8.19), (8.26), and (8.25), respec-

tively, for a range of water table depths up to 10 m. Figure 8.15c indicates that saltation (and hence dust emission) is essentially absent for water tables within 4 m of the surface for this set of parameters. As the water table depth increases, saltation flux also increases, asymptotically approaching a maximum value of nearly $6\,g/m^2/s$. For water tables deeper that 10 m, the surface is dry enough that soil moisture plays a relatively minor role. The shape of the saltation curve depends primarily on the value of the wilting point moisture θ_w. Figure 8.15d, for example, illustrates flux curves for representative end-member values of 0.2 and 0.3. A larger wilting-point moisture means a smaller role for soil moisture in suppressing emissions. The $\theta_w = 0.3$ case, therefore, has saltation initiated at a shallower water table depth (2 m instead of 4 m), and rises more rapidly to its asymptotic value as a function of water table depth. In contrast, saltation curves are relatively insensitive to the values of b, α and λ. Varying each of these parameters by 10%, for example, results in flux changes of only a few percent for any particular water table depth.

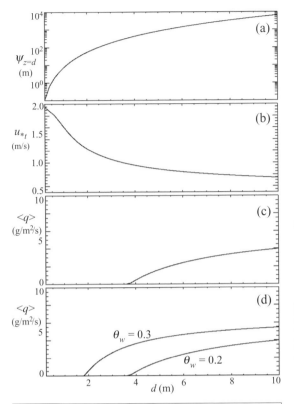

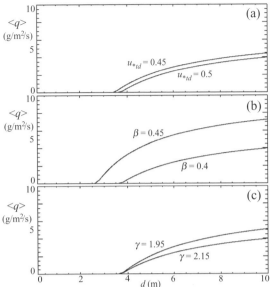

Fig 8.16 Sensitivity of saltation fluxes to changes in (a) dry threshold friction velocity, (b) Weibull scale factor, and (c) Weibull shape factor, for a range of water table depths. From Pelletier (2006). Reproduced with permission of Elsevier Limited.

Fig 8.15 Solutions to (a) Eq. (8.19), (b) Eq. (8.26), and (c) Eq. (8.25) for the model parameters. (d) Sensitivity to the value of the wilting-point moisture. Modified from Pelletier (2006). Reproduced with permission of Elsevier Limited.

Figure 8.16 illustrates the sensitivity of the saltation curve to changes in the dry threshold friction velocity and the mean and variability of wind speeds. Although likely future changes in regional wind speeds cannot be easily quantified, many global climate models suggest that wind speeds will become more variable in the future. This type of scenario should be considered as a possibility within future air-quality management plans. Figure 8.16a illustrates the saltation flux as a function of water-table depth corresponding to a 10% decrease in the dry threshold friction velocity. This 10% decrease causes a proportionate increase in saltation flux for all water table depths. A 10% increase in the Weibull scale factor β (closely related to mean wind speed), shown in Figure 8.16b, results in a near doubling of saltation flux for most water-table depths. This result is not surprising given the nonlinear, threshold

dependence on saltation flux on friction velocity. A 10% change in the Weibull shape factor γ leads to a small but significant flux response ($\approx 20\%$) for deep water tables.

Field measurements of saltation activity indicate that threshold friction velocities are variable in time (e.g. Gillette *et al.*, 1980, 1997). This variability can occur for many reasons, including limited availability of transportable material, short-term bursts in near-surface turbulence, temporal changes in microsurficial characteristics (i.e. formation and disturbance of protective crusts), and the selective entrainment of fine particles and subsequent surface armoring with coarse lag deposits. In the absence of detailed quantitative observations of controlling surficial characteristics, a stochastic model component is necessary to represent this variability. Here we generalize the model equations to include a range of threshold wind velocities.

Figures 8.17a–8.17c illustrate the variability in threshold velocities in the CLIM-MET station data. In this figure, soil moisture is plotted as a function of wind speed, with saltating conditions

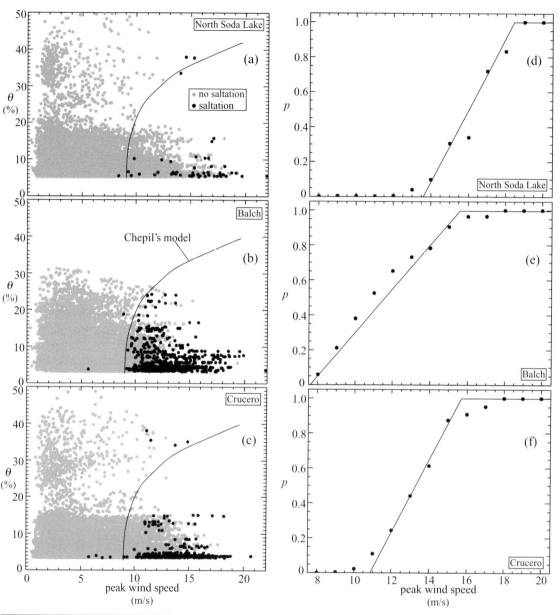

Fig 8.17 (a)–(c) Control on saltation activity by wind speed and soil moisture at (a) North Soda Lake, (b) Balch, and (c) Crucero CLIM-MET stations. Saltation activity is indicated with black dots; gray dots indicate no saltation. Chepil's quadratic relationship between soil moisture and threshold friction velocity is also shown. (d)–(f) The frequency of saltation activity at a given wind speed shows a linear increase from a minimum threshold velocity to a maximum threshold velocity. Modified from Pelletier (2006). Reproduced with permission of Elsevier Limited.

indicated by black dots and non-saltating conditions indicated by gray dots. These data illustrate that high wind speeds occasionally fail to produce saltating conditions. Conversely, low wind speeds can sometimes trigger saltation. Subsetting the CLIM-MET data at time intervals of a few months does not significantly reduce this overlap. The range of threshold friction velocities at these locations, therefore, appears to be related to microsurficial changes that occur over intra-annual time scales.

The CLIM-MET station data support a linear relationship between excess friction velocity and the probability of saltation, as shown in Figures 8.17d–8.17f. Mathematically, this can be expressed as a linear increase in the probability of saltation p from a minimum value of 0 at $u = u_{*td,min}$ to a maximum value of 1 at $u_{*td,max}$:

$$p = \frac{u_* - u_{*td,min}}{u_{*td,max} - u_{*td,min}} \tag{8.29}$$

where $u_{*td,min}$ and $u_{*td,max}$ define a range of threshold friction velocities. For a surface that experiences minor microsurficial variability through time, the difference between $u_{*td,min}$ and $u_{*td,max}$ will be small. Conversely, large surficial changes will result in a correspondingly large range of values. Among the three CLIM-MET stations, North Soda Lake and Crucero (Figures 8.17d and 8.17f) exhibit a relatively small range of threshold velocities, while Balch (Figure 8.17e) exhibits a relatively large range.

The range of threshold friction velocities can be included in the model by integrating Eq. (8.29) in a piecewise manner:

$$<q> = \frac{\rho}{g} \left(\int_{u_{*td,min}}^{u_{*td,max}} du_* \, p(u_*) f_w(u_*) u_* (u_*^2 - u_{*t}^2) \right.$$
$$\left. + \int_{u_{*td,max}}^{\infty} du_* \, f_w(u_*) u_* (u_*^2 - u_{*t}^2) \right) \tag{8.30}$$

to obtain

$$<q> \approx \frac{\rho}{g} \frac{\beta}{\gamma^2} \left(6\beta^2 - u_{*td,max}^2 \right.$$
$$\left. + 2(u_{*td,max} - u_{*td,min})^2 \right) \tag{8.31}$$

The incomplete gamma functions have been approximated as unity in Eq. (8.31) to yield an expression analogous to Eq. (8.28). The additional terms in Eq. (8.28) were found to have a minor effect on the dependence of saltation on water-table depth (e.g. Figures 8.15 and 8.16) because the additional terms in Eq. (8.31) are related directly to wind speed rather than soil moisture. Equation (8.31) does, however, provide a more accurate estimation for the absolute value of the saltation flux than Eq. (8.28), because it explicitly represents the range of threshold wind velocities observed in CLIM-MET station data.

8.8 | The frequency-size distribution of landslides

Landslides occur in regions of steep slopes and are often triggered by intense rainfall, rapid snowmelt, and ground motion from earthquakes and volcanic eruptions. Given the spatial and temporal complexity of these triggering mechanisms and the underlying topography, it is not surprising that stochastic models play a prominent role in our understanding of landslide populations. Although our ability to forecast landslides is limited, we can gain understanding of landslide population statistics by combining deterministic models of landslide occurrence with stochastic models of the underlying spatial and temporal quantities such as topographic slope and soil moisture. This section illustrates that approach.

In this section we investigate the frequency-size distribution of landslides in areas where different triggering mechanisms dominate. We argue that fractional Gaussian noises are useful for modeling the variations in soil moisture and topography that trigger landslide instability. A number of authors have examined the frequency-size distribution of landslides (Whitehouse and Griffiths, 1983; Ohmori and Hirano, 1988; Hovius et al., 1997). These authors have consistently noted that the distribution is power-law above some threshold size. This is precisely analogous to the Gutenberg–Richter law for earthquakes, which states that the cumulative frequency of events with a seismic moment greater than M_0 is a power-law function of M_0.

Studies suggest that a threshold dependence on soil moisture is appropriate for slope instablity since landslides often occur when the product of rainfall duration and intensity exceeds a threshold value (Caine, 1980; Wieczoreck, 1987). In situ monitoring of pore pressure has shown that increases in pore pressure resulting from heavy precipitation are coincident with landslides (Johnson and Sitar, 1990). Several authors have shown that variations in the frequency of occurrence of landslides respond to climatic changes, with higher rates of landslides associated with wetter climates (Grove, 1972; Innes, 1983; Pitts, 1983; Brooks and Richards, 1994). Further evidence for correlations between landslides and soil moisture is the observation that soil drainage is the most successful method of landslide prevention in the United States. Seismic triggering of landslides has also been studied by many authors (Keefer, 1984; Jibson, 1996). Correlations between landslides and earthquakes can be inferred in a number of ways. Many earthquakes have been known to trigger large numbers of landslides and landslide densities have a strong correlation with active seismic belts.

The Washita experimental station has provided very useful information on the evolution of soil moisture in space and time. By characterizing the spatial variability of soil moisture using observed data sets, we can construct statistical models that match the observed statistics. For this purpose, microwave remotely-sensed soil moisture data collected from June 10 to June 18, 1992 at the Washita experimental watershed can be used (Jackson, 1993). The watershed received heavy rainfall preceding the experiment but did not receive rainfall during the experiment. The data are gridded estimates of soil moisture in the top five centimeters of soil calculated using the algorithm of Jackson and Le Vine (1996). Each pixel represents an area of 200 m × 200 m and the total area considered is 45.6 km by 18.6 km. The soil moisture values do not correlate with relief in the watershed, which is very small. Figure 8.18a shows a grayscale image of the soil moisture at Washita on June 17, 1992. White spaces indicate areas where the watershed is interrupted by roads or lakes. Figure 8.18b is a synthetic two-dimensional fractional Gaussian noise with $\beta = 1.8$ constructed using the Fourier-filtering method. The mean and variance of the synthetic moisture field have been chosen to match those of the observed image. The synthetic image reproduces the correlated structure of the real soil moisture image.

Entekhabi and Rodriguez-Iturbe (1994) proposed a partial differential equation for the dynamics of the soil moisture field $s(\vec{x}, t)$:

$$\frac{\partial s}{\partial t} = D \nabla^2 s - \eta s + \xi(\vec{x}, t) \qquad (8.32)$$

Fig 8.18 (a) Grayscale map of microwave remotely sensed estimates of soil moisture on June 17, 1992 at the Washita experimental watershed. (b) Grayscale map of a synthetic soil moisture field with the power spectrum $S(k) \propto k^{-1.8}$ with the same mean and variance as the remotely sensed data. Modified from Pelletier et al. (1997). Reproduced with permission of Elsevier Limited.

In this equation, the evapotranspiration rate η is assumed to be constant in space and time. Soil moisture disperses diffusively in the soil, and rainfall input $\xi(\vec{x}, t)$ is modeled as a random function in space and time. Variations in rainfall from place to place cause spatial variations in soil moisture that are damped by the effects of diffusion and evapotranspiration. Without spatial variations in rainfall input, there are no variations in soil moisture according to this model. As pointed out by Rodriguez-Iturbe et al. (1995), the variability at the small scale of the Washita watershed is not likely to be the result of spatial variations in rainfall.

An alternative approach to this problem which will generate spatial variations in soil moisture at these small scales assumes that the evapotranspiration rate is a random function in space and time. This models spatial and temporal variations in evapotranspiration resulting from variable atmospheric conditions and heterogeneity in soil, topography, and vegetation characteristics. The resulting equation is

$$\frac{\partial s}{\partial t} = D \nabla^2 s - \eta(x, y, t) s \qquad (8.33)$$

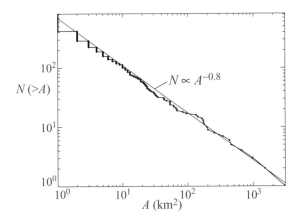

Fig 8.19 Cumulative frequency-size distribution of patches of soil moisture larger than a threshold value for the synthetic soil moisture field with power spectrum $S(k) \propto k^{-1.8}$. Modified from Pelletier et al. (1997). Reproduced with permission of Elsevier Limited.

This equation is a variant of the Kardar–Parisi–Zhang equation (Kardar et al., 1986), which has received a great deal of attention in the physics literature. This equation is nonlinear and can only be solved numerically.

Amar and Family (1989) have solved this equation and have found that the solutions have transects with power spectra that have a power-law dependence on wave number with exponent $\beta = 1.8$. The similarity between the field generated by Eq. (8.33) and the soil moisture observed in the Washita images suggests that Eq. (8.33) may be capturing much of the essential dynamics of soil moisture at these scales.

Rodriguez-Iturbe et al. (1995) quantified the scale-invariance of soil moisture variations by computing the cumulative frequency-size distribution of patches of soil with moisture levels higher than a prescribed value. They found that the cumulative frequency-size distribution had a power-law function on area with an exponent of approximately −0.8. The number of soil patches N depended on the area according to

$$N(> A) \propto A^{-0.8} \tag{8.34}$$

This is a power-law or fractal relation. We have determined the equivalent distribution for the synthetic soil moisture field illustrated in Figure 8.18b. The result is plotted in Figure 8.19. The

same distribution that was observed in real soil moisture fields is obtained.

Landslides occur when the shear stress exceeds a threshold value given approximately by the Mohr–Coulomb failure criterion (Terzaghi, 1962):

$$\tau_f = \tau_0 + (\sigma - u)\tan\phi \tag{8.35}$$

where τ_0 is the cohesive strength of the soil, σ is the normal stress on the slip plane, u is the pore pressure, and ϕ is the angle of internal friction. Landslides are initiated in places where τ_f is greater than a threshold value. The movement of the soil at the point of instability increases the shear stress in adjacent points on the hillslope causing failure of a connected domain with shear stress larger than the threshold value. To model the landslide instability and, in particular, the frequency-size distribution of landslides, it is therefore necessary to model the spatial variations of τ_0, σ, u, and ϕ. The variables τ_0 and u are primarily dependent on soil moisture for a homogeneous lithology and a slip plane of constant depth. The dependence of each of these variables on soil moisture has been approximated using power-law functions (Johnson, 1984). The shear stress and normal stress are linearly proportional to soil moisture through the weight of water in the soil. The shear stress and normal stress are also trigonometric functions of the local slope. Based on the results given earlier, the soil moisture will be modeled as a two-dimensional fractional Brownian walk with $\beta = 1.8$. Power spectral analyses of one-dimensional transects of topography have shown topography to have a power-law power spectrum with $\beta = 2$ (e.g. Huang and Turcotte, 1989). This behavior is applicable only over a certain range of scales, however. At scales smaller than the inverse of the drainage density, the topography is controlled by hillslope processes and does not exhibit fractal behavior. To represent the small-scale smoothness of hillslope topography as well as the fractal behavior at larger scales, we can first construct a two-dimensional surface with one-dimensional power spectrum $S(k) \propto k^{-2}$ with the Fourier-filtering method. At small scales, however, the synthetic topography is linearly interpolated. A shaded-relief example of the model

(a)

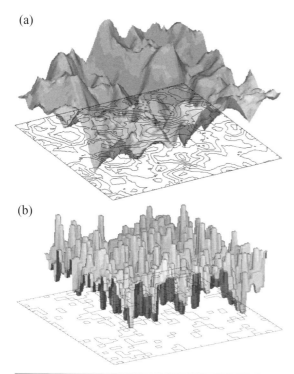

(b)

topography is illustrated in Figure 8.20a along with its contour plot. The plot has 128×128 grid points with interpolation below a scale of 8 pixels. The result of the interpolation is clearly identified as piecewise linear segments in the transects along the boundaries of the plot. The slope corresponding to this model of topography is illustrated in Figure 8.20b. Below a scale of 8 pixels, the slopes are constant. Above this scale, the slopes are a two-dimensional Gaussian white noise. This follows from the fact that our model for topography at large scales, a Brownian walk with $S(k) \propto k^{-2}$, can be defined as the summation of a Gaussian white noise time series. White noise means that adjacent values are totally uncorrelated. The contour map of the slope function

shows that there are no contours smaller than 8×8 pixels.

The shear stress necessary for failure is a complex function of soil moisture and slope. However, to show how landslide areas may be associated with areas of simultaneously high levels of soil moisture and steep slopes, we assume a threshold shear stress criterion proportional to the product of the soil moisture and the slope. In addition, we assume that slope and soil moisture are uncorrelated. A grid of synthetic soil moisture and topography of 512×512 grid points was constructed according to the models described above. The domains where the product of the soil moisture and the topography were above a threshold value are shown in Figure 8.21a. The threshold value was chosen such that only a small fraction of the region was above the threshold. Figure 8.21b shows the cumulative frequency-size distribution of the regions above threshold, our model landslides. It can be seen that at large areas a power-law distribution with an exponent of -1.6 exists. Actual landslide distributions, discussed below, show a similar trend. Distributions obtained with different realizations of the model yielded power-law exponents of 1.6 ± 0.1 when fit to the landslides above $A = 10$. This form was independent of the threshold value chosen for landslide initiation, as long as the threshold was chosen such that only a small fraction of the lattice is above threshold. The exponent of this distribution is more negative than that of the soil moisture patches of Figure 8.19. This results from the less correlated slope field "breaking up" some of the large soil moisture patches. The effect of smooth topography at the hillslope scales is to create a rolloff in the frequency-area distribution for small landslide areas. This is a consequence of the fact that fractal scaling breaks down at hillslope scales, and hence slopes tend to fail as a unit rather than triggering many small, isolated landslides on the same slope. The effect of strong ground motion from earthquakes is to lower the shear stress necessary for failure. This does not alter the frequency-size distribution of landslides according to this model since the form of the distribution is independent of the value of threshold.

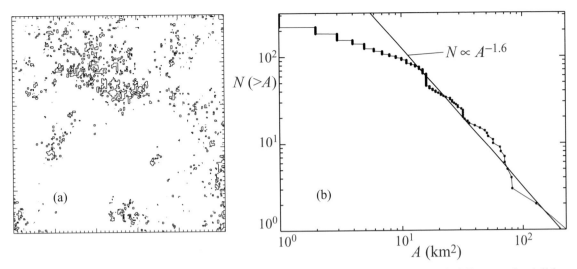

$N(>A)$

$N \propto A^{-1.6}$

(a)

(b)

$A \ (km^2)$

Fig 8.21 (a) Contour map of the product of the synthetic soil moisture field with the synthetic slope function. Areas inside the contour loops represent model landslides. (b) Cumulative frequency-size distribution of model landslides. The distribution compares favorably to the distributions of real landslides. Modified from Pelletier et al. (1997). Reproduced with permission of Elsevier Limited.

Do actual landslide distributions produce a similar distribution? Whitehouse and Griffiths (1983), Ohmori and Hirano (1988), and Hovius et al. (1997), among others, have all presented evidence that landslide frequency-size distributions are power-law functions of area for large landslide areas. Figure 8.22 presents cumulative frequency-size distributions of landslide area from three areas first analyzed in Pelletier et al. (1997). In Figure 8.22a, the number of landslides with an area greater than A are plotted as a function of A for seven lithologic units from a dataset of 3 424 landslides with areas larger than $10^4 \, m^2$ in the Akaishi Ranges, central Japan (Ohmori and Sugai, 1995). For large landslides, the distribution is well characterized by a power law with an exponent of approximately -2, that is

$$N(> A) \propto A^{-2} \tag{8.36}$$

In this region landslides occur as a result of both heavy rainfall and strong seismicity. Figure 8.22b presents two sets of landslide areas mapped in adjacent watersheds in the Yungas region of the Eastern Cordillera, Bolivia. All of these landslides were triggered by heavy rainfall. Large landslides in the two areas match power-law distributions from Eq. (8.36) with exponents -1.6 and -2. Figure 8.22c presents the cumulative frequency-area distribution of more than 11 000 landslides triggered by the 1994 Northridge earthquake over an area of 10 000 km^2 (Harp and Jibson, 1995). Working with high-resolution aerial photography acquired the day after the earthquake, Harp and Jibson estimate that their catalog of landslides is complete for landslides larger that 25 m^2. At smaller sizes, a significant number of landslides may have been missed. Large landslides in this dataset have a power-law dependence on area with an exponent of approximately -1.6. As in the Japan and Bolivia datasets there is a roll off in the power-law distribution for small areas. These results suggest that cumulative frequency-size distributions are remarkably similar despite different triggering mechanisms, and that they match the predictions predicted by a stochastic model for landslides triggered by soil moisture and topography.

Some studies have suggested that rolloffs at the small end of landslide distributions are the result of an incomplete catalog (Stark and Hovius, 2001). While it is certainly true that power-law relationships describe only the upper tail of the landslide size distribution, it is not true that breaks in scaling at the lower end of the distribution are primarily due to incomplete sampling. Consider the Northridge dataset shown in Figure 8.22c. If power-law scaling held down to

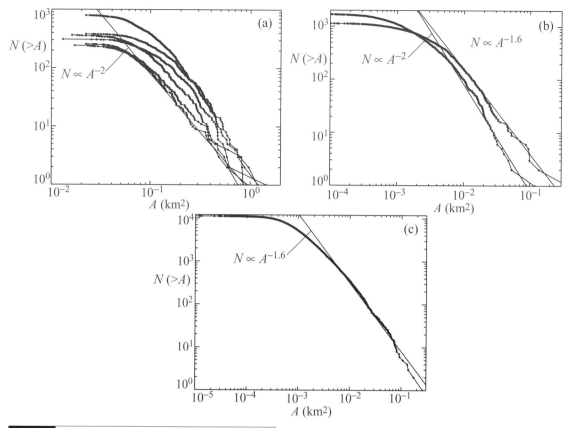

Fig 8.22 (a) Cumulative frequency-size distribution of landslides in six lithologic zones in Japan. The distributions are well characterized by $N(> A) \propto A^{-2}$ above approximately 0.1 km^2. (b) Cumulative frequency-size distribution of landslides in two areas in Bolivia. (c) Cumulative frequency-size distribution of landslides triggered by the 1994 Northridge, California earthquake. This dataset is characterized by $N(> A) \propto A^{-1.6}$ above approximately 0.01 km^2.

landslide areas as small as 10^{-4} km^2, for example, the power-law extrapolation in Figure 8.22c would predict approximately 1 million landslides larger than 10^{-4} km^2. Working with detailed aerial photographs, Harp and Jibson found that they could resolve landslides as small as 25 m^2. Therefore, if the break in power-law scaling is due to undersampling, this implies that Harp and Jibson missed approximately 990 000 landslides at least four times larger than the minimum area they could resolve in their images. The fact that so few small landslides occur in the Northridge case relative to the power-law trend for larger

landslides clearly shows that there is some physical mechanism for limiting these small landslides. In the model of this section, the break in topographic scaling at the hillslope scale is the cause of the rolloff in the landslide size distribution. The planarity of slopes at small scales means that when slides are triggered, large sections of the slope tend to fail, with fewer isolated sections of slope failure relative to a power-law distribution.

8.9 | Coherence resonance and the timing of ice ages

The behavior of Earth's climate system in the last 2 Myr is a consequence of both deterministic and stochastic forces. Significant periodicities exist in time series of Late-Pleistocene global climate (i.e. proxies for temperature and ice volume) near 29 and 41 kyr. These periodicities are controlled by the precession and obliquity of Earth's orbit.

There is little agreement, however, on which processes and feedbacks control other characteristics of Late-Pleistocene climate, including the 100-kyr cycle and the asymmetry of glacial-interglacial transitions. In this section we explore a simple model of Earth's climate system in the Pleistocene as an example of how stochastic processes can amplify nonlinear system behavior. Although not strictly limited to surface processes, the climate system over these time scales is strongly controlled by both ice sheets and the lithospheric response to ice-sheet loading. As such, the climate system involves key aspects of surface processes we have discussed in earlier chapters.

Many models have been introduced over the past few decades that reproduce one or more of the key features of Earth's climate in the Pleistocene. Several of the most successful models have focused on nonlinear feedbacks between continental ice-sheet growth, radiation balance, and lithospheric deflection (e.g. Oerlemans, 1980b; Pollard, 1983). In this section we will explore a simple model of Late-Pleistocene climate that reproduces the dominant 100-kyr oscillation of ice ages. The mechanism for producing this oscillation process is "coherence resonance." Coherence resonance has been investigated by statistical physicists for several years (e.g. Masoller, 2002). The necessary components of a coherently-resonating system are (1) a system with two stable states, (2) a delayed feedback, and (3) random variability (Tsimring and Pikovsky, 2001). A number of physical, chemical, and biological systems have these components and exhibit coherence resonance. A simple example of a system with two stable states is a ball that rolls back and forth between two valleys when pushed over a hill. Random noise may be introduced into this system to produce random "kicks" that push the ball over the hill and into the other valley if enough kicks work in the same direction. If the kicks are random, they generally tend to cancel each other out to keep the ball trapped in one valley. From time to time, however, a series of kicks in the same direction will occur that move the ball over the hill. Delayed feedback occurs if an additional force is exerted on the ball that is not random, but depends on the elevation of the ball at a previous time. The basic idea is that coherence resonance can make a randomly kicked ball move back and forth periodically between the two hills if a delayed feedback is introduced into the system. In this climate system, the growth and decay of ice sheets and the motion of the lithosphere beneath them provide that delayed feedback.

There are two time scales in this model system: the delay time and the average time that the ball spends in each valley before switching over to the other valley. This latter time scale is controlled by the height of the hill and the magnitude of the random variability. If these two time scales are close in value, a resonance occurs in which the random switching between valleys is made more regular by the delayed feedback, which encourages a repetition in the history of the system. For example, if a series of random kicks move the ball over the hill, the delay feedback will act to push the ball up over the hill in the same manner at a later time by providing systematic kicks in the uphill direction. This behavior is strongly self-reinforcing: the more regularly a ball makes its way up over the hill, the more regularly will the delay feedback work to systematically push it up the hill at the later time.

The standard equation that displays coherence resonance is given by

$$\frac{\Delta T_n}{\Delta t} = T_n - T_n^3 + \epsilon T_{n-\tau} + D \eta_n \qquad (8.37)$$

where T is the system variable (e.g. elevation in the case of a rolling ball, or temperature in the climate system), n is the time step, τ is the delay time, and η is a white noise with standard deviation D. The first two terms on the right side of Eq. (8.37) represent bistability in the system. The delay feedback and noise are represented by the third and fourth terms, respectively. Systems with both positive and negative feedback may occur depending on whether ϵ is positive or negative. Coherence resonance occurs in both cases, but with a different periodicity.

An example of model output from Eq. (8.37) is plotted in Figure 8.23a, with the corresponding power spectrum plotted in Figure 8.23b. Model parameters for this example are $\tau = 600$, $\epsilon = -0.1$, and $D = 0.1$. The power spectrum in Figure 8.23b was computed by averaging the spectra of 1000 independent model time series. The

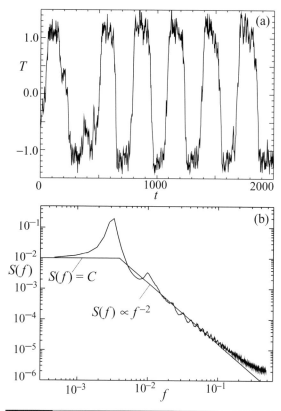

spectrum in Figure 8.23b is constant for low frequencies and proportional to f^{-2} above a threshold frequency. This is called a *Lorentzian* spectrum and is a common feature of stochastic models with a restoring or negative feedback. A Lorentzian spectral signature was documented in Late-Pleistocene paleoclimatic time-series data by Komintz and Pisias (1979), suggesting a stochastic element to the climate system limited by a negative feedback mechanism at long time scales. This observation was the basis for many stochastic climate models including Hasselman (1971), North *et al.* (1981), Nicolis and Nicolis (1982), and Pelletier (1997), among others.

The climate model we consider here is a single equation for global temperature analogous to Eq. (8.37). The Earth maintains a mean global temperature of approximately $15\,^\circ\text{C}$ through a balance between incoming solar radiation and long-wavelength outgoing radiation. Long-wavelength outgoing radiation is characterized by the Stefan–Boltzmann Law, in which outgoing radiated energy depends on the fourth power of absolute temperature. For small temperature fluctuations around an equilibrium temperature, the T^4 Boltzmann dependence may be linearized to yield the finite-difference equation (North *et al.*, 1981)

$$C \frac{\Delta T_n}{\Delta t} = -B T_n \qquad (8.38)$$

where n is an index of the time step, C is the heat capacity per unit surface area of the atmosphere-ocean-cryosphere system, B is the coefficient of temperature dependence of outgoing radiation, and T_n is the temperature difference from equilibrium. This equation characterizes a negative feedback: global warming (cooling) resulting from short-term climatic fluctuations results in more (less) outgoing radiation. C may be estimated from the specific heat of water, $4.2 \times 10^3\,\text{J/kg/}^\circ\text{C}$, and the mass of the oceans, $1.4 \times 10^{21}\,\text{kg}$, and the surface area of the Earth, $5.1 \times 10^{14}\,\text{m}^2$, to obtain $C = 1.2 \times 10^{10}\,\text{J/}^\circ\text{C}$, assuming that the heat capacity of the climate system is dominated by the oceans. The value of B is constrained from satellite measurements to be $2.1\,\text{J/s/m}^2/^\circ\text{C}$ (North, 1991). The ratio of B to C is defined to be c_1 in this model and defines a radiative-damping time scale for the climate

time series in Figure 8.23a has two stable states at $T = +1$ and $T = -1$. The system switches between these states with a period equal to $\tau/2$, *despite the absence of any periodic terms in the model equation.* In Figure 8.23 we have illustrated the negative-feedback case by choosing $\epsilon < 0$. Positive feedback produces a similar output, but with a dominant period equal to τ rather than $\tau/2$ (Tsimring and Pikovsky, 2001). Less-dominant periodicities also occur at odd harmonics of the dominant periodicity (i.e. periods of $\tau/6$, $\tau/10$, $\tau/14$, etc.) in Figure 8.23b. The "background"

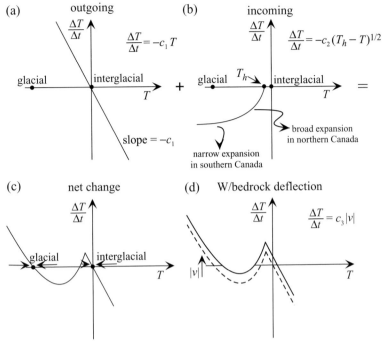

Fig 8.24 Schematic diagram of processes included in our radiation-balance model. Each graph is a sketch of the functional relationship between temperature change, $\Delta T/\Delta t$, and global temperature, T, for each process. (a) A high global temperature results in more outgoing long-wavelength radiation (and relative cooling). This effect is proportional to the temperature difference from equilibrium where the constant of proportionality, c_1, is equal to the radiation damping time, constrained to be $c_1 = 180$ yr as described in the text. (b) Lower global temperatures result in increased albedo (and relative cooling) with a function that depends on the ice sheet distribution in Figure 8.25. (c) Long-wavelength outgoing radiation and the ice-albedo feedback taken together yield a system with two fixed points, corresponding to glacial and interglacial states. (d) Bedrock subsidence and rebound modify the effects of the ice-albedo feedback by shifting the curve towards greater relative warming by an amount proportional to $|v|$. From Pelletier (2003).

system equal to 180 yr (the inverse of c_1). The simple linear relationship between temperature change and mean global temperature in Eq. (8.38) is plotted in Figure 8.24a.

The negative feedback of Eq. (8.38) is balanced by natural variability in Earth's radiation balance driven by short-term (decadal and centennial) fluctuations in atmospheric temperature. These fluctuations create random variability in long-wavelength outgoing radiation. This variability may be modeled by introducing a random noise in Eq. (8.38) to obtain (Hasselman, 1971; North et al., 1981)

$$\frac{\Delta T_n}{\Delta t} = -c_1 T_n + D \eta_n \tag{8.39}$$

where η is a white noise with standard deviation equal to D. Short-term internal temperature variations generate significant low-frequency temperature variations because the climate system *integrates* the short-term random variations through their effect on the outgoing radiation flux. The result of this process is Brownian-walk behavior for global temperature. Equation (8.39) has the behavior of a *damped* Brownian walk, however, because the equation also includes a negative feedback term.

The growth of ice sheets provides a positive feedback that dominates the negative feedback of outgoing radiation if ice sheets grow nonlinearly with global temperature. The effects of the ice–albedo feedback on global climate may be modeled by including a term in Eq. (8.38), $a(T)$, which reflects the dependence of Earth's albedo on global temperature, to obtain

$$\frac{\Delta T_n}{\Delta t} = -c_1 T_n - c_2 a(T_n) \tag{8.40}$$

We assume that the modern interglacial state is a steady state of the climate system. This provides the constraint $a(0) = 0$ and fixes the temperature of any other stable states of the system to be with respect to an origin at the modern interglacial temperature.

Stable states of Eq. (8.40) are determined by setting the right hand side equal to zero. $T = 0$, the modern interglacial, is a fixed point because we defined $a(0) = 0$. Whether or not another fixed point exists depends on how $a(T)$ increases with decreasing T. If the Earth's albedo increases greater than linearly with T, the positive ice–albedo feedback dominates the system. A dominant ice–albedo feedback leads to an ice-covered Earth if $a(T)$ increases greater than linearly for all T. However, if the ice–albedo feedback increases less than linearly for *any* T, the combination of outgoing radiation and the ice–albedo feedback will be negative or self-limiting for those values of T.

Determining the form of $a(T)$ is complicated by uncertainties in the geographic distribution of ice as a function of global temperature. Most numerical ice-sheet-climate models determine $a(T)$ by ice-sheet dynamics in their full complexity, including internal flow, basal sliding, and mass wasting. The complexity of these models results in dozens of free parameters for the motion of continental ice sheets. These processes, especially basal sliding, are poorly constrained. The distribution of sea ice as a function of global temperature is also highly uncertain. As an alternative to modeling ice-sheet dynamics, the extent of land ice through time during the last deglaciation may be inferred from the pattern of postglacial rebound (Peltier, 1994). Peltier's reconstruction provides ice extents at 1-kyr intervals from the Last Glacial Maximum (LGM) to the present. His reconstruction provides data on ice-sheet topography as well as extent, but only the data on ice-sheet extent are necessary to constrain variations in albedo. We can estimate the relationship between global albedo and temperature from the LGM to the present using Peltier's ice extents and the Vostok time series. First, a global-temperature time series can be constructed by averaging the Vostok time series in 1-kyr intervals centered on each integer value of kyr. The time-averaged

Vostok data provide the independent data to determine $a(T)$. The albedo data were obtained by multiplying each $1° \times 1°$ grid square in Peltier's data by its area and the amount of incident solar radiation it receives to obtain the relative change in albedo as a function of global temperature:

$$a(T) = \sum_{i=-180}^{180} \sum_{j=-90}^{90} a_{i,j}(T) \cos^2(j) \qquad (8.41)$$

where $a_{i,j} = 0.25$ if the grid square is ice-free and $a_{i,j} = 0.85$ if the grid square is ice-covered.

Equation (8.41) provides an estimate of the relative change in Earth's albedo due to the presence of Late-Pleistocene continental ice sheets. The albedo difference between modern and LGM conditions calculated in Eq. (8.41) equals 10% of the modern albedo assuming that ice-free areas are also continuously snow free as well. 10% is an overestimate, however, because many areas assumed to be ice-free in Eq. (8.41) have seasonal snow and ice cover. Accounting for modern seasonal snow and sea-ice cover, Oerlemans (1980a) estimated that the albedo difference between the LGM and the present was approximately 5%. We have used this value to scale the albedo curve to a maximum change of 5% between the present and the LGM. In other words, we have used Oerlemans' albedo estimate for the total change between the LGM and the present, and Peltier's reconstructions to estimate the relative change through time. The result is plotted in Figure 8.25.

The data of Figure 8.25 are well approximated by a square-root function:

$$a(T) = \begin{cases} (T_h - T)^{\frac{1}{2}} & \text{if } T < T_h \\ 0 & \text{if } T \geq T_h \end{cases} \qquad (8.42)$$

where $T_h = -1°C$ (relative to the modern temperature) is the temperature at which subpolar ice-sheet growth is initiated in Hudson Bay. The value of the coefficient of $a(T)$, c_2, is given by

$$c_2 = \frac{Q}{C} \frac{a(T_g) - a(T_h)}{(T_h - T_g)^{\frac{1}{2}}} \qquad (8.43)$$

where T_g is the temperature of the full glacial climate state. Q, the incoming solar energy, is equal to 340 J/s/m^2 (North, 1991). Equation (8.43) yields $c_2 = 5.4 \times 10^{-10} °C^{1/2}/s$. Scaled to the radiative time scale of 180 yr this is $c_2 = 3.0 °C^{1/2}$. The relationship between incoming energy and

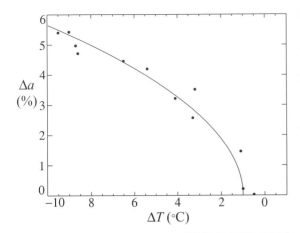

Fig 8.25 Observed relationship between albedo and global temperature based on Peltier's (1994) gravitationally self-consistent sea-level inversion and Vostok temperatures for the same time periods. The data closely approximate a square-root function between albedo and global temperature ($a(T) = c_2(T_h - T)^{\frac{1}{2}}$ with $T_h = -1$). The shape of this curve largely reflects the sensitivity of the Laurentide ice sheet coverage to global temperatures as the ice sheet retreated westward and southward from Hudson Bay during the later stages of retreat. Earlier stages of retreat were not as sensitive because ablation resulted in limited retreat due to the tapered ice sheet geometry in southern Canada. From Pelletier (2003).

global temperature is plotted in Figure 8.24b. The combination of the ice–albedo feedback and long-wavelength outgoing radiation are illustrated in Figure 8.24c. The combined effect of these processes results in two fixed points in the model system corresponding to the full glacial and interglacial states.

Bistability in the climate system is based on a balance between the negative feedback of long-wavelength outgoing radiation and the positive ice–albedo feedback. The negative feedback of outgoing radiation is dominant in an Earth without extensive subpolar ice sheets. When large continental ice sheets expand in North America and Northern Europe, however, temperature drops result in an albedo increase large enough to dominate the negative feedback of outgoing radiation. The net positive feedback driven by ice-sheet expansion sends the climate system into a self-enhancing feedback of colder temperatures and larger ice sheets. This feedback, however,

continues only as long as ice sheets expand non-linearly with global temperature. The fact that ice sheets grow preferentially on the continents eventually weakens the positive feedback as the North American ice sheets expand into a tapering continent. The result is a second stable state characterized by an equilibrium between the ice--albedo feedback and the negative feedback of outgoing radiation. The square-root form may be approximately understood by considering the constraints on the North American ice sheets (including the Laurentide and former ice sheets) as they expand from their initial location in Hudson Bay. When the North American ice sheet is expanding southward and westward across Canada during its initial growth, the Earth's albedo is sensitive to small changes in global temperature. The result is a strong positive feedback between cooler temperatures and larger ice sheets. The ice sheet eventually reaches its maximum westward extent in northern Canada as the global temperature further drops, however. If we assume that ice sheets expand preferentially on land, the ice--albedo feedback weakens at this stage because a further decrease in global temperature results in smaller changes in ice-sheet extent because expansion is limited to a southerly direction and a tapering North America. The combined effect of long-wavelength outgoing radiation and the ice–albedo feedback is positive and of large magnitude for small ice sheets, but becomes progressively weaker as ice sheets grow in extent and are eventually geographically limited in where they can expand. As global temperature continues to drop, a balance is eventually achieved between the long-wavelength outgoing radiation and the increase in Earth's albedo. This is the full-glacial climate.

Figure 8.24c illustrates the radiation balance of the Earth as a function of global temperature for the combination of long-wavelength outgoing radiation and the ice–albedo feedback. The fixed points are obtained by setting Eq. (8.43) equal to zero to obtain

$$T = c_2(-1 - T)^{\frac{1}{2}} \qquad (8.44)$$

with $c_2 = 3.0\,°C^{1/2}$. This quadratic equation has two roots at $T = -1.1$ and $T = -8.0\,°C$. The root at $T = -8.0\,°C$ is a stable fixed point; cooling below this point leads to a net warming and a return to the fixed point. The root at $T = -1.1\,^\circ$C is an unstable fixed point. Equation (8.44) predicts two stable states of the Late-Pleistocene climate system: an interglacial climate with $T = 0\,^\circ$C and a full-glacial climate with $T = -8\,^\circ$C. This analysis is independent of the system dynamics and depends only on equilibria of Eq. (8.40).

The flexural isostatic deflection of the lithosphere beneath an ice sheet may have at least two effects on the geometry of the ice sheet. First, lithospheric subsidence may lower accumulation rates by decreasing the ice-sheet elevation. This "load-accumulation" feedback is perhaps the most well-studied effect of lithospheric deflection on climate (North, 1991). The strength of this feedback, however, is limited by another effect of lithospheric subsidence: changes in lateral ice-flow velocities. An ice sheet may be considered to be a conveyor belt in which ice flows from zones of accumulation to zones of ablation. In steady-state, this flow occurs without a change in surface topography or ice thickness. In fact, an increase in accumulation may be completely accommodated by an increase in lateral ice-flow velocities such that no change in ice volume occurs. This possibility highlights the difficulty of modeling ice-sheet response to the small changes in insolation experienced during the Pleistocene. In particular, a small change in basal-sliding parameters may greatly change the sensitivity of ice sheets to external forcing. Basal-sliding parameters are poorly known, introducing great uncertainty into forward models of climate based on ice-sheet dynamics.

Although the absolute values of basal-sliding parameters are poorly known, the *relative* effects of lithospheric deflection on ice-sheet sliding may be determined more precisely. Lithospheric subsidence perturbs a steady-state ice sheet by increasing the basal slope of the ice sheet, reducing the gravitational force available to drive lateral ice flow. This flow reduction thickens the ice even under conditions of uniform mass balance. Conversely, lithospheric rebound enhances lateral ice-flow velocities and results in ice-sheet thinning. The effects of lithospheric subsidence and rebound may be illustrated by considering the steady-state geometry of an ice sheet slid-

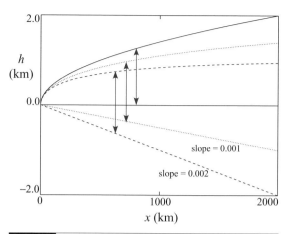

Fig 8.26 Steady-state ice-surface profiles of a threshold-sliding ice sheet with different scenarios of bedrock topography, calculated from Eqs. (8.45) and (8.46). The elevation of the ice surface and bedrock topography are shown for flat bedrock (solid line), and uniformly sloping bedrock (dipping away from the margin) with a slope of $\alpha = 0.002$ (small-dashed line), and $\alpha = 0.004$ (long-dashed line). As the bedrock depression deepens, the ice surface elevation decreases, but not as much as the bedrock. The result is a counterflow within the ice sheet that causes thickening or thinning of the ice sheet depending on whether the lithosphere is subsiding or rebounding. From Pelletier (2003).

ing on its bed. On a flat bed, the profile of a threshold-sliding ice sheet is given by

$$h = \sqrt{2h_o(L - x)} \qquad (8.45)$$

and the thickness for an ice sheet on a sloping base of gradient α is

$$x - L = \frac{h}{\alpha} + \frac{h_o}{\alpha^2} \ln|\alpha h - h_o| \qquad (8.46)$$

where L is the length of the ice sheet and h_o is the maximum thickness of the ice sheet on a flat surface. h_o is a function of the sliding shear stress if we assume a sliding rather than a plastic ice sheet. To illustrate the effects of lithospheric deflection on the ice-sheet thickness, we have plotted ice-sheet profiles from Eqs. (8.45) and (8.46) for a flat surface and for surfaces with $\alpha = 0.002$ and 0.004 in Figure 8.26. These profiles illustrate that as the lithosphere subsides, an ice sheet will thicken because lateral ice flow is inhibited by the increase in basal slope. Conversely, lithospheric rebound leads to ice-sheet thinning.

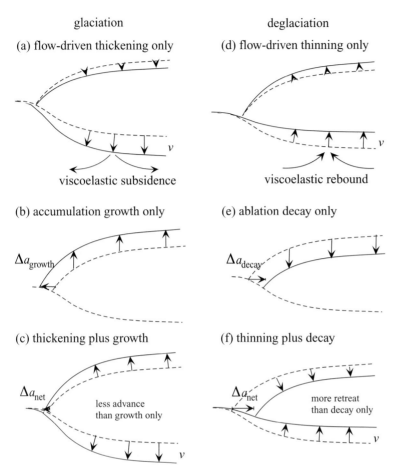

glaciation

(a) flow-driven thickening only

viscoelastic subsidence

(b) accumulation growth only

Δa_{growth}

(c) thickening plus growth

Δa_{net}

less advance than growth only

v

deglaciation

(d) flow-driven thinning only

viscoelastic rebound

(e) ablation decay only

Δa_{decay}

(f) thinning plus decay

Δa_{net}

more retreat than decay only

v

Fig 8.27 Schematic illustration of the coupling between ice sheet adjustment by lithospheric deflection and ice-sheet advance or retreat. The initial bedrock and ice-sheet geometry is shown (dashed line) along with the final geometry (solid line) in each case. During glaciation: (a) Bedrock subsidence considered alone leads to ice-sheet thickening. (b) Accumulation growth considered alone leads to uniform thickening (assuming uniform accumulation rate along the ice sheet) with a corresponding advance. (c) Both processes taken together act destructively because some accumulation growth is canceled out by subsidence, reducing the effective albedo change. During deglaciation: (d) Bedrock rebound alone leads to ice-sheet thinning. (e) Ablation alone leads to uniform thinning of the ice sheet with a corresponding retreat. (f) Together these processes act constructively to modify albedo because ablation results in a greater retreat for a thinner ice sheet. Modified from Pelletier (2003).

Ice-sheet thickening and thinning, in turn, modify the effects of changes in global temperature to retard glaciations and accelerate deglaciations. This coupling may be called the "load-advance" feedback and is illustrated in Figure 8.27. Figures 8.27a–8.27c illustrate how a decrease in lateral ice-flow velocities results in a reduced rate of ice-sheet advance. Conversely, Figures 8.27d–8.27f illustrate how ice-sheet retreat is enhanced by an increase in lateral flow driven by lithospheric rebound. In both cases, lithospheric adjustment acts to increase the albedo relative to cases with a fixed lithosphere. Figure 8.27a illustrates the effect of subsidence on ice-sheet geometry: a reduction in lateral ice-flow velocities results in ice-sheet thickening as shown in Figure 8.26. In Figure 8.27b the effects of accumulation growth are shown in the absence of subsidence. Accumulation is assumed to occur uniformly along the ice-sheet profile. The effects of subsidence-induced thickening and accumulation growth taken together are illustrated in Figure 8.27c. The combination results in a smaller ice-sheet advance relative to the case with accumulation alone. In cases in which accumulation is not uniform (i.e. accumulation increases with distance from the margin) the limiting effect of subsidence on ice-sheet advance is even greater. Figures 8.27d–8.27f illustrate the load-advance feedback during deglaciation. Figure 8.27d illustrates lithospheric rebound and resultant ice-sheet thinning. Figure 8.27e illustrates the ice-sheet retreat resulting from ablation. Finally, the combination of thinning and ablation are shown in Figure 8.27f. This figure illustrates how ice-sheet thinning enhances retreat.

The effects of lithospheric subsidence and rebound on the motion of an ice margin are dependent on the vertical *velocity* of the lithosphere, $|v|$, not on its elevation. This is a key difference

Table 8.1	Values of the parameters in the reference model.				
	c_1 (s^{-1})	c_2 (°C$^{\frac{1}{2}}$/s)	c_3	c_4 (°C/s)	τ (kyr)
original	1.8×10^{-10}	5.4×10^{-10}	N/A	N/A	10
scaled to c_1	1	3.0	25.0	0.9	56

between the load-accumulation feedback and the load-advance feedback. In the load-advance feedback, the retarding and amplifying effects of Figures 8.27c and 8.27f are observed only when the lithosphere is actively subsiding or rebounding. With the load-accumulation feedback, in contrast, the reduction in accumulation rate that accompanies subsidence is a function of the bedrock *elevation*, not its velocity. This is an important distinction, because asymmetric glacial-interglacial transitions are only reproduced in a model with velocity-dependent feedbacks.

To model the load-advance feedback, we include a term in the model equation (Eq. (8.40)) proportional to the magnitude of the velocity, $|v|$, given by

$$v_n = \frac{\Delta B_n}{\Delta t} \tag{8.47}$$

where

$$B_n = \sum_{n'=0}^{n'=n} \frac{T_h - T}{\tau} e^{-(n-n')/\tau} \tag{8.48}$$

and τ is the Maxwell-relaxation time of the viscoelastic crust (Watts, 2001) and B is the bedrock depth.

The model equation, including long-wavelength outgoing radiation, the ice–albedo feedback, and lithospheric deflection, is

$$\frac{\Delta T_n}{\Delta t} = -c_1 T_n + c_2 (T_h - T_n)^{\frac{1}{2}} + c_3 |v|_n + c_4 \eta_n \tag{8.49}$$

To simplify, we rescale Eq. (8.49) to make $c_1 = 1$:

$$\frac{\Delta T_n}{\Delta t} = -T_n + c_2 (T_h - T_n)^{\frac{1}{2}} + c_3 |v|_n + c_4 \eta_n \tag{8.50}$$

Model time steps are equal to the radiative-damping time of 180 yr in Eq. (8.50). Equation (8.50) is analogous to the coherence resonance

Eq. (8.37). The first two terms on the right side of (8.50) represent the bistability in the system. The third and fourth terms represent the delay feedback and random variability, respectively. The values of c_1 through c_4 and τ used in the reference model run of Eq. (8.50) are given in Table 8.1. The values of c_1 and c_2 have been determined from empirical data in the preceding sections. The value of c_4 may be constrained using the Vostok time series. To do this, we first averaged the Vostok time series in equal intervals of the model scaling time of 180 years. Second, we computed the standard deviation of ΔT_n from the Vostok data. The result is $c_4 = 0.9$ °C and quantifies the magnitude of short-term global temperature changes. This value is consistent with other available estimates of natural century-scale global-temperature variations. The Maxwell relaxation time of the mantle was chosen to be 10 kyr, a value consistent with postglacial rebound. The viscous response time of the mantle actually varies with the size of the load so a constant value will only be an approximation. The remaining model coefficient, c_3, is the only free parameter in the model. This coefficient characterizes the strength of the load-advance feedback, which depends on the geometries and sliding rates of the ice sheets. We have not attempted to constrain this parameter with empirical data because of the complexity of the processes involved. Instead, the model was run with a range of parameters to qualitatively identify the system behavior. Coherence resonance occurs over a broad range of values for c_3. $c_3 = 25$ was chosen as the reference parameter value for the numerical model experiments because it leads to both coherence resonance and behavior similar to the Vostok time series, including asymmetric transitions.

The behavior of Eq. (8.50) is illustrated in Figure 8.28b with the reference values of Table 8.1.

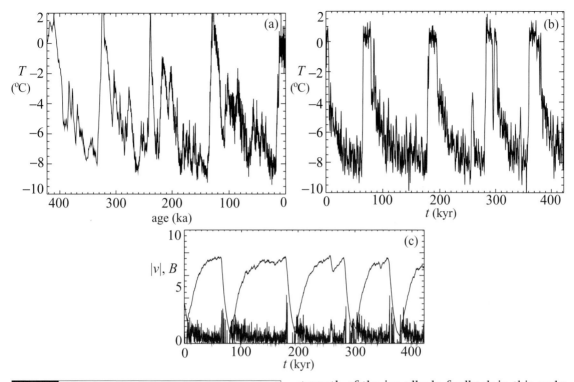

Fig 8.28 (a) Time series of atmospheric temperature variations inferred from deuterium concentrations at Vostok, Antarctica (Petit, 1999). (b) Time series of model temperature variations with $c_2 = 3.0$, $c_3 = 25.0$, $c_4 = 0.9$, and $\tau = 10$ kyr. (c) Time series of bedrock elevation and velocity in the model example of (b). From Pelletier (2003).

Variations in bedrock depth B and the magnitude of bedrock velocity $|v|$ are shown in Figure 8.28c for this run. The Vostok time series is plotted in Figure 8.28a for comparison. Note that the Vostok data is plotted here with time, not age, increasing toward the right in order to simplify comparison with the model time series.

The model time series reproduces behavior very similar to the Vostok data, including the same temperature range (with full glacial climates 8 °C cooler than interglacials), a 100-kyr periodicity, and asymmetric glacial–interglacial transitions. More specifically, glaciations in both the model and observed data are relatively rapid at first but slow prior to a very rapid deglaciation. This pattern may be understood in terms of the strength of radiation feedbacks and the variations in bedrock velocity through time. Glaciations begin rapidly at first, reflecting the strength of the ice–albedo feedback in this early stage. The glaciation slows as it proceeds, reflecting a decrease in the ice–albedo feedback that limits ice advance. Deglaciation is triggered by a warming trend that is long enough to initiate lithospheric rebound. The length of the warming trend is important because lithospheric motion lags behind ice loading. Therefore, a *sustained* warming trend is required to initiate deglaciation through the combined effects of ice sheet melting and a lithospheric rebound. During deglaciation, the rebounding lithosphere acts to thin the ice sheet and amplify the ice–albedo feedback in an additional feedback mechanism. In this additional mechanism, faster ice-sheet melting results in faster lithospheric rebound and an enhanced ice–albedo feedback. The ice–albedo feedback, in turn, drives rapid ice-sheet melting in a positive feedback. It should be noted that even though the relaxation time of the mantle is 10 kyr, the dominant oscillation is close to 100 kyr.

The similarity between the model behavior and the Vostok time series may be documented more thoroughly by comparing the histograms and power spectra of both data sets. The power

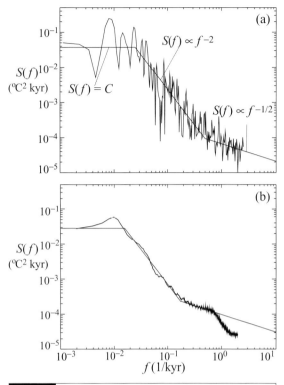

Fig 8.29 (a) Lomb periodogram, $S(f)$, of Vostok time series data plotted as a function of frequency f. Note logarithmic scales on both axes. Spectrum has a dominant frequency corresponding to a period near 100 kyr, a constant background power spectrum between time scales of 1 Ma and 40 kyr, an f^{-2} spectrum between 40 kyr and 2 kyr, and an $f^{-1/2}$ spectrum at higher frequencies. (b) Average power spectrum of 1000 samples of the model time series, with same scale as (a). The spectrum has a dominant periodicity at 120 kyr superimposed on a background spectrum similar to (a) (except at the highest frequencies). In addition, small-amplitude periodicities appear above the background at odd harmonics of the dominant periodicity. From Pelletier (2003).

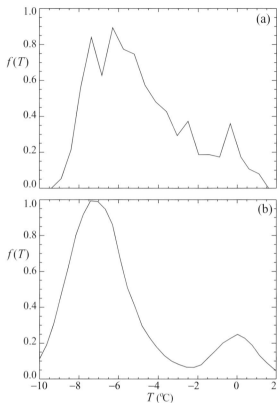

Fig 8.30 Histograms of (a) Vostok time series data, uniformly sampled at 200 yr intervals, and (b) model time series. These curves reflect the fraction of time the climate is in the interglacial state, the glacial state, or fluctuating in between. The time spent near the glacial state exceeds the time spent near the interglacial state by a ratio of about 5:1. From Pelletier (2003).

spectra of the Vostok and model time series are plotted in Figures 8.29a and 8.29b, respectively. We estimated the power spectrum of the Vostok time series using the Lomb periodogram (Press *et al.*, 1992). This technique is commonly used for nonuniformly sampled time-series data. The power spectrum of the model time series was estimated by averaging the spectra of 1000 independent samples to obtain a precise spectrum. Both spectra exhibit a dominant 100-kyr periodicity

superimposed on a Lorenztian spectrum. Notably, both spectra also exhibit a high-frequency transition to $f^{-1/2}$ behavior. At the highest frequencies, power in the model spectrum drops as an artifact of the simulation method. Specifically, the viscoelastic load was averaged at time scales less than 1 kyr to speed up the code for long runs.

Histograms of the Vostok and model data are plotted in Figures 8.30a and 8.30b, respectively. Histograms of 1000 model samples have been averaged to obtain Figure 8.30b. Both histograms are bimodal, reflecting the stable states of the system. The asymmetry of the model and observed data are reflected in the larger peak near the full glacial climate.

Exercises

8.1 Using the code in Appendix 7, generate a log-normal fractional Gaussian noise with $\beta = 1/2$, a mean value of 1 acre-ft/s, and a coefficient of variation of 1.0. Use this series as a model for the discharge into a reservoir. Using Excel, model the volume of the reservoir as a function of time, assuming that each year the reservoir loses 1 acre-ft/s to evaporation. In a simulation of 1000 yr, determine the duration of the longest drought (defined as consecutive years below 50% of normal volume).

8.2 Develop a code that implements Masek and Turcotte's DLA model for drainage network evolution. However, instead of growing the network each time a random walker lands on a site whose neighbor is part of the network, introduce a threshold condition that requires that N walkers visit a site before the site becomes part of the network. Qualitatively compare the drainage networks produced with $N = 10$ with those produced with $N = 1$.

8.3 The spreading of contaminants in underground aquifers can be modeled as a diffusion process. Such a model, however, does not incorporate the effects of random spatial variations in porosity within the aquifer. As an alternative to the diffusion model, construct a 2D random walk model of contaminant transport in aquifers. Define the central grid point to be the contaminant source. Simulate the variation in concentration from the source using random walkers. Assuming that the contaminant migrates in a random direction 10 m each year, what is the most likely time following the leak that the contaminant will first arrive a distance of 10 km from the source? Consider the worst possible scenario. What is the minimum time required for the contaminant to travel 10 km from the source? Imagine that the aquifer has a net transport direction to the south. Model this effect by making the probability that the walker will migrate south twice as large as the probability that it will migrate north. What is the relative contamination risk north and south of the source, expressed as the ratio of the most likely arrival times 10 km north and south of the contaminant source?

Appendix 1

Codes for solving the diffusion equation

A1.1 | Preliminaries

The appendices of this book provide sample code for implementing many of the methods we have described. The codes are written in the standard C language and make extensive use of Numerical Recipes (Press *et al.*, 1992) routines. In addition to Numerical Recipes, the codes make use of the standard C libraries `malloc` (memory allocation, useful for constructing vectors and matrices), `math` (standard math functions), `stdio` (standard input and output of data), and `stdlib` (miscellaneous standard functions). In order to inform the C compiler that we will be using these functions, the first few lines of every program that uses code from this book should include:

```
#include<malloc.h>
#include<math.h>
#include<stdio.h>
#include<stdlib.h>
#include"nr.h"
#include"nrutil.h"
```

Most C compilers come standard with the first four libraries. The final two libraries, `nr.h` and `nrutil.h` (Numerical Recipes and its utilities), should be copied into the local directory where the code will be compiled or appropriately linked from another location.

The codes in this book have been successfully compiled using the UNIX cc compiler and the Microsoft Visual C++ compiler. Every effort has been made to ensure that they are error free, but no guarantees are made. Moreover, these codes will generally need to be modified to work with other data sets, boundary conditions, etc. In addition, codes sometimes need to be modified on different systems and as systems change over time. As such, code used from this book should be thoroughly tested by the user for accuracy and internal consistency when applied to specific problems.

The codes in this book use `float` data types, which provide accuracy to six decimal places for individual operations. For greater accuracy, data structures of type `double` should be used. Numerical Recipes includes a version of each of their routines for double precision. For example, if a user is implementing a code from the book that calls the `qsimp` function and he or she requires optimal accuracy, all `float` variables should be changed to type `double` and the routine `dqsimp` should be called instead of `qsimp`.

Output from these codes is usually sent to an ASCII text file in two-line format: x h <return> (for 1D models) and as a series of rows and columns for 2D models (with one grid point on each line).

In 2D cases, all grid output proceeds as rows first, columns second, beginning with the upper left corner of the grid, moving from left to right and top to bottom. Many standard commercial and freeware programs are available for visualizing 1D and 2D output. For 1D output, the results can be imported from an ASCII text file into Microsoft Excel (http://www.microsoft.com), for example. For visualizing 2D grids, the RiverTools software package (http://www.rivix.com) provides a good commercial option while 3DEM provides a good freeware option (http://www.visualizationsoftware.com). Both RiverTools and 3DEM can import and export raster grids as ASCII text files as well as a variety of standard Geographic Information System (GIS) file formats. Routines that create graphical output in the PostScript language can also be written directly in the C language. Gershenfeld (1999) provides a number of very useful visualization routines written in C.

The majority of methods in this book require that we refer to neighboring grid points when performing grid operations. Many of the codes use a simple labeling scheme presented below. In this scheme, the vectors iup[i] and idown[i] are defined to be i+1 and i-1, respectively, for all values of i except on the boundaries of the grid where i+1 and i-1 are not defined. The values of iup[i] and idown[i] at the boundaries depend on the type of boundary conditions. For periodic boundary conditions, the value of iup[lattice_size_x] should be 1 and the value of idown[1] should be lattice_size_x. For fixed boundary conditions, the value of iup[lattice_size_x] should be lattice_size_x and the value of idown[1] should be 1. This labeling scheme has two advantages. First, it enables the boundary conditions to be switched from periodic to fixed by changing just a couple of lines of code. Second, it eliminates the need to include conditional statements. If, for example, we used i+1 and i-1 to refer to grid neighbors, we would need to include conditional statements such as

```
if ((i>1)&&(i<lattice_size_x)&&(j>1)&&(j<lattice_size_y))
 h[i][j]+=timestep*D/(delta*delta)*
  (hold[i+1][j]+hold[i-1][j]+hold[i][j+1]+hold[i][j-1]-4*hold[i][j]);
```

throughout the code in order to avoid a segmentation fault on the second line.

The following code is a standard routine for initializing grid-neighbor labels for a 2D grid with fixed boundary conditions. Many codes in this book use this function, but we will list it only once. Some of the modules that appear in early appendices are used in later appendices. In subsequent usage of setupgridneighbors,

```
int lattice_size_x,lattice_size_y,*iup,*idown,*jup,*jdown;

void setupgridneighbors()
{    int i,j;

     idown=ivector(1,lattice_size_x);
     iup=ivector(1,lattice_size_x);
     jup=ivector(1,lattice_size_y);
     jdown=ivector(1,lattice_size_y);
     for (i=1;i<=lattice_size_x;i++)
      {idown[i]=i-1;
       iup[i]=i+1;}
     idown[1]=1;
     iup[lattice_size_x]=lattice_size_x;
     for (j=1;j<=lattice_size_y;j++)
      {jdown[j]=j-1;
```

```
      jup[j]=j+1;}
    jdown[1]=1;
    jup[lattice_size_y]=lattice_size_y;
}
```

If periodic boundary conditions are needed, setupgridneighbors can be modified as follows:

```
int lattice_size_x,lattice_size_y,*iup,*idown,*jup,*jdown;

void setupgridneighborsperiodic()
{    int i,j;

    idown=ivector(1,lattice_size_x);
    iup=ivector(1,lattice_size_x);
    jup=ivector(1,lattice_size_y);
    jdown=ivector(1,lattice_size_y);
    for (i=1;i<=lattice_size_x;i++)
      {idown[i]=i-1;
       iup[i]=i+1;}
    idown[1]=lattice_size_x;
    iup[lattice_size_x]=1;
    for (j=1;j<=lattice_size_y;j++)
      {jdown[j]=j-1;
       jup[j]=j+1;}
    jdown[1]=lattice_size_y;
    jup[lattice_size_y]=1;
}
```

Another common trick we will use involves time stepping. In the case of the linear diffusion equation with constant diffusivity, the stability criterion is $\Delta t < (\Delta x)^2/(2D)$. In nonlinear and/or coupled equations in which the diffusivity changes through time as the solution evolves, the stability criterion is sometimes difficult to determine. Stability can be enforced by varying the time step to limit the maximum change in the variable or variables that are being modeled. This approach is crude but it works. The downside to this approach is that it requires some trial and error in order to determine how big a step is allowed before the solution becomes unstable. In the example code below, maxchange is set equal to the largest change allowed in any value of topo[i][j] during a single time step (determined by trial and error). If the largest change exceeds that maximum value, the values of topo revert to the values at the last time step topoold and the time step is reduced by a factor of 2. If the largest change is smaller than one tenth of the maximum value then the time step is increased by a factor of 1.2. In this way, the time step is increased and decreased dynamically in order to maintain similar maximum changes in the function or functions we are modeling. We will use this scheme in a number of example codes in this book.

```
max=0;
for (i=1;i<lattice_size_x;i++)
 {deltah=timestep/deltax*(flux[i]-flux[iup[i]]);
  topo[i]+=deltah;
  if (fabs(deltah)>max) max=fabs(deltah);}
elapsedtime+=timestep;
```

```
if (max>maxchange)
 {elapsedtime-=timestep;
  timestep/=2.0;
  for (i=1;i<=lattice_size_x;i++)
   topo[i]=topoold[i];}
else
  {if (max<0.1*maxchange) timestep*=1.2;}
```

A1.2 | FTCS method for 2D diffusion equation

The sample code below solved the 2D diffusion equation using the FTCS method. The code accepts an input grid of type float, size 300×300, and resolution 10 m/pixel from the local directory, calculates the solution for $t =$ duration, and then outputs the result to an ASCII file in the local directory.

```
int lattice_size_x,lattice_size_y,*iup,*idown,*jup,*jdown;

main()
{    FILE *fp1,*fp2;
     float delta,**topo,**topoold,D,duration,timestep;
     int i,j,t,nsteps;

     fp1=fopen("inputdem","r");
     fp2=fopen("outputdem","w");
     lattice_size_x=300;
     lattice_size_y=300;
     delta=10.0;       /* m */
     D=1.0;            /* m^2/kyr */
     duration=100.0; /* kyr */
     setupgridneighbors();
     topo=matrix(1,lattice_size_x,1,lattice_size_y);
     topoold=matrix(1,lattice_size_x,1,lattice_size_y);
     for (j=1;j<=lattice_size_y;j++)
      for (i=1;i<=lattice_size_x;i++)
       {fscanf(fp1,"%f",&topo[i][j]);
        topoold[i][j]=topo[i][j];}
     timestep=0.5*delta*delta/(2*D);
     nsteps=(int)(duration/timestep);
     for (t=1;t<=nsteps;t++)
      {for (j=1;j<=lattice_size_y;j++)
        for (i=1;i<=lattice_size_x;i++)
          topo[i][j]+=timestep*D/(delta*delta)*(topoold[iup[i]][j]+topoold[idown[i]][j]
            +topoold[i][jup[j]]+topoold[i][jdown[j]]-4*topoold[i][j]);
       for (j=1;j<=lattice_size_y;j++)
        for (i=1;i<=lattice_size_x;i++)
          topoold[i][j]=topo[i][j];}
     for (j=1;j<=lattice_size_y;j++)
      for (i=1;i<=lattice_size_x;i++)
```

```
            fprintf(fp2,"%f\n",topo[i][j]);
        fclose(fp1);
        fclose(fp2);
}
```

As a concrete example, we consider the case of terrace diffusion described in Section 2.3.4. The following code can be used to reproduce Figure 2.23b, given the input file `rillmask`, available from the CUP website. This code also makes use of an "efficiency mask" that identifies the grid points and their neighbors that are actively undergoing erosion. This speeds up the code early in the simulation when erosion is localized close to the rills.

```
#define PI 3.141592653589793

main()
{    FILE *fp1,*fp2;
     int length_x,length_y,delta,lattice_size_x,lattice_size_y,**rills,**efficiencymask,i,j;
     float deltah,simulationlength,elapsedtime,timestep,dissectionrelief,kappa,depositslope;
     float depositaspect,northwestdatum,**topo,**topoold,**initialtopo;

     fp1=fopen("rillmask","r");
     fp2=fopen("terracediffusion2d","w");
     length_x=300;             /* m */
     length_y=300;
     delta=1.0;                /* m */
     dissectionrelief=10.0;
     kappa=1.0;                /* m^2/kyr */
     depositslope=0.06;        /* m/m */
     depositaspect=300*PI/180; /* surface drains SE; degrees converted to radians */
     northwestdatum=0.0;       /* elevations are relative to NE corner = 0 m */
     lattice_size_x=length_x/delta;
     lattice_size_y=length_y/delta;
     topo=matrix(1,lattice_size_x,1,lattice_size_y);
     topoold=matrix(1,lattice_size_x,1,lattice_size_y);
     initialtopo=matrix(1,lattice_size_x,1,lattice_size_y);
     rills=imatrix(1,lattice_size_x,1,lattice_size_y);
     efficiencymask=imatrix(1,lattice_size_x,1,lattice_size_y);
     timestep=0.5*delta*delta/(2*kappa);
     simulationlength=50;      /* kyr */
     for (j=1;j<=lattice_size_y;j++)
      for (i=1;i<=lattice_size_x;i++)
       {fscanf(fp1,"%d",&rills[i][j]);
        if (rills[i][j]>0) rills[i][j]=0; else rills[i][j]=1;
        efficiencymask[i][j]=1;
        if (i<lattice_size_x) if (rills[i+1][j]==1) efficiencymask[i][j]=0;
        if (i>1) if (rills[i-1][j]==1) efficiencymask[i][j]=0;
        if (j<lattice_size_y) if (rills[i][j+1]==1) efficiencymask[i][j]=0;
        if (j>1) if (rills[i][j-1]==1) efficiencymask[i][j]=0;
        topo[i][j]=northwestdatum-(i-1)*delta*depositslope*cos(depositaspect)+
          (j-1)*delta*depositslope*sin(depositaspect);
```

```
     if (rills[i][j]==1) topo[i][j]-=dissectionrelief;
     topoold[i][j]=topo[i][j];
     initialtopo[i][j]=topo[i][j];}
  elapsedtime=0.0;
  while (elapsedtime<=simulationlength)
   {printf("%f %f\n",elapsedtime,timestep);
    for (j=2;j<=lattice_size_y-1;j++)
     for (i=2;i<=lattice_size_x-1;i++)
      if (efficiencymask[i][j]==0)
       {if (rills[i][j]==0)
         deltah=timestep*kappa/(delta*delta)*(topoold[i+1][j]+topoold[i-1][j]+
           topoold[i][j+1]+topoold[i][j-1]-4*topoold[i][j]);
        else
         deltah=0;
        topo[i][j]+=deltah;
        if (fabs(topo[i][j]-initialtopo[i][j])>0.01*dissectionrelief)
         {efficiencymask[i+1][j]=0;
          efficiencymask[i-1][j]=0;
          efficiencymask[i][j+1]=0;
          efficiencymask[i][j-1]=0;}}
    elapsedtime+=timestep;
    i=1;
    for (j=1;j<=lattice_size_y;j++)
     topo[i][j]=topo[i+1][j]+depositslope*cos(depositaspect)*delta;
    i=lattice_size_x;
    for (j=1;j<=lattice_size_y;j++)
     topo[i][j]=topo[i-1][j]-depositslope*cos(depositaspect)*delta;
    j=1;
    for (i=1;i<=lattice_size_x;i++)
     topo[i][j]=topo[i][j+1]-depositslope*sin(depositaspect)*delta;
    j=lattice_size_y;
    for (i=1;i<=lattice_size_x;i++)
     topo[i][j]=topo[i][j-1]+depositslope*sin(depositaspect)*delta;
    for (j=1;j<=lattice_size_y;j++)
     for (i=1;i<=lattice_size_x;i++)
      topoold[i][j]=topo[i][j];}
  for (j=1;j<=lattice_size_y;j++)
   for (i=1;i<=lattice_size_x;i++)
    fprintf(fp2,"%f\n",topo[i][j]);
  fclose(fp1);
  fclose(fp2);
}
```

A1.3 | FTCS method for 1D nonlinear diffusion equation

The code presented below is used to solve for the evolution of a 1D hillslope according to the nonlinear diffusion equation (described in Section 2.3.3) subject to instantaneous base level drop.

```
main()
{   FILE *fp1;
    int *iup,*idown,printed,outputinterval,lattice_size_x,i;
    float kappa,Sc,slope,den,*flux,duration,deltax,deltah,elapsedtime,max,timestep;
    float *topo,*topoold,maxchange;

    fp1=fopen("nonlinbasedrop","w");
    lattice_size_x=102;
    deltax=1.0; /* m */
    kappa=1.0;  /* m^2/kyr */
    Sc=1.0;     /* m/m */
    iup=ivector(1,lattice_size_x);
    idown=ivector(1,lattice_size_x);
    topo=vector(1,lattice_size_x);
    topoold=vector(1,lattice_size_x);
    flux=vector(1,lattice_size_x);
    duration=1000.0; /* kyr */
    outputinterval=1000.0;
    maxchange=0.001;
    for (i=1;i<=lattice_size_x;i++)
     {topo[i]=1.0;
      if ((i==lattice_size_x)||(i==lattice_size_x-1)||(i==lattice_size_x-2)) topo[i]=0.0;
      topoold[i]=topo[i];
      flux[i]=0;
      iup[i]=i+1;idown[i]=i-1;}
    iup[lattice_size_x]=lattice_size_x;idown[1]=1;
    timestep=0.0001*deltax*deltax;
    elapsedtime=0.0;printed=0;
    while (elapsedtime<duration)
     {printed=elapsedtime/outputinterval;
      for (i=1;i<=lattice_size_x;i++)
        {slope=(topoold[idown[i]]-topoold[i])/deltax;
         den=1-fabs(slope)*fabs(slope)/(Sc*Sc);
         if (den<0.01) den=0.01;
         flux[i]=kappa*slope/den;
         if (i==1) flux[i]=0.0;
         if (i==lattice_size_x) flux[i]=flux[idown[i]];
         if (flux[i]<flux[idown[i]]) flux[i]=flux[idown[i]];}
      max=0;
      for (i=1;i<lattice_size_x;i++)
        {deltah=timestep/deltax*(flux[i]-flux[iup[i]]);
         topo[i]+=deltah;
         if (fabs(deltah)>max) max=fabs(deltah);}
      elapsedtime+=timestep;
      if (max>maxchange)
        {elapsedtime-=timestep;
         timestep/=2.0;
         for (i=1;i<lattice_size_x;i++)
          topo[i]=topoold[i];}
```

```
       else
        {if (max<0.1*maxchange) timestep*=1.2;}
      for (i=1;i<lattice_size_x;i++)
        topoold[i]=topo[i];
      if ((int)(elapsedtime/outputinterval)>printed)
        {printf("%f %f\n",elapsedtime,timestep);
        for (i=1;i<=lattice_size_x;i++)
          fprintf(fp1,"%d %f\n",i,topo[i]);
        printed=(int)(elapsedtime/outputinterval);}}
    fclose(fp1);
}
```

A1.4 | ADI method for solving the 2D diffusion equation

The following code accepts a 30 m/pixel US Geological Survey DEM covering Hanaupah Canyon from file hanaupahtopo and uses the ADI technique to solve the diffusion equation on hillslopes. In the hillslopediffusion subroutine, the diffusion equation is solved for each row and each column in an alternating fashion using the tridag routine of Numerical Recipes, which inverts a tridiagonal matrix. Hillslopes are defined to be pixels with a contributing area less than 0.1 km^2. Channel pixels act as fixed boundary conditions for the channel pixels. In this example, 10 Myr of diffusive evolution is simulated with a diffusivity value of $D = 10 \text{ m}^2/\text{kyr}$ using time steps of 1 Myr. The file hanaupahtopo can be downloaded from the CUP website. The code below makes use of two functions: fillinpitsandflats and mfdflowroute introduced in Appendix 2.

```
float D,delta,thresholdarea,*ax,*bx,*cx,*ux,*rx,*ay,*by,*cy,*uy,*ry;
float **topoold,*topovec,**topo,**flow,**flow1,**flow2,**flow3,**flow4,**flow5;
float **flow6,**flow7,**flow8;
int *topovecind,lattice_size_x,lattice_size_y,*iup,*idown,*jup,*jdown;

void hillslopediffusion()
{
    int i,j,count;
    float term1;

    count=0;
    while (count<5)
     {count++;
      for (i=1;i<=lattice_size_x;i++)
       for (j=1;j<=lattice_size_y;j++)
        topoold[i][j]=topo[i][j];
      for (i=1;i<=lattice_size_x;i++)
       {for (j=1;j<=lattice_size_y;j++)
         {term1=D*timestep/(delta*delta);
          if (flow[i][j]<thresholdarea)
           {ay[j]=-term1;
            cy[j]=-term1;
            by[j]=4*term1+1;
            ry[j]=term1*(topo[iup[i]][j]+topo[idown[i]][j])+topoold[i][j];}
```

```
              else
               {by[j]=1;
                ay[j]=0;
                cy[j]=0;
                ry[j]=topoold[i][j];}
              if (j==1)
               {by[j]=1;
                cy[j]=0;
                ry[j]=topoold[i][j];}
              if (j==lattice_size_y)
               {by[j]=1;
                ay[j]=0;
                ry[j]=topoold[i][j];}}
            tridag(ay,by,cy,ry,uy,lattice_size_y);
            for (j=1;j<=lattice_size_y;j++)
             topo[i][j]=uy[j];}
        for (i=1;i<=lattice_size_x;i++)
         for (j=1;j<=lattice_size_y;j++)
          topoold[i][j]=topo[i][j];
        for (j=1;j<=lattice_size_y;j++)
         {for (i=1;i<=lattice_size_x;i++)
            {term1=D*timestep/(delta*delta);
             if (flow[i][j]<thresholdarea)
              {ax[i]=-term1;
               cx[i]=-term1;
               bx[i]=4*term1+1;
               rx[i]=term1*(topo[i][jup[j]]+topo[i][jdown[j]])+topoold[i][j];}
             else
              {bx[i]=1;
               ax[i]=0;
               cx[i]=0;
               rx[i]=topoold[i][j];}
             if (i==1)
              {bx[i]=1;
               cx[i]=0;
               rx[i]=topoold[i][j];}
             if (i==lattice_size_x)
              {bx[i]=1;
               ax[i]=0;
               rx[i]=topoold[i][j];}}
            tridag(ax,bx,cx,rx,ux,lattice_size_x);
            for (i=1;i<=lattice_size_x;i++)
             topo[i][j]=ux[i];}}
}

main()
{    FILE *fp1,*fp2;
     float time,timestep,duration;
     int i,j,t,dum;
```

```
      fp1=fopen("hanaupahtopo","r");
      fp2=fopen("hanaupahtopodiffuse","w");
      lattice_size_x=640;
      lattice_size_y=335;
      delta=30.0;              /* m */
      D=10.0;                  /* m^2/kyr */
      thresholdarea=0.1;       /* km */
      duration=10000.0;        /* kyr */
      timestep=1000.0;         /* kyr */
      setupgridneighbors();
      /* grids needed for matrix inversion */
      ax=vector(1,lattice_size_x);
      ay=vector(1,lattice_size_y);
      bx=vector(1,lattice_size_x);
      by=vector(1,lattice_size_y);
      cx=vector(1,lattice_size_x);
      cy=vector(1,lattice_size_y);
      ux=vector(1,lattice_size_x);
      uy=vector(1,lattice_size_y);
      rx=vector(1,lattice_size_x);
      ry=vector(1,lattice_size_y);
      /* other grids */
      topo=matrix(1,lattice_size_x,1,lattice_size_y);
      topovec=vector(1,lattice_size_x*lattice_size_y);
      topovecind=ivector(1,lattice_size_x*lattice_size_y);
      topoold=matrix(1,lattice_size_x,1,lattice_size_y);
      flow=matrix(1,lattice_size_x,1,lattice_size_y);
      flow1=matrix(1,lattice_size_x,1,lattice_size_y);
      flow2=matrix(1,lattice_size_x,1,lattice_size_y);
      flow3=matrix(1,lattice_size_x,1,lattice_size_y);
      flow4=matrix(1,lattice_size_x,1,lattice_size_y);
      flow5=matrix(1,lattice_size_x,1,lattice_size_y);
      flow6=matrix(1,lattice_size_x,1,lattice_size_y);
      flow7=matrix(1,lattice_size_x,1,lattice_size_y);
      flow8=matrix(1,lattice_size_x,1,lattice_size_y);
      for (j=1;j<=lattice_size_y;j++)
       for (i=1;i<=lattice_size_x;i++)
        {flow[i][j]=(delta/1000)*(delta/1000);  /* km */
         fscanf(fp1,"%d",&dum);
topo[i][j]=dum;}
      for (j=1;j<=lattice_size_y;j++)
       for (i=1;i<=lattice_size_x;i++)
        fillinpitsandflats(i,j);
      for (j=1;j<=lattice_size_y;j++)
       for (i=1;i<=lattice_size_x;i++)
        topovec[(j-1)*lattice_size_x+i]=topo[i][j];
      indexx(lattice_size_x*lattice_size_y,topovec,topovecind);
      t=lattice_size_x*lattice_size_y+1;
      while (t>1)
```

```
        {t--;
         i=(topovecind[t])%lattice_size_x;
         if (i==0) i=lattice_size_x;
         j=(topovecind[t])/lattice_size_x+1;
         if (i==lattice_size_x) j--;
         mfdflowroute(i,j);}
       time=0;
       while (time<duration)
        {time+=timestep;
         hillslopediffusion();}
       for (j=1;j<=lattice_size_y;j++)
        for (i=1;i<=lattice_size_x;i++)
         fprintf(fp2,"%f\n",topo[i][j]);
       fclose(fp1);
       fclose(fp2);
}
```

A1.5 | Numerical integration of Fourier–Bessel terms

The following code accepts a list of the first 300 zeros of the Bessel function $J_0(r)$ from the local directory and performs numerical integration to calculate the value of the terms in Eq. (2.57) for the case of a volcanic cone with $r_f = 200$ m, $r_r = 500$ m, and $r_c = 1500$ m.

```
int nsum;
float *zeros;

float func(x)
float x;
{
        return x*x*bessj0(zeros[nsum]*x);
}


main()
{    FILE *fp0,*fp1;
     double a,zero=0.0,s1,s2,s3,rf,rr,rc;

     fp0=fopen("besselzeros","r");
     fp1=fopen("besselzerosintegral","w");
     zeros=vector(1,300);
     a=3000.0;
     rf=200.0/a;
     rr=500.0/a;
     rc=1500.0/a;
     for (nsum=1;nsum<=300;nsum++)
      {fscanf(fp0,"%f",&zeros[nsum]);
       s1=qsimp(func,zero,rf);
       s2=qsimp(func,zero,rr);
       s3=qsimp(func,zero,rc);
```

```
    fprintf(fp1,"%f %12.10f %12.10f %12.10f\n",zeros[nsum],s1,s2,s3);}
    fclose(fp1);
    fclose(fp2);
}
```

The following code accepts output from the code above, computes the results of the Bessel series in Eq. (2.57), and prints the results to an output file.

```
main()
{
    FILE *fp1,*fp2;
    float rf,rr,rc,*f,kappaT,deltax,a,*prefact,*topo,*J0rr,*J0rc,*J1rr,*J1rc;
    float *J0rf,*J1rf,*Intrf,*J0r0,*J1r0,*Intrr,*Intrc;
    int lattice_size_x,i,nsum;

    fp1=fopen("besselzerosintegral","r");
    fp2=fopen("volcanicconeoutput","w");
    kappaT=0.0001;  /* dimensionless  scaled to a^2 */
    lattice_size_x=300;
    deltax=10.0;
    a=lattice_size_x*deltax;
    rf=200.0/a;
    rr=500.0/a;
    rc=1500.0/a;
    topo=vector(1,lattice_size_x);
    f=vector(1,lattice_size_x);
    zeros=vector(1,300);
    J0rc=vector(1,300);
    J1rc=vector(1,300);
    Intrc=vector(1,300);
    J0rr=vector(1,300);
    J1rr=vector(1,300);
    Intrr=vector(1,300);
    J0rf=vector(1,300);
    J1rf=vector(1,300);
    Intrf=vector(1,300);
    prefact=vector(1,300);
    for (nsum=1;nsum<=300;nsum++)
     {fscanf(fp1,"%f %f %f %f",&zeros[nsum],&Intrf[nsum],&Intrr[nsum],&Intrc[nsum]);
      J0rc[nsum]=bessj0(rc*zeros[nsum]);
      J1rc[nsum]=bessj1(rc*zeros[nsum]);
      J0rr[nsum]=bessj0(rr*zeros[nsum]);
      J1rr[nsum]=bessj1(rr*zeros[nsum]);
      J0rf[nsum]=bessj0(rf*zeros[nsum]);
      J1rf[nsum]=bessj1(rf*zeros[nsum]);}
    for (i=1;i<=lattice_size_x;i++)
     {if (i*deltax/a<rf)
       f[i]=1+(rf-rr)/(rc-rr);
      else if (i*deltax/a<rr)
       f[i]=1+(i*deltax/a-rr)/(rc-rr);
```

```
         else if (i*deltax/a<rc)
          f[i]=1+(rr-i*deltax/a)/(rc-rr);
         else f[i]=0;
         /* print initial condition first */
         fprintf(fp2,"%f %12.10f\n",i*deltax,f[i]);}
    for (i=1;i<=lattice_size_x;i++)
      topo[i]=0;
    for (nsum=1;nsum<=300;nsum++)
      prefact[nsum]=2.0/zeros[nsum]*exp(-kappaT*zeros[nsum]*zeros[nsum])/
       (bessj1(zeros[nsum])*bessj1(zeros[nsum]));
    for (i=1;i<=lattice_size_x;i++)
      {for (nsum=1;nsum<=300;nsum++)
        topo[i]+=prefact[nsum]*bessj0(zeros[nsum]*i*deltax/a)*(rf*rf/(rc-rr)*
        J1rf[nsum]-2*rr*rr/(rc-rr)*J1rr[nsum]+(1+rr/(rc-rr))*
        rc*J1rc[nsum]+2*zeros[nsum]/(rc-rr)*Intrr[nsum]-zeros[nsum]/
        (rc-rr)*(Intrf[nsum]+Intrc[nsum]));
       fprintf(fp2,"%f %12.10f\n",i*deltax,topo[i]);}
    fclose(fp1);
    fclose(fp2);
}
```

Appendix 2

Codes for flow routing

A2.1 | Filling in pits and flats in a DEM

The following code is used to eliminate pits and flats from a DEM. The output from this type of processing is sometimes called a hydrologically corrected DEM. The code is currently constructed to accept a DEM from the local directory named inputdem that has 300 columns and 300 rows. In this routine, all of the grid points are tested to determine whether they have a downslope neighbor. If they do not have a downslope neighbor then they must be a flat or the bottom of a pit. In such cases, the local elevation value is incremented by a small value fillincrement and the neighbors are searched, recursively, for additional pits and flats. Any additional pits and flats that are found also have their elevation values incremented. This procedure stops when all of the pixels in the pit or flat have a downslope neighbor (i.e. the areas drains). For very large pits or flats, it is possible for this routine to run out of memory by calling itself too many times. In such cases, the value of fillincrement can be raised, and/or a counter can be set up fillinpitsandflats that limits the number of times that the routine can call itself before returning to the main program.

```
#define fillincrement 0.01

float **topo,
int lattice_size_x,lattice_size_y,*iup,*idown,*jup,*jdown;

void fillinpitsandflats(i,j)
int i,j;
{   float min;

    min=topo[i][j];
    if (topo[iup[i]][j]<min) min=topo[iup[i]][j];
    if (topo[idown[i]][j]<min) min=topo[idown[i]][j];
    if (topo[i][jup[j]]<min) min=topo[i][jup[j]];
    if (topo[i][jdown[j]]<min) min=topo[i][jdown[j]];
    if (topo[iup[i]][jup[j]]<min) min=topo[iup[i]][jup[j]];
    if (topo[idown[i]][jup[j]]<min) min=topo[idown[i]][jup[j]];
    if (topo[idown[i]][jdown[j]]<min) min=topo[idown[i]][jdown[j]];
    if (topo[iup[i]][jdown[j]]<min) min=topo[iup[i]][jdown[j]];
    if ((topo[i][j]<=min)&&(i>1)&&(j>1)&&(i<lattice_size_x)&&(j<lattice_size_y))
```

```
        {topo[i][j]=min+fillincrement;
         fillinpitsandflats(i,j);
         fillinpitsandflats(iup[i],j);
         fillinpitsandflats(idown[i],j);
         fillinpitsandflats(i,jup[j]);
         fillinpitsandflats(i,jdown[j]);
         fillinpitsandflats(iup[i],jup[j]);
         fillinpitsandflats(idown[i],jup[j]);
         fillinpitsandflats(idown[i],jdown[j]);
         fillinpitsandflats(iup[i],jdown[j]);}
}

main ()
{    FILE *fp1,*fp2;
     int i,j;

     fp1=fopen("inputdem","r");
     fp2=fopen("filleddem","w");
     setupgridneighbors();
     lattice_size_x=300;
     lattice_size_y=300;
     topo=matrix(1,lattice_size_x,1,lattice_size_y);
     for (j=1;j<=lattice_size_y;j++)
      for (i=1;i<=lattice_size_x;i++)
        fscanf(fp1,"%f ",&topo[i][j]);
     for (j=1;j<=lattice_size_y;j++)
      for (i=1;i<=lattice_size_x;i++)
        fillinpitsandflats(i,j);
     for (j=1;j<=lattice_size_y;j++)
      for (i=1;i<=lattice_size_x;i++)
        fprintf(fp2,"%f\n",topo[i][j]);
     fclose(fp1);
     fclose(fp2);
}
```

A2.2 | MFD flow routing method

The following code calculates the contributing area of a 30 m/pixel DEM using MFD flow routing assuming a local input file named inputdem.

```
#define oneoversqrt2 0.707106781187

float **topo,**flow,**flow1,**flow2,**flow3,**flow4,**flow5,**flow6,**flow7,**flow8;
int lattice_size_x,lattice_size_y,*iup,*idown,*jup,*jdown;
```

```
void mfdflowroute(i,j)
int i,j;
{    float tot;

    tot=0;
    if (topo[i][j]>topo[iup[i]][j])
     tot+=pow(topo[i][j]-topo[iup[i]][j],1.1);
    if (topo[i][j]>topo[idown[i]][j])
     tot+=pow(topo[i][j]-topo[idown[i]][j],1.1);
    if (topo[i][j]>topo[i][jup[j]])
     tot+=pow(topo[i][j]-topo[i][jup[j]],1.1);
    if (topo[i][j]>topo[i][jdown[j]])
     tot+=pow(topo[i][j]-topo[i][jdown[j]],1.1);
    if (topo[i][j]>topo[iup[i]][jup[j]])
     tot+=pow((topo[i][j]-topo[iup[i]][jup[j]])*oneoversqrt2,1.1);
    if (topo[i][j]>topo[iup[i]][jdown[j]])
     tot+=pow((topo[i][j]-topo[iup[i]][jdown[j]])*oneoversqrt2,1.1);
    if (topo[i][j]>topo[idown[i]][jup[j]])
     tot+=pow((topo[i][j]-topo[idown[i]][jup[j]])*oneoversqrt2,1.1);
    if (topo[i][j]>topo[idown[i]][jdown[j]])
     tot+=pow((topo[i][j]-topo[idown[i]][jdown[j]])*oneoversqrt2,1.1);
    if (topo[i][j]>topo[iup[i]][j])
     flow1[i][j]=pow(topo[i][j]-topo[iup[i]][j],1.1)/tot;
      else flow1[i][j]=0;
    if (topo[i][j]>topo[idown[i]][j])
     flow2[i][j]=pow(topo[i][j]-topo[idown[i]][j],1.1)/tot;
      else flow2[i][j]=0;
    if (topo[i][j]>topo[i][jup[j]])
     flow3[i][j]=pow(topo[i][j]-topo[i][jup[j]],1.1)/tot;
      else flow3[i][j]=0;
    if (topo[i][j]>topo[i][jdown[j]])
     flow4[i][j]=pow(topo[i][j]-topo[i][jdown[j]],1.1)/tot;
      else flow4[i][j]=0;
    if (topo[i][j]>topo[iup[i]][jup[j]])
     flow5[i][j]=pow((topo[i][j]-topo[iup[i]][jup[j]])*oneoversqrt2,1.1)/tot;
      else flow5[i][j]=0;
    if (topo[i][j]>topo[iup[i]][jdown[j]])
     flow6[i][j]=pow((topo[i][j]-topo[iup[i]][jdown[j]])*oneoversqrt2,1.1)/tot;
      else flow6[i][j]=0;
    if (topo[i][j]>topo[idown[i]][jup[j]])
     flow7[i][j]=pow((topo[i][j]-topo[idown[i]][jup[j]])*oneoversqrt2,1.1)/tot;
      else flow7[i][j]=0;
    if (topo[i][j]>topo[idown[i]][jdown[j]])
     flow8[i][j]=pow((topo[i][j]-topo[idown[i]][jdown[j]])*oneoversqrt2,1.1)/tot;
      else flow8[i][j]=0;
    flow[iup[i]][j]+=flow[i][j]*flow1[i][j];
    flow[idown[i]][j]+=flow[i][j]*flow2[i][j];
    flow[i][jup[j]]+=flow[i][j]*flow3[i][j];
    flow[i][jdown[j]]+=flow[i][j]*flow4[i][j];
```

```
      flow[iup[i]][jup[j]]+=flow[i][j]*flow5[i][j];
      flow[iup[i]][jdown[j]]+=flow[i][j]*flow6[i][j];
      flow[idown[i]][jup[j]]+=flow[i][j]*flow7[i][j];
      flow[idown[i]][jdown[j]]+=flow[i][j]*flow8[i][j];
}


main ()
{    FILE *fp1,*fp2;
     int i,j,t,*topovecind;
     float *topovec,delta;

     fp1=fopen("inputdem","r");
     fp2=fopen("MFDcontribarea","w");
     setupgridneighbors();
     lattice_size_x=300;
     lattice_size_y=300;
     delta=30.0;           /* m */
     topo=matrix(1,lattice_size_x,1,lattice_size_y);
     topovec=vector(1,lattice_size_x*lattice_size_y);
     topovecind=ivector(1,lattice_size_x*lattice_size_y);
     flow=matrix(1,lattice_size_x,1,lattice_size_y);
     flow1=matrix(1,lattice_size_x,1,lattice_size_y);
     flow2=matrix(1,lattice_size_x,1,lattice_size_y);
     flow3=matrix(1,lattice_size_x,1,lattice_size_y);
     flow4=matrix(1,lattice_size_x,1,lattice_size_y);
     flow5=matrix(1,lattice_size_x,1,lattice_size_y);
     flow6=matrix(1,lattice_size_x,1,lattice_size_y);
     flow7=matrix(1,lattice_size_x,1,lattice_size_y);
     flow8=matrix(1,lattice_size_x,1,lattice_size_y);
     for (j=1;j<=lattice_size_y;j++)
      for (i=1;i<=lattice_size_x;i++)
       {fscanf(fp1,"%f ",&topo[i][j]);
         flow[i][j]=delta*delta;}
     for (j=1;j<=lattice_size_y;j++)
      for (i=1;i<=lattice_size_x;i++)
       fillinpitsandflats(i,j);
     for (j=1;j<=lattice_size_y;j++)
      for (i=1;i<=lattice_size_x;i++)
       topovec[(j-1)*lattice_size_x+i]=topo[i][j];
     indexx(lattice_size_x*lattice_size_y,topovec,topovecind);
     t=lattice_size_x*lattice_size_y+1;
     while (t>1)
      {t--;
       i=(topovecind[t])%lattice_size_x;
       if (i==0) i=lattice_size_x;
       j=(topovecind[t])/lattice_size_x+1;
       if (i==lattice_size_x) j--;
       mfdflowroute(i,j);}
     for (j=1;j<=lattice_size_y;j++)
```

```
      for (i=1;i<=lattice_size_x;i++)
        fprintf(fp2,"%f\n",flow[i][j]);
    fclose(fp1);
    fclose(fp2);
}
```

A2.3 | Successive flow routing with MFD method

The following code performs a successive flow routing analysis on an input DEM. The user is prompted
for the names of the input and output DEM files, the size of the DEM, and the prescribed runoff
depth. The routine uses Manning's equation to convert a prescribed discharge into an equivalent
depth, assuming steady flow. As such, the local flow depths depend sensitively on the local bed
slope. For some applications of this code it may be necessary to first smooth the input DEM in order
to remove topographic irregularities. In the example below it is assumed that the smallest bed slope
was 1% (any smaller value is made equal to 1%).

```
#define niterations 100
#define oneoversqrt2 0.707106781186

float deltax,**topo,**toposave,*topovec,**depth,**slope;
float **flow,**flow1,**flow2,**flow3,**flow4,**flow5,**flow6,**flow7,**flow8;
int lattice_size_x,lattice_size_y,*iup,*idown,*jup,*jdown;

void calculatealongchannelslope(i,j)
int i,j;
{    float down;

     down=0;
     if (topo[iup[i]][j]-topo[i][j]<down) down=topo[iup[i]][j]-topo[i][j];
     if (topo[idown[i]][j]-topo[i][j]<down) down=topo[idown[i]][j]-topo[i][j];
     if (topo[i][jup[j]]-topo[i][j]<down) down=topo[i][jup[j]]-topo[i][j];
     if (topo[i][jdown[j]]-topo[i][j]<down) down=topo[i][jdown[j]]-topo[i][j];
     if ((topo[iup[i]][jup[j]]-topo[i][j])*oneoversqrt2<down)
       down=(topo[iup[i]][jup[j]]-topo[i][j])*oneoversqrt2;
     if ((topo[idown[i]][jup[j]]-topo[i][j])*oneoversqrt2<down)
       down=(topo[idown[i]][jup[j]]-topo[i][j])*oneoversqrt2;
     if ((topo[iup[i]][jdown[j]]-topo[i][j])*oneoversqrt2<down)
       down=(topo[iup[i]][jdown[j]]-topo[i][j])*oneoversqrt2;
     if ((topo[idown[i]][jdown[j]]-topo[i][j])*oneoversqrt2<down)
       down=(topo[idown[i]][jdown[j]]-topo[i][j])*oneoversqrt2;
     slope[i][j]=fabs(down)/deltax;
}

main()
{    FILE *fp1,*fp2;
     float runoff,manningsn,*topovec;
     char inputfile[100],outputfile[100];
     int i,j,k,t,*topovecind;
```

```
scanf("%s",&inputfile);
scanf("%s",&outputfile);
scanf("%d",&lattice_size_x);
scanf("%d",&lattice_size_y);
scanf("%f",&deltax);  /* m */
scanf("%f",&runoff);  /* m */
scanf("%f",&manningsn);
fp1=fopen(inputfile,"r");
fp2=fopen(outputfile,"w");
setupgridneighbors();
topo=matrix(1,lattice_size_x,1,lattice_size_y);
toposave=matrix(1,lattice_size_x,1,lattice_size_y);
topovec=vector(1,lattice_size_x*lattice_size_y);
topovecind=ivector(1,lattice_size_x*lattice_size_y);
depth=matrix(1,lattice_size_x,1,lattice_size_y);
slope=matrix(1,lattice_size_x,1,lattice_size_y);
flow=matrix(1,lattice_size_x,1,lattice_size_y);
flow1=matrix(1,lattice_size_x,1,lattice_size_y);
flow2=matrix(1,lattice_size_x,1,lattice_size_y);
flow3=matrix(1,lattice_size_x,1,lattice_size_y);
flow4=matrix(1,lattice_size_x,1,lattice_size_y);
flow5=matrix(1,lattice_size_x,1,lattice_size_y);
flow6=matrix(1,lattice_size_x,1,lattice_size_y);
flow7=matrix(1,lattice_size_x,1,lattice_size_y);
flow8=matrix(1,lattice_size_x,1,lattice_size_y);
for (j=1;j<=lattice_size_y;j++)
 for (i=1;i<=lattice_size_x;i++)
  fscanf(fp1,"%f",&topo[i][j]);
for (j=1;j<=lattice_size_y;j++)
 for (i=1;i<=lattice_size_x;i++)
  fillinpitsandflats(i,j);
for (j=1;j<=lattice_size_y;j++)
 for (i=1;i<=lattice_size_x;i++)
  {toposave[i][j]=topo[i][j];
   calculatechannelslope(i,j);
   if (slope[i][j]<0.01) slope[i][j]=0.01;}
for (k=1;k<=niterations;k++)
 {printf("iteration = %d/%d\n",k,niterations);
  for (j=1;j<=lattice_size_y;j++)
   for (i=1;i<=lattice_size_x;i++)
    {flow[i][j]=pow(runoff,1.66667)*sqrt(slope[i][j])*deltax/manningsn;
     topovec[(j-1)*lattice_size_x+i]=topo[i][j];}
     indexx(lattice_size_x*lattice_size_y,topovec,topovecind);
     t=lattice_size_x*lattice_size_y+1;
     while (t>1)
      {t--;
       i=(topovecind[t])%lattice_size_x;
       if (i==0) i=lattice_size_x;
       j=(topovecind[t])/lattice_size_x+1;
```

```
            if (i==lattice_size_x) j--;
            mfdflowroute(i,j);
            depth[i][j]=pow(flow[i][j]*manningsn/
              (deltax*sqrt(slope[i][j])),0.6);}
       for (j=1;j<=lattice_size_y;j++)
        for (i=1;i<=lattice_size_x;i++)
         if ((toposave[i][j]+depth[i][j])>topo[i][j])
topo[i][j]=toposave[i][j]+depth[i][j]/niterations;
        for (j=1;j<=lattice_size_y;j++)
         for (i=1;i<=lattice_size_x;i++)
          fillinpitsandflats(i,j);}
    for (j=1;j<=lattice_size_y;j++)
     for (i=1;i<=lattice_size_x;i++)
      fprintf(fp2,"%f\n",topo[i][j]-toposave[i][j]);
}
```

Appendix 3

Codes for solving the advection equation

A3.1 | Coupled 1D bedrock-alluvial channel evolution

The following code models the coupled 1D evolution of a bedrock-alluvial channel longitudinal profile described in Section 4.5.

```
main()
{    FILE *fp1,*fp2;
     float transport,c,D,*hold,time,factor,max,totsed,deltah,*h,sum;
     int bedrock,xc,i,lattice_size,check;

     fp1=fopen("appalachiansspace1","w");
     fp2=fopen("appalachianstime1","w");
     c=0.1;
     D=1.0;
     lattice_size=128; /* delta= 1 km, so basin is 128 km in length */
     bedrock=16;       /* bedrock-alluvial transition 16 km from divide */
     h=vector(1,lattice_size);
     hold=vector(1,lattice_size);
     xc=bedrock;
     for (i=1;i<=lattice_size;i++)
      if (i>bedrock)
       {h[i]=0.0;
        hold[i]=0.0;
        /* plot the initial condition first */
        fprintf(fp1,"%f %f\n",i/(float)(lattice_size),h[i]);}
         else  {h[i]=1.0;
                hold[i]=1.0;
                fprintf(fp1,"%f %f\n",i/(float)(lattice_size),h[i]);}
     factor=1.0;
     time=0.0;
     check=10000;
     while (time<100000.0)
      {if (time>check)
        {sum=0;
```

```
    check+=10000;
    for (i=1;i<=lattice_size;i++)
     {sum+=h[i];
       fprintf(fp1,"%f %f\n",i/(float)(lattice_size),h[i]);}
     fprintf(fp2,"%f %f\n",time,sum/lattice_size);
     fprintf(fp1,"%f %f\n",xc/(float)(lattice_size),h[xc]);}
   max=0;
   totsed=0;
   deltah=c*factor*(1/(float)(lattice_size))*(h[1]-h[2]);
   if (fabs(deltah)>max) max=fabs(deltah);
   h[1]-=deltah;
   totsed=deltah;
   for (i=2;i<=lattice_size-1;i++)
    {deltah=c*factor*(i/(float)(lattice_size))*(hold[i]-hold[i+1]);
     totsed+=deltah;
     transport=D*factor*(i/(float)(lattice_size))*(hold[i]-hold[i+1]);
     if ((transport>(totsed+0.01))&&(i<=bedrock))
       {xc=i;
        if (fabs(deltah)>max) max=fabs(deltah);
        h[i]-=deltah;}
     else
       {deltah=D*factor*((i-1)/(float)(lattice_size))*(hold[i-1]-hold[i])-
          D*factor*(i/(float)(lattice_size))*(hold[i]-hold[i+1]);
        if (fabs(deltah)>max) max=fabs(deltah);
        h[i]+=deltah;}}
   if (max<0.001)
     {for (i=1;i<=lattice_size;i++)
        hold[i]=h[i];
        time+=factor;}
   else
     {h[i]=hold[i];
      factor/=3;}
   if (max<0.0001) factor*=3;}
  fclose(fp1);
  fclose(fp2);
}
```

A3.2 | Modeling the development of topographic steady state in the stream-power model

The following code implements the stream power model on a landscape with $m = 1/2$ and $n = 1$ for a uniformly uplifting square mountain block. In order to apply the stream-power model, we must create some initial landscape. In this example, the initial landscape is created by first generating a random white noise with very minor relief and then diffusing that landscape as it is uplifted by 1 m. This process creates a low-relief landscape that drains towards the edges but also has microtopographic variability that establishes initial drainage pathways in an unstructured manner (Figure A3.1a). After the initial landscape is created, the block is uplifted at a rate $U = 1$ m/kyr and the stream-power

Fig A3.1   Output of the code in Section A3.2 that implements the stream-power model in a uniformly uplifting mountain block. (a) Early in the simulation, drainages grow headward into a diffusing plateau with initially random topography. (b) After development of steady state.

model with $K_w = 1\,\mathrm{kyr}^{-1}$ is imposed using upwind differencing. On hillslopes, a simple avalanching rule is used: any slope above $30°$ sheds mass to restore the $30°$ angle of stability. The $30°$ stability criterion is imposed in routine `avalanche` as a threshold relief between adjacent pixels.

```
#define sqrt2 1.414213562373
#define oneoversqrt2 0.707106781186
#define fillincrement 0.001

float **flow1,**flow2,**flow3,**flow4,**flow5,**flow6,**flow7,**flow8,**flow;
float **topo,**topoold,**topo2,**slope,deltax,*ax,*ay,*bx,*by,*cx,*cy,*ux,*uy;
float *rx,*ry,U,K,D,**slope,timestep,*topovec,thresh;
int *topovecind,lattice_size_x,lattice_size_y,*iup,*idown,*jup,*jdown;

void hillslopediffusioninit()
{    int i,j,count;
     float term1;

     ax=vector(1,lattice_size_x);
     ay=vector(1,lattice_size_y);
     bx=vector(1,lattice_size_x);
     by=vector(1,lattice_size_y);
     cx=vector(1,lattice_size_x);
     cy=vector(1,lattice_size_y);
     ux=vector(1,lattice_size_x);
     uy=vector(1,lattice_size_y);
```

```
rx=vector(1,lattice_size_x);
ry=vector(1,lattice_size_y);
D=10000000.0;
count=0;
term1=D/(deltax*deltax);
for (i=1;i<=lattice_size_x;i++)
 {ax[i]=-term1;
  cx[i]=-term1;
  bx[i]=4*term1+1;
  if (i==1)
   {bx[i]=1;
    cx[i]=0;}
  if (i==lattice_size_x)
   {bx[i]=1;
    ax[i]=0;}}
for (j=1;j<=lattice_size_y;j++)
 {ay[j]=-term1;
  cy[j]=-term1;
  by[j]=4*term1+1;
  if (j==1)
   {by[j]=1;
    cy[j]=0;}
  if (j==lattice_size_y)
   {by[j]=1;
    ay[j]=0;}}
while (count<5)
 {count++;
  for (i=1;i<=lattice_size_x;i++)
   for (j=1;j<=lattice_size_y;j++)
    topo2[i][j]=topo[i][j];
  for (i=1;i<=lattice_size_x;i++)
   {for (j=1;j<=lattice_size_y;j++)
     {ry[j]=term1*(topo[iup[i]][j]+topo[idown[i]][j])+topoold[i][j];
      if (j==1)
       ry[j]=topoold[i][j];
      if (j==lattice_size_y)
       ry[j]=topoold[i][j];}
    tridag(ay,by,cy,ry,uy,lattice_size_y);
    for (j=1;j<=lattice_size_y;j++)
     topo[i][j]=uy[j];}
  for (i=1;i<=lattice_size_x;i++)
   for (j=1;j<=lattice_size_y;j++)
    topo2[i][j]=topo[i][j];
  for (j=1;j<=lattice_size_y;j++)
   {for (i=1;i<=lattice_size_x;i++)
     {rx[i]=term1*(topo[i][jup[j]]+topo[i][jdown[j]])+topoold[i][j];
      if (i==1)
       rx[i]=topoold[i][j];
      if (i==lattice_size_x)
```

```
            rx[i]=topoold[i][j];}
         tridag(ax,bx,cx,rx,ux,lattice_size_x);
         for (i=1;i<=lattice_size_x;i++)
          topo[i][j]=ux[i];}}
}


void avalanche(i,j)
int i,j;
{
     if (topo[iup[i]][j]-topo[i][j]>thresh)
      topo[iup[i]][j]=topo[i][j]+thresh;
     if (topo[idown[i]][j]-topo[i][j]>thresh)
      topo[idown[i]][j]=topo[i][j]+thresh;
     if (topo[i][jup[j]]-topo[i][j]>thresh)
      topo[i][jup[j]]=topo[i][j]+thresh;
     if (topo[i][jdown[j]]-topo[i][j]>thresh)
      topo[i][jdown[j]]=topo[i][j]+thresh;
     if (topo[iup[i]][jup[j]]-topo[i][j]>(thresh*sqrt2))
      topo[iup[i]][jup[j]]=topo[i][j]+thresh*sqrt2;
     if (topo[iup[i]][jdown[j]]-topo[i][j]>(thresh*sqrt2))
      topo[iup[i]][jdown[j]]=topo[i][j]+thresh*sqrt2;
     if (topo[idown[i]][jup[j]]-topo[i][j]>(thresh*sqrt2))
      topo[idown[i]][jup[j]]=topo[i][j]+thresh*sqrt2;
     if (topo[idown[i]][jdown[j]]-topo[i][j]>(thresh*sqrt2))
      topo[idown[i]][jdown[j]]=topo[i][j]+thresh*sqrt2;
}


main()
{    FILE *fp1;
     float deltah,time,max,duration;
     int printinterval,idum,i,j,t,step;

     fp1=fopen("streampowertopo","w");
     lattice_size_x=250;
     lattice_size_y=250;
     U=1;                /* m/kyr */
     K=0.05;             /* kyr^-1 */
     printinterval=100;
     deltax=200.0;       /* m */
     thresh=0.58*deltax; /* 30 deg */
     timestep=1;         /* kyr */
     duration=200;
     setupgridneighbors();
     topo=matrix(1,lattice_size_x,1,lattice_size_y);
     topo2=matrix(1,lattice_size_x,1,lattice_size_y);
     topoold=matrix(1,lattice_size_x,1,lattice_size_y);
     slope= matrix(1,lattice_size_x,1,lattice_size_y);
     flow=matrix(1,lattice_size_x,1,lattice_size_y);
     flow1=matrix(1,lattice_size_x,1,lattice_size_y);
```

```
flow2=matrix(1,lattice_size_x,1,lattice_size_y);
flow3=matrix(1,lattice_size_x,1,lattice_size_y);
flow4=matrix(1,lattice_size_x,1,lattice_size_y);
flow5=matrix(1,lattice_size_x,1,lattice_size_y);
flow6=matrix(1,lattice_size_x,1,lattice_size_y);
flow7=matrix(1,lattice_size_x,1,lattice_size_y);
flow8=matrix(1,lattice_size_x,1,lattice_size_y);
topovec=vector(1,lattice_size_x*lattice_size_y);
topovecind=ivector(1,lattice_size_x*lattice_size_y);
for (i=1;i<=lattice_size_x;i++)
 for (j=1;j<=lattice_size_y;j++)
  {topo[i][j]=0.5*gasdev(&idum);
   topoold[i][j]=topo[i][j];
   flow[i][j]=1;}
/*construct diffusional landscape for initial flow routing */
for (step=1;step<=10;step++)
 {hillslopediffusioninit();
  for (i=2;i<=lattice_size_x-1;i++)
   for (j=2;j<=lattice_size_y-1;j++)
    {topo[i][j]+=0.1;
     topoold[i][j]+=0.1;}}
time=0;
while (time<duration)
 {/*perform landsliding*/
  for (j=1;j<=lattice_size_y;j++)
   for (i=1;i<=lattice_size_x;i++)
    topovec[(j-1)*lattice_size_x+i]=topo[i][j];
  indexx(lattice_size_x*lattice_size_y,topovec,topovecind);
  t=0;
  while (t<lattice_size_x*lattice_size_y)
   {t++;
    i=(topovecind[t])%lattice_size_x;
    if (i==0) i=lattice_size_x;
    j=(topovecind[t])/lattice_size_x+1;
    if (i==lattice_size_x) j--;
    avalanche(i,j);}
  for (j=1;j<=lattice_size_y;j++)
   for (i=1;i<=lattice_size_x;i++)
    topoold[i][j]=topo[i][j];
  for (j=1;j<=lattice_size_y;j++)
   for (i=1;i<=lattice_size_x;i++)
    fillinpitsandflats(i,j);
  for (j=1;j<=lattice_size_y;j++)
   for (i=1;i<=lattice_size_x;i++)
    {flow[i][j]=1;
     topovec[(j-1)*lattice_size_x+i]=topo[i][j];}
  indexx(lattice_size_x*lattice_size_y,topovec,topovecind);
  t=lattice_size_x*lattice_size_y+1;
  while (t>1)
```

```
      {t--;
       i=(topovecind[t])%lattice_size_x;
       if (i==0) i=lattice_size_x;
       j=(topovecind[t])/lattice_size_x+1;
       if (i==lattice_size_x) j--;
       mfdflowroute(i,j);}
     for (i=2;i<=lattice_size_x-1;i++)
      for (j=2;j<=lattice_size_y-1;j++)
       {topo[i][j]+=U*timestep;
        topoold[i][j]+=U*timestep;}
     /*perform upwind erosion*/
     max=0;
     for (i=2;i<=lattice_size_x-1;i++)
      for (j=2;j<=lattice_size_y-1;j++)
       {calculatealongchannelslope(i,j);
        deltah=timestep*K*sqrt(flow[i][j])*deltax*slope[i][j];
        topo[i][j]-=deltah;
        if (topo[i][j]<0) topo[i][j]=0;
        if (K*sqrt(flow[i][j])>max) max=K*sqrt(flow[i][j]);}
     time+=timestep;
     if (max>0.3*deltax/timestep)
      {time-=timestep;
       timestep/=2.0;
       for (i=2;i<=lattice_size_x-1;i++)
        for (j=2;j<=lattice_size_y-1;j++)
         topo[i][j]=topoold[i][j]-U*timestep;}
      else
      {if (max<0.03*deltax/timestep) timestep*=1.2;
       for (j=1;j<=lattice_size_y;j++)
        for (i=1;i<=lattice_size_x;i++)
         topoold[i][j]=topo[i][j];}
     if (time>printinterval)
      {printinterval+=250;
       for (i=1;i<=lattice_size_x;i++)
        for (j=1;j<=lattice_size_y;j++)
         {fprintf(fp1,"%f\n",topo[i][j]);}}}
     fclose(fp1);
   }
```

A3.3 | Knickpoint propagation in the 2D sediment-flux-driven bedrock erosion model

The following code solves for the fluvial bedrock incision of an uplifted square block according to the sediment-flux-driven bedrock erosion model. In the model, tectonic uplift at a rate of $U = 1\,\text{m/kyr}$ occurs for 1 Myr. After that initial period, isostatic rebound is calculated using the 2D Fourier-filtering

Fig A3.2 Grayscale maps of example output of the code in Section A3.3. This code implements the sediment-flux driven model in a mountain block that is uniformly uplifted for 1 Myr and then undergoes erosional decay with isostatic rebound. (a) Topography, and instantaneous (b) erosion rate, (c) sediment flux, and (d) flexural rebound at the end of a 40 Myr simulation.

method (see Chapter 5). The code accepts an input DEM named `sedfluxdriventopoinitial` (available from the CUP website). The initial topography is divided by a factor of five to create a low-relief surface with a drainage network geometry already formed. Figure A3.2 illustrates example output from the model.

```
#define sqrt2 1.414213562373
#define oneoversqrt2 0.707106781186

float delrho,alpha,*load,**slope,**erode,S,averosionrate,avreboundrate,avelevation;
float oneoverdeltax,oneoverdeltax2,**U,K,D,X,duration,timestep;
float *topovec,thresh,**deltah,**channel,**area,**sedflux,**sedfluxold,**flow1;
float **flow2,**flow3,**flow4,**flow5,**flow6,**flow7,**flow8,**flow,**topo,**topoold;
float **topo2,deltax,*ax,*ay,*bx,*by,*cx,*cy,*ux,*uy,*rx,*ry;
int *nn,*topovecind,lattice_size_x,lattice_size_y,*iup,*idown,*jup,*jdown;

void computeflexure()
{   int i,j,index;
    float fact;

    for (j=1;j<=lattice_size_y;j++)
      for (i=1;i<=lattice_size_x;i++)
       {load[2*(i-1)*lattice_size_y+2*j-1]=erode[i][j];
        load[2*(i-1)*lattice_size_y+2*j]=0.0;}
    fourn(load,nn,2,1);
    load[1]*=1/delrho;
    load[2]*=1/delrho;
    for (j=1;j<=lattice_size_y/2;j++)
      {fact=1/(delrho+pow(alpha*PI*j/lattice_size_y,4.0));
       load[2*j+1]*=fact;
       load[2*j+2]*=fact;
       load[2*lattice_size_y-2*j+1]*=fact;
       load[2*lattice_size_y-2*j+2]*=fact;}
    for (i=1;i<=lattice_size_x/2;i++)
      {fact=1/(delrho+pow(alpha*PI*i/lattice_size_x,4.0));
       load[2*i*lattice_size_y+1]*=fact;
       load[2*i*lattice_size_y+2]*=fact;
       load[2*lattice_size_x*lattice_size_y-2*i*lattice_size_y+1]*=fact;
       load[2*lattice_size_x*lattice_size_y-2*i*lattice_size_y+2]*=fact;}
    for (i=1;i<=lattice_size_x/2;i++)
      for (j=1;j<=lattice_size_y/2;j++)
       {fact=1/(delrho+pow(alpha*sqrt((PI*PI*i*i)/(lattice_size_x*lattice_size_x)
       +(PI*PI*j*j)/(lattice_size_y*lattice_size_y)),4.0));
        index=2*i*lattice_size_y+2*j+1;
        load[index]*=fact;
        load[index+1]*=fact;
        index=2*i*lattice_size_y+2*lattice_size_y-2*j+1;
        load[index]*=fact;
        load[index+1]*=fact;
        index=2*lattice_size_x*lattice_size_y-2*(i-1)*lattice_size_y-2*(lattice_size_y-j)+1;
        load[index]*=fact;
        load[index+1]*=fact;
        index=2*lattice_size_x*lattice_size_y-2*lattice_size_y-
          2*(i-1)*lattice_size_y+2*(lattice_size_y-j)+1;
        load[index]*=fact;
```

```
          load[index+1]*=fact;}
      fourn(load,nn,2,-1);
      for (j=2;j<=lattice_size_y-1;j++)
        for (i=2;i<=lattice_size_x-1;i++)
          {U[i][j]=(load[2*(iup[i]-1)*lattice_size_y+2*j-1]+
            load[2*(idown[i]-1)*lattice_size_y+2*j-1]+
           load[2*(iup[i]-1)*lattice_size_y+2*jdown[j]-1]+load[2*(i-
             1)*lattice_size_y+2*jup[j]-1])/(4*4*lattice_size_x*lattice_size_y);}
}


void setupmatrices()
{    int i,j;

      ax=vector(1,lattice_size_x);
      ay=vector(1,lattice_size_y);
      bx=vector(1,lattice_size_x);
      by=vector(1,lattice_size_y);
      cx=vector(1,lattice_size_x);
      cy=vector(1,lattice_size_y);
      ux=vector(1,lattice_size_x);
      uy=vector(1,lattice_size_y);
      rx=vector(1,lattice_size_x);
      ry=vector(1,lattice_size_y);
      topo=matrix(1,lattice_size_x,1,lattice_size_y);
      topo2=matrix(1,lattice_size_x,1,lattice_size_y);
      topoold=matrix(1,lattice_size_x,1,lattice_size_y);
      sedflux=matrix(1,lattice_size_x,1,lattice_size_y);
      sedfluxold=matrix(1,lattice_size_x,1,lattice_size_y);
      area=matrix(1,lattice_size_x,1,lattice_size_y);
      slope=matrix(1,lattice_size_x,1,lattice_size_y);
      deltah=matrix(1,lattice_size_x,1,lattice_size_y);
      load=vector(1,2*lattice_size_x*lattice_size_y);
      erode=matrix(1,lattice_size_x,1,lattice_size_y);
      U=matrix(1,lattice_size_x,1,lattice_size_y);
      channel=matrix(1,lattice_size_x,1,lattice_size_y);
      flow=matrix(1,lattice_size_x,1,lattice_size_y);
      flow1=matrix(1,lattice_size_x,1,lattice_size_y);
      flow2=matrix(1,lattice_size_x,1,lattice_size_y);
      flow3=matrix(1,lattice_size_x,1,lattice_size_y);
      flow4=matrix(1,lattice_size_x,1,lattice_size_y);
      flow5=matrix(1,lattice_size_x,1,lattice_size_y);
      flow6=matrix(1,lattice_size_x,1,lattice_size_y);
      flow7=matrix(1,lattice_size_x,1,lattice_size_y);
      flow8=matrix(1,lattice_size_x,1,lattice_size_y);
      topovec=vector(1,lattice_size_x*lattice_size_y);
      topovecind=ivector(1,lattice_size_x*lattice_size_y);
}


main()
```

```
{    FILE *fp0,*fp1,*fp2,*fp3,*fp4,*fp5;
     float change,maxelevation,capacity,time,max,hillslopeerosionrate;
     int printinterval,i,j,t;

     fp0=fopen("sedfluxdriventopoinitial","r");
     fp1=fopen("sedfluxdriventopo","w");
     fp2=fopen("sedfluxdrivenerosion","w");
     fp3=fopen("sedfluxdrivensedflux","w");
     fp4=fopen("sedfluxdrivenrebound","w");
     lattice_size_x=256;
     lattice_size_y=256;
     deltax=500;   /* m */
     delrho=0.28;  /* (rho_m-rho_c)/rho_m */
     oneoverdeltax=1.0/deltax;
     oneoverdeltax2=1.0/(deltax*deltax);
     timestep=1.0; /* kyr */
     alpha=100000; /* m */
     hillslopeerosionrate=0.01; /* m/kyr */
     K=0.0005;       /* m^1/2/kyr */
     D=1.0;          /* m^2/kyr */
     X=0.005;        /* m^-1 */
     alpha/=deltax;
     setupmatrices();
     setupgridneighbors();
     for (j=1;j<=lattice_size_y;j++)
      for (i=1;i<=lattice_size_x;i++)
       {fscanf(fp0,"%f",&topo[i][j]);topo[i][j]/=5;
        if ((i==1)||(j==1)||(i==lattice_size_x)||(j==lattice_size_y)) topo[i][j]=0;
        topoold[i][j]=topo[i][j];
        flow[i][j]=deltax*deltax;
        area[i][j]=0;
        U[i][j]=0;
        erode[i][j]=0;
        channel[i][j]=0;
        sedflux[i][j]=0;
        sedfluxold[i][j]=0;}
     time=0;
     nn=ivector(1,2);
     nn[1]=lattice_size_x;
     nn[2]=lattice_size_y;
     duration=40000;         /* 40 Myr */
     printinterval=40000;
     while (time<duration)
      {for (j=1;j<=lattice_size_y;j++)
        for (i=1;i<=lattice_size_x;i++)
         topoold[i][j]=topo[i][j];
       /*compute uplift rate*/
       if (time>1000) computeflexure();  /* compute isostatic uplift */
       else
```

```
{/* or prescribe a tectonic uplift rate */
 for (j=2;j<=lattice_size_y-1;j++)
  for (i=2;i<=lattice_size_x-1;i++)
   U[i][j]=1.0;  /* m/kyr */}
avelevation=0;avreboundrate=0;maxelevation=0;
for (j=2;j<=lattice_size_y-1;j++)
 for (i=2;i<=lattice_size_x-1;i++)
  {avelevation+=topo[i][j];
   if (topo[i][j]>maxelevation) maxelevation=topo[i][j];
   avreboundrate+=U[i][j];
   topoold[i][j]+=U[i][j]*timestep;
   topo[i][j]+=U[i][j]*timestep;
   deltah[i][j]=0;
   channel[i][j]=0;}
avelevation/=(lattice_size_x-2)*(lattice_size_y-2);
avreboundrate/=(lattice_size_x-2)*(lattice_size_y-2);
printf("%f %f %f %f %f\n",time,avelevation,averosionrate,avreboundrate,maxelevation);
/* hillslope erosion can occur by prescribing a uniform rate, as done here
or using the ADI technique to solve the diffusion equation on hillslopes */
for (j=2;j<=lattice_size_y-1;j++)
 for (i=2;i<=lattice_size_x-1;i++)
  if (channel[i][j]==0) topo[i][j]-=hillslopeerosionrate*timestep;
for (j=2;j<=lattice_size_y-1;j++)
 for (i=2;i<=lattice_size_x-1;i++)
  erode[i][j]=(topoold[i][j]-topo[i][j])/timestep;
for (j=1;j<=lattice_size_y;j++)
 for (i=1;i<=lattice_size_x;i++)
  fillinpitsandflats(i,j);
/*route water from highest gridpoint to lowest*/
for (j=1;j<=lattice_size_y;j++)
 for (i=1;i<=lattice_size_x;i++)
  flow[i][j]=deltax*deltax;
for (j=1;j<=lattice_size_y;j++)
 for (i=1;i<=lattice_size_x;i++)
  topovec[(j-1)*lattice_size_x+i]=topo[i][j];
indexx(lattice_size_x*lattice_size_y,topovec,topovecind);
t=lattice_size_x*lattice_size_y+1;
while (t>1)
 {t--;
  i=(topovecind[t])%lattice_size_x;
  if (i==0) i=lattice_size_x;
  j=(topovecind[t])/lattice_size_x+1;
  if (i==lattice_size_x) j--;
  mfdflowroute(i,j);}
for (i=2;i<=lattice_size_x-1;i++)
 for (j=2;j<=lattice_size_y-1;j++)
  area[i][j]=flow[i][j];
/*perform upwind differencing */
max=0;
```

```
for (i=2;i<=lattice_size_x-1;i++)
 for (j=2;j<=lattice_size_y-1;j++)
  {calculatealongchannelslope(i,j);
    capacity=slope[i][j]*sqrt(area[i][j]);
    if (capacity>X)
     {change=timestep*K*sqrt(fabs(sedflux[i][j]))*deltax*slope[i][j];
       deltah[i][j]+=change;
       erode[i][j]+=change/timestep;
       if (deltah[i][j]<0) deltah[i][j]=0;
       channel[i][j]=1;}
    topo[i][j]-=deltah[i][j];
    if (topo[i][j]<0) topo[i][j]=0;
    if (K*sqrt(fabs(sedflux[i][j]))*deltax>max)
     max=K*sqrt(fabs(sedflux[i][j]))*deltax;}
for (j=1;j<=lattice_size_y;j++)
 for (i=1;i<=lattice_size_x;i++)
  fillinpitsandflats(i,j);
averosionrate=0;
for (i=2;i<=lattice_size_x-1;i++)
 for (j=2;j<=lattice_size_y-1;j++)
  {flow[i][j]=erode[i][j];
    averosionrate+=erode[i][j];}
averosionrate/=(lattice_size_x-2)*(lattice_size_y-2);
for (j=1;j<=lattice_size_y;j++)
 for (i=1;i<=lattice_size_x;i++)
  topovec[(j-1)*lattice_size_x+i]=topo[i][j];
indexx(lattice_size_x*lattice_size_y,topovec,topovecind);
t=lattice_size_x*lattice_size_y+1;
while (t>1)
 {t--;
  i=(topovecind[t])%lattice_size_x;
  if (i==0) i=lattice_size_x;
  j=(topovecind[t])/lattice_size_x+1;
  if (i==lattice_size_x) j--;
  mfdflowroute(i,j);}
for (i=2;i<=lattice_size_x-1;i++)
 for (j=2;j<=lattice_size_y-1;j++)
  {sedfluxold[i][j]=sedflux[i][j];
    sedflux[i][j]=flow[i][j];}
time+=timestep;
if (max>5.0*deltax/timestep)
 {time-=timestep;
  timestep/=2.0;
  for (i=2;i<=lattice_size_x-1;i++)
   for (j=2;j<=lattice_size_y-1;j++)
    {sedflux[i][j]=sedfluxold[i][j];
      topo[i][j]=topoold[i][j]-U[i][j]*timestep;}}
 else
  {if (max<0.5*deltax/timestep) timestep*=1.2;}
```

```
        if (time>=printinterval)
         {printinterval+=40000;
          for (i=1;i<=lattice_size_x;i++)
           for (j=1;j<=lattice_size_y;j++)
            {fprintf(fp1,"%f\n",topo[i][j]);
             fprintf(fp2,"%f\n",erode[i][j]);
             fprintf(fp3,"%f\n",sedflux[i][j]);
             fprintf(fp4,"%f\n",U[i][j]);}}}
   fclose(fp0);
   fclose(fp1);
   fclose(fp2);
   fclose(fp3);
   fclose(fp4);
   fclose(fp5);
}
```

Appendix 4

Codes for solving the flexure equation

A4.1 | Fourier filtering in 1D

The following code uses the Numerical Recipes routine `realft` to Fourier-transform the loading function input from file `load1dandes` (two-column format with elevation as the first column and distance along the transect as the second column; file available from the CUP website), filter it using the kernel of the flexure equation, and inverse-Fourier-transform it back to real space. The profile is read into the first half of a vector of length 4096; the second half of the vector is zeroed (to avoid periodic boundary effects). In this example, the loading function is constructed from a topographic cross section through the Andes Mountains with a resolution of 0.52 km/pixel. In the code, only elevation values greater that 1000 m a.s.l. contribute to the load; elevation values lower than 1000 m are made equal to zero before the Fourier-filtering procedure begins.

```
main()
{    float deltax,alpha,*w,dum,delrho, L,k;
     int lattice_size_x,i;
     FILE *fp1,*fp2;

     fp1=fopen("load1dandes","r");
     fp2=fopen("load1ddeflect50","w");
     lattice_size_x=4096;
     deltax=0.52; /* km */
     L=deltax*2*lattice_size_x;
     alpha=50.0; /* (D/(rho_c*g))^0.25 */
     delrho=0.27; /* (rho_m-rho_c)/rho_m */
     w=vector(1,2*lattice_size_x);
     for (i=1;i<=2*lattice_size_x;i++)
       {if (i<=lattice_size_x/2) fscanf(fp1,"%f %f",&w[i],&dum);
          else w[i]=0;
        if (fabs(w[i])<1000) w[i]=0.0;}
     realft(w,2*lattice_size_x,1);
     w[1]=w[1]/delrho;
     k=PI/deltax;
     w[2]=w[2]/(delrho+pow(k*alpha,4.0));
for (i=3;i<=2*lattice_size_x;i++)
     {k=((i-1)/2)*PI/L;
```

```
    w[i]=w[i]/(delrho+pow(k*alpha,4.0));}
    realft(w,2*lattice_size_x,-1);
    for (i=1;i<=2*lattice_size_x;i++)
     fprintf(fp2,"%f %f\n",i*deltax,w[i]/(lattice_size_x));
    fclose(fp1);
    fclose(fp2);
}
```

A4.2 | Integral solution in 2D

The integral solution method uses point source solutions to the flexure equation. The point source solution has the form of the Kelvin function $kei(r)$. The functions given below provide approximate expressions for the kei function in terms of series expansions for the ber and bei functions.

```
#define PI 3.141592653589793

float ber(r)
float r;
{    float x,z;

     x = pow(pow(r/8.0,2),2);
     z = 1.0 +   x * (-64.0 + x * (113.77777774 +
       x * (-32.36345652 + x * (  2.64191397 +
       x * ( -0.08349609 + x * (  0.00122552 +
       x * -0.00000901))))));
     return z;
}

float bei(r)
float r;
{    float x,y,z;
     y = pow(r/8.0,2);
     x = pow(y,2);
     z = y *        (16.0 + x * (-113.77777774 +
       x * (72.81777742 + x * ( -10.56765779 +
       x * ( 0.52185615 + x * (  -0.01103667 +
       x * 0.00011346))))));
     return z;
}

float kei(r)
float r;
{    float x,y,z;

     if (r == 0) r = 1e-50;
     y = pow(r/8.0,2);
     x = pow(y,2);
     z = (-log(0.50 * r) * bei(r)) - (0.25*PI*ber(r)) +
```

```
        y * (    6.76454936 + x * (-142.91827687 +
        x * (124.23569650 + x * (  -21.30060904 +
        x * (    1.17509064 + x * (   -0.02695875 +
        x * 0.00029532))))));
        return z;
}
```

A4.3 | Fourier filtering in 2D

The following code uses the Numerical Recipes routine `fourn` to Fourier-transform the loading function input from file `load2dandes` (in two-column format with elevation as the first column and distance as the second column; available from the CUP website), filter it using the kernel of the flexure equation, and inverse-Fourier-transform it back to real space. In this example, the loading function is constructed from a DEM of the central Andes Mountains with a resolution of 1.1 km/pixel. In the code, only elevation values greater that 1000 m a.s.l. contribute to the load; elevation values lower than 1000 m are made equal to zero before the Fourier-filtering procedure begins.

```
main()
{    float dum,delta,alpha,fact,*w,delrho;
     int lattice_size_x,lattice_size_y,*nn,i,j;
     FILE *fp1,*fp2;

     fp1=fopen("load2dandes","r");
     fp2=fopen("load2dandesdeflect50","w");
     lattice_size_x=2048;
     lattice_size_y=4096;
     delta=1.1;    /* km */
     alpha=50.0; /* (D/(rho_c*g))^0.25 */
     delrho=0.27; /* (rho_m-rho_c)/rho_m */
     nn=ivector(1,2);
     nn[1]=lattice_size_x;
     nn[2]=lattice_size_y;
     w=vector(1,2*lattice_size_x*lattice_size_y);
     for (j=1;j<=lattice_size_y;j++)
      for (i=1;i<=lattice_size_x;i++)
        {fscanf(fp1,"%f",&dum);
         if (dum<1000) w[2*(i-1)*lattice_size_y+2*j-1]=0;
           else w[2*(i-1)*lattice_size_y+2*j-1]=dum;
         w[2*(i-1)*lattice_size_y+2*j]=0.0;}
     fourn(w,nn,2,1);
     w[1]*=1/delrho;
     w[2]*=1/delrho;
     for (j=1;j<=lattice_size_y/2;j++)
      {fact=1/(delrho+pow(alpha*j*PI/lattice_size_y,4.0));
       w[2*j+1]*=fact;
       w[2*j+2]*=fact;
       w[2*lattice_size_y-2*j+1]*=fact;
```

```
      w[2*lattice_size_y-2*j+2]*=fact;}
   for (i=1;i<=lattice_size_x/2;i++)
    {fact=1/(delrho+pow(alpha*i*PI/lattice_size_x,4.0));
     w[2*i*lattice_size_y+1]*=fact;
     w[2*i*lattice_size_y+2]*=fact;
     w[2*lattice_size_x*lattice_size_y-2*i*lattice_size_y+1]*=fact;
     w[2*lattice_size_x*lattice_size_y-2*i*lattice_size_y+2]*=fact;}
   for (i=1;i<=lattice_size_x/2;i++)
    for (j=1;j<=lattice_size_y/2;j++)
     {fact=1/(delrho+pow(alpha*sqrt((PI*PI*i*i)/(lattice_size_x*lattice_size_x)+
       (PI*PI*j*j)/(lattice_size_y*lattice_size_y)),4.0));
      w[2*i*lattice_size_y+2*j+1]*=fact;
      w[2*i*lattice_size_y+2*j+2]*=fact;
      w[2*i*lattice_size_y+2*lattice_size_y-2*j+1]*=fact;
      w[2*i*lattice_size_y+2*lattice_size_y-2*j+2]*=fact;
      w[2*lattice_size_x*lattice_size_y-2*(i-1)*lattice_size_y-
        2*(lattice_size_y-j)+1]*=fact;
      w[2*lattice_size_x*lattice_size_y-2*(i-1)*lattice_size_y-
        2*(lattice_size_y-j)+2]*=fact;
      w[2*lattice_size_x*lattice_size_y-2*lattice_size_y-
        2*(i-1)*lattice_size_y+2*(lattice_size_y-j)+1]*=fact;
 w[2*lattice_size_x*lattice_size_y-2*lattice_size_y-
        2*(i-1)*lattice_size_y+2*(lattice_size_y-j)+2]*=fact;}
   fourn(w,nn,2,-1);
   for (j=1;j<=lattice_size_y;j++)
    for (i=1;i<=lattice_size_x;i++)
     fprintf(fp2,"%f\n",w[2*(i-1)*lattice_size_y+2*j-1]
      /(lattice_size_x*lattice_size_y));
   fclose(fp1);
   fclose(fp2);
}
```

A4.4 | ADI technique applied to glacial isostatic response modeling

The following code uses the ADI technique to solve the flexure equation with spatially variable rigidity. The code accepts two files: westtelcomb3km (a 3 km/pixel grid of elastic thickness values) and westelamask3km (a mask of equilibrium lines rectified to the same coordinate system as the elastic thickness grid). These files can be downloaded from the CUP website. The ADI method is repeated 5000 times in this example (the method converges slowly). In practice, the number of iterations should be determined by the required level of accuracy.

```
#define PI 3.141592653589793
#define adiiterations 5000

int idum,*iup,*idown,*jup,*jdown,lattice_size_y,lattice_size_x;
float **D,delrho,deltax,deltay,**load,**deflect,**deflect2,te;
```

```
void solveimpl()
{
    int i,j,count;
    float d,**ax,**ay,*xx,*xy,*rx,*ry,**alx,**aly,tot;
    unsigned long *indx,*indy;

    indx=lvector(1,lattice_size_x);
    indy=lvector(1,lattice_size_y);
    ax=matrix(1,lattice_size_x,1,5);
    ay=matrix(1,lattice_size_y,1,5);
    xx=vector(1,lattice_size_x);
    xy=vector(1,lattice_size_y);
    rx=vector(1,lattice_size_x);
    ry=vector(1,lattice_size_y);
    alx=matrix(1,lattice_size_x,1,2);
    aly=matrix(1,lattice_size_y,1,2);
    for (count=1;count<=adiiterations;count++)
     {for (j=1;j<=lattice_size_y;j++)
       for (i=1;i<=lattice_size_x;i++)
        deflect2[i][j]=deflect[i][j];
      tot=0;
      for (j=1;j<=lattice_size_y;j++) {
       for (i=1;i<=lattice_size_x;i++)
        if ((i>2)&&(i<lattice_size_x-1))
        {rx[i]=-load[i][j];
         ax[i][1]=D[i][j];
         ax[i][2]=-4*D[i][j];
         ax[i][3]=12*D[i][j]+delrho;
         rx[i]+=-D[i][j]*(deflect2[i][jup[jup[j]]]+deflect2[i][jdown[jdown[j]]]-
          4*(deflect2[i][jup[j]]+deflect2[i][jdown[j]]));
         ax[i][4]=-4*D[i][j];
         ax[i][5]=D[i][j];}
        else
        {rx[i]=0;
         ax[i][1]=0;
         ax[i][2]=0;
         ax[i][3]=1;
         ax[i][4]=0;
         ax[i][5]=0;}
       bandec(ax,lattice_size_x,2,2,alx,indx,&d);
       banbks(ax,lattice_size_x,2,2,alx,indx,rx);
       for (i=1;i<=lattice_size_x;i++)
        deflect[i][j]=rx[i];}
      for (j=1;j<=lattice_size_y;j++)
       for (i=1;i<=lattice_size_x;i++)
        deflect2[i][j]=deflect[i][j];
      for (i=1;i<=lattice_size_x;i++) {
       for (j=1;j<=lattice_size_y;j++)
        if ((j>2)&&(j<lattice_size_y-1))
```

```
         {ry[j]=-load[i][j];
          ay[j][1]=D[i][j];
          ay[j][2]=-4*D[i][j];
          ay[j][3]=12*D[i][j]+delrho;
          ry[j]+=-D[i][j]*(deflect2[iup[iup[i]]][j]+deflect2[idown[idown[i]]][j]-
           4*(deflect2[iup[i]][j]+deflect2[idown[i]][j]));
          ay[j][4]=-4*D[i][j];
          ay[j][5]=D[i][j];}
         else
          {ry[j]=0;
          ay[j][1]=0;
          ay[j][2]=0;
          ay[j][3]=1;
          ay[j][4]=0;
          ay[j][5]=0;}
       bandec(ay,lattice_size_y,2,2,aly,indy,&d);
       banbks(ay,lattice_size_y,2,2,aly,indy,ry);
       for (j=1;j<=lattice_size_y;j++)
        deflect[i][j]=ry[j];}
     }
}


main()
{    FILE *fp1,*fp2,*fp3;
     int i,j;

     fp1=fopen("westtelcomb3km","r");
     fp2=fopen("westsnowlinemask3km","r");
     fp3=fopen("westdeflectcombnew","w");
     lattice_size_x=626;
     lattice_size_y=642;
     deltax=3;    /* km */
     deltay=3;    /* km */
     delrho=6000; /* kg/m^3 */
     deflect=matrix(1,lattice_size_x,1,lattice_size_y);
     deflect2=matrix(1,lattice_size_x,1,lattice_size_y);
     load=matrix(1,lattice_size_x,1,lattice_size_y);
     D=matrix(1,lattice_size_x,1,lattice_size_y);
     setupgridneighborsperiodic();
     for (j=1;j<=lattice_size_y;j++)
      for (i=1;i<=lattice_size_x;i++)
       {fscanf(fp1,"%f",&te);
        if ((i<5)||(i>lattice_size_x-5)||(j<5)||(j>lattice_size_y-5)) te=0;
        D[i][j]=te*te*te*70000000/(deltax*deltax*deltax*deltax)/
         (12*(1-0.25*0.25));
        fscanf(fp2,"%f",&load[i][j]);
        load[i][j]*=0.82;  /* rho_c/rho_m */
        if ((i<5)||(i>lattice_size_x-5)||(j<5)||(j>lattice_size_y-5)) load[i][j]=0;
        deflect[i][j]=load[i][j];}
```

```
        solveimpl();
        for (j=1;j<=lattice_size_y;j++)
         for (i=1;i<=lattice_size_x;i++)
          fprintf(fp3,"%f\n",deflect[i][j]);
        fclose(fp1);
        fclose(fp2);
        fclose(fp3);
}
```

Appendix 5

Codes for modeling non-Newtonian flows

A5.1 │ 2D radially symmetric lava flow simulation

The following code implements the 2D radially symmetric gravity flow model with temperature-dependent viscosity described in Section 6.3.

```
main()
{    FILE *fp1;
     int lattice_size_x,*iup,*idown,i;
     float deltar,nu,ro,rn,*height,*heightold,*temp,*tempold,*flux;
     float simulationlength,outputinterval,timestep,elapsedtime,maxdelt,maxdelh,del;

     fp1=fopen("viscousflownu100","w");
     lattice_size_x=260;
     deltar=0.02;
     nu=100.0;
     ro=1.0;
     rn=1.1;
     iup=ivector(1,lattice_size_x);
     idown=ivector(1,lattice_size_x);
     height=vector(1,lattice_size_x);
     heightold=vector(1,lattice_size_x);
     temp=vector(1,lattice_size_x);
     tempold=vector(1,lattice_size_x);
     flux=vector(1,lattice_size_x);
     simulationlength=10.0;
     outputinterval=0.1;
     for (i=1;i<=lattice_size_x;i++)
      {temp[i]=exp(-pow(i*deltar/rn,20));
       tempold[i]=temp[i];
       if (i*deltar<=ro) height[i]=pow(4.0/(1+nu)*log(rn/ro),0.25);
        else if (i*deltar<=rn) height[i]=
         pow(4.0/(1+nu)*log(rn/(i*deltar)),0.25); else height[i]=0.001;
       heightold[i]=height[i];
       flux[i]=0;
```

```
    iup[i]=i+1;idown[i]=i-1;}
  iup[lattice_size_x]=lattice_size_x;idown[1]=1;
  timestep=0.00001;
  elapsedtime=0.0;
  while (elapsedtime<simulationlength)
   {maxdelt=0;maxdelh=0;
    flux[1]=1;
    for (i=1;i<=lattice_size_x;i++)
     if (i*deltar<ro) flux[i]=1;
      else flux[i]=i*(1+nu*tempold[i])*pow(0.5*(heightold[idown[i]]+heightold[i]),3)*
       (heightold[idown[i]]-heightold[i]);
    for (i=1;i<=lattice_size_x-1;i++)
     {del=timestep/(i*deltar*deltar)*(flux[i]-flux[iup[i]]);
      height[i]+=del;
      if (fabs(del)>maxdelh) maxdelh=del;
      del=0;
      if (heightold[i]>0)
       {del=timestep*(flux[i]*(tempold[idown[i]]-tempold[i])/
         (heightold[i]*i*deltar*deltar)-
           tempold[i]/(heightold[i]*heightold[i]));
   if (i*deltar<ro) del=0;
      temp[i]+=del;} else del=0;
      if (fabs(del)>maxdelt) maxdelt=fabs(del);
      if (temp[i]<0) temp[i]=0;
      if (temp[i]>1) temp[i]=1;}
    for (i=1;i<=lattice_size_x;i++)
     if (i*deltar<ro) height[i]=height[(int)(ro/deltar)];
    elapsedtime+=timestep;
    if ((maxdelt>0.1)||(maxdelh>0.1))
      {elapsedtime-=timestep;
       timestep/=2.0;
       for (i=1;i<=lattice_size_x;i++)
        {height[i]=heightold[i];
          temp[i]=tempold[i];}}
     else
      if ((maxdelt<0.01)&&(maxdelh<0.01)) timestep*=1.2;
    for (i=1;i<=lattice_size_x;i++)
      {heightold[i]=height[i];
tempold[i]=temp[i];}}
    for (i=1;i<=lattice_size_x;i++)
     fprintf(fp1,"%d %f %f\n",i,height[i],temp[i]);
    fclose(fp1);
}
```

A5.2 | Sandpile method for ice-sheet and glacier reconstruction

The following code implements the sandpile method of ice-sheet or glacier reconstruction using the Greenland ice sheet as an example. The code accepts the input files greenlandsmallbed (a 10 km/pixel

DEM of the bed topography) and greenlandsmallmask (a 10 km/pixel mask grid defining locations of ice coverage) from the local directory and outputs the file greenlandsmallsurf. The routine checkmin is just a simple version of fillinpitsandflats that only applies to the masked area (i.e. the area with ice cover).

```
#define fillincrement 0.1

float taugam,**topo,block,*heightvec,**mask,**height,delta;
int *heightvecind,*iup,*jup,*idown,*jdown,lattice_size_x,lattice_size_y,change;

void addiflessthanthreshold(i,j)
int i,j;
{    float slope,slopex,slopey;

    height[i][j]+=block;
    slopex=height[i][j]-height[iup[i]][j];
    if (height[i][j]-height[idown[i]][j]>slopex) slopex=height[i][j]-height[idown[i]][j];
    slopey=height[i][j]-height[i][jup[j]];
    if (height[i][j]-height[i][jdown[j]]>slopey) slopey=height[i][j]-height[i][jdown[j]];
    slope=sqrt(slopex*slopex+slopey*slopey)/delta;
    /* this version implements slope-dependent threshold
            tau(S)=15S^0.55 */
    if (pow(slope,0.45)*(height[i][j]-topo[i][j])>taugam*15)
     height[i][j]-=block;
    else change=1;
}


void checkmin(i,j)
int i,j;
{    float min;

    min=10000;
    if (mask[i][j]>0.1) {
    if (height[iup[i]][j]<min) min=height[iup[i]][j];
    if (height[idown[i]][j]<min) min=height[idown[i]][j];
    if (height[i][jup[j]]<min) min=height[i][jup[j]];
    if (height[i][jdown[j]]<min) min=height[i][jdown[j]];
    if (height[i][j]<min) {height[i][j]=min+0.1;
                        checkmin(iup[i],j);
                        checkmin(idown[i],j);
                        checkmin(i,jup[j]);
                        checkmin(i,jdown[j]);} }
}

main()
{    FILE *fp1,*fp2,*fp3;
    int i,j,t;

    fp1=fopen("greenlandsmallbed","r");
    fp2=fopen("greenlandsmallmask","r");
```

```
      fp3=fopen("greenlandsmallsurf","w");
      lattice_size_x=150;
      lattice_size_y=280;
      delta=10000.0;                   /* km */
      taugam=100000.0/(920.0*9.8);    /* tau = 10^5 Pa */
      height=matrix(1,lattice_size_x,1,lattice_size_y);
      mask=matrix(1,lattice_size_x,1,lattice_size_y);
      topo=matrix(1,lattice_size_x,1,lattice_size_y);
      setupgridneighbors();
      heightvec=vector(1,lattice_size_x*lattice_size_y);
      heightvecind=ivector(1,lattice_size_x*lattice_size_y);
      for (j=1;j<=lattice_size_y;j++)
       for (i=1;i<=lattice_size_x;i++)
        {fscanf(fp1,"%f",&topo[i][j]);
         fscanf(fp2,"%f",&mask[i][j]);
         height[i][j]=topo[i][j];}
      block=100;              /* start at 100 m block size */
      while (block>0.9)      /* stop when block size < 1 m */
       {change=1;
        while (change>0)
         {change=0;
          for (j=1;j<=lattice_size_y;j++)
           for (i=1;i<=lattice_size_x;i++)
            heightvec[(j-1)*lattice_size_x+i]=height[i][j];
          indexx(lattice_size_x*lattice_size_y,heightvec,heightvecind);
          t=0;
          while (t<lattice_size_x*lattice_size_y)
           {t++;
            i=(heightvecind[t])%lattice_size_x;
            if (i==0) i=lattice_size_x;
            j=(heightvecind[t])/lattice_size_x+1;
            if (mask[i][j]>0.1)
              addiflessthanthreshold(i,j);}
          for (j=1;j<=lattice_size_y;j++)
           for (i=1;i<=lattice_size_x;i++)
            checkmin(i,j);}
         block=block/3.33;}
       for (j=1;j<=lattice_size_y;j++)
        for (i=1;i<=lattice_size_x;i++)
         {if (height[i][j]>0)
           fprintf(fp3,"%f\n",height[i][j]);
      else fprintf(fp3,"0\n");}
      fclose(fp1);
      fclose(fp2);
      fclose(fp3);
}
```

Appendix 6

Codes for modeling instabilities

A6.1 | Werner's eolian dune model

The following code implements Werner's eolian dune model described in Section 7.4 with parameters $l = 5$, $p_s = 0.6$, $p_{ns} = 0.4$, and a shadow-zone angle equal to half of the angle of repose. The code was written by the author based on the description in Werner (1995). The initial condition is a flat bed with three sand slabs at each grid point.

```
int **height,*iup,*jup,*idown,*jdown,lattice_size_x,lattice_size_y;
float **mask,thresh;

void avalanchedown(i,j)
int i,j;
{       float hshadow;
        int i2,violate,min,mini,minj;

        violate=0;min=height[i][j];mini=i;minj=j;
        if (height[i][j]-height[iup[i]][j]>thresh)
         {violate=1;
          if (height[iup[i]][j]<min)
           {min=height[iup[i]][j];mini=iup[i];minj=j;}}
        if (height[i][j]-height[idown[i]][j]>thresh)
         {violate=1;
          if (height[idown[i]][j]<min)
           {min=height[idown[i]][j];mini=idown[i];minj=j;}}
        if (height[i][j]-height[i][jup[j]]>thresh)
         {violate=1;
          if (height[i][jup[j]]<min)
           {min=height[i][jup[j]];mini=i;minj=jup[j];}}
        if (height[i][j]-height[i][jdown[j]]>thresh)
         {violate=1;
          if (height[i][jdown[j]]<min)
           {min=height[i][jdown[j]];mini=i;minj=jdown[j];}}
        if (violate==1)
         {height[i][j]--;
```

```
                  height[mini][minj]++;
                  avalanchedown(mini,minj);}
               else
                {hshadow=height[i][j]-0.5;
                 i2=iup[i];
                 while (height[i2][j]<hshadow)
                  {if (mask[i2][j]<hshadow) mask[i2][j]=hshadow;
                   i2=iup[i2];hshadow-=0.5;}}
        }


void avalancheup(i,j)
int i,j;
{      float hshadow;
       int i2,violate,min,mini,minj;

       violate=0;min=height[i][j];mini=i;minj=j;
       if (height[iup[i]][j]-height[i][j]>thresh)
        {violate=1;
         if (height[iup[i]][j]>min)
           {min=height[iup[i]][j];mini=iup[i];minj=j;}}
       if (height[idown[i]][j]-height[i][j]>thresh)
        {violate=1;
         if (height[idown[i]][j]>min)
           {min=height[idown[i]][j];mini=idown[i];minj=j;}}
       if (height[i][jup[j]]-height[i][j]>thresh)
        {violate=1;
         if (height[i][jup[j]]>min)
           {min=height[i][jup[j]];mini=i;minj=jup[j];}}
       if (height[i][jdown[j]]-hcight[i][j]>thresh)
        {violate=1;
         if (height[i][jdown[j]]>min)
           {min=height[i][jdown[j]];mini=i;minj=jdown[j];}}
       if (violate==1)
        {height[i][j]++;
         height[mini][minj]--;
         avalancheup(mini,minj);}
       else
        {hshadow=height[i][j]-0.5;
         i2=iup[i];
         while ((hshadow<mask[i2][j])&&(hshadow>0))
          {if (height[i2][j]>=hshadow) mask[i2][j]=0; else mask[i2][j]=hshadow;
           i2=iup[i2];hshadow-=0.5;}}
}


main()
{   FILE *fp1,*fp2;
    int idum,t,l,ijump,jjump,i,j,duration;
    float psand,pbed,p;
```

```
fp1=fopen("wernermodeltopo","w");
fp2=fopen("wernermodelshadowmask","w");
idum=-56;
psand=0.6;
pbed=0.4;
thresh=1;
l=5;
lattice_size_x=300;
lattice_size_y=300;
duration=100;
height=imatrix(1,lattice_size_x,1,lattice_size_y);
mask=matrix(1,lattice_size_x,1,lattice_size_y);
setupgridneighborsperiodic();
for (i=1;i<=lattice_size_x;i++)
 for (j=1;j<=lattice_size_y;j++)
  {height[i][j]=3;
   mask[i][j]=0;}
for (t=1;t<=duration*lattice_size_x*lattice_size_y;t++)
 {if (t%(lattice_size_x*lattice_size_y)==0) printf("%d\n",t);
  ijump=(int)(ran3(&idum)*lattice_size_x)+1;
  while (ijump>lattice_size_x) ijump=(int)(ran3(&idum)*lattice_size_x)+1;
  jjump=(int)(ran3(&idum)*lattice_size_y)+1;
  while (jjump>lattice_size_y) jjump=(int)(ran3(&idum)*lattice_size_y)+1;
  while ((height[ijump][jjump]==0)||(mask[ijump][jjump]>0.1))
   {ijump=(int)(ran3(&idum)*lattice_size_x)+1;
    while (ijump>lattice_size_x) ijump=(int)(ran3(&idum)*lattice_size_x)+1;
    jjump=(int)(ran3(&idum)*lattice_size_y)+1;
    while (jjump>lattice_size_y)
     jjump=(int)(ran3(&idum)*lattice_size_y)+1;}
  height[ijump][jjump]--;
  avalancheup(ijump,jjump);
  ijump=(ijump+1)%lattice_size_x+1;
  if (mask[ijump][jjump]>0.1) p=1;
    else if (height[ijump][jjump]>0) p=psand; else p=pbed;
  while (ran3(&idum)>p)
   {ijump=(ijump+1)%lattice_size_x+1;
    if (height[ijump][jjump]>0) p=psand; else p=pbed;}
  height[ijump][jjump]++;
  avalanchedown(ijump,jjump);}
for (j=1;j<=lattice_size_y;j++)
 for (i=1;i<=lattice_size_x;i++)
  {fprintf(fp1,"%d\n",height[i][j]);
   fprintf(fp2,"%f\n",mask[i][j]);}
fclose(fp1);
fclose(fp2);
}
```

A6.2 | Oscillations in arid alluvial channels

The following code solves the equations for arid channel oscillations described in Section 7.5.

```
#define PI 3.141592653589793

main()
{   FILE *fp1;
    int idum,*iup,*idown,count,printed,outputinterval,length_y,lattice_size_x,i;
    float courantwave,courantdiff,maxdelw,del,*h2,*w2,h0,*flux,*deltaw,*deltah;
    float w0,simulationlength,deltax,elapsedtime,timestep,channelslope;
    float *height,*heightold,*width,*widthold,maxheightchange;

    fp1=fopen("channeloscillations","w");
    lattice_size_x=100;
    deltax=100;              /* m */
    idum=-76;
    channelslope=0.01;       /* m/m */
    w0=100;                  /* m */
    h0=1;                    /* m */
    iup=ivector(1,lattice_size_x);
    idown=ivector(1,lattice_size_x);
    height=vector(1,lattice_size_x);
    heightold=vector(1,lattice_size_x);
    deltah=vector(1,lattice_size_x);
    width=vector(1,lattice_size_x);
    widthold=vector(1,lattice_size_x);
    deltaw=vector(1,lattice_size_x);
    flux=vector(1,lattice_size_x);
    h2=vector(1,lattice_size_x);
    w2=vector(1,lattice_size_x);
    simulationlength=5000000.0; /* yr */
    outputinterval=100000.0;
    for (i=1;i<=lattice_size_x;i++)
     {width[i]=w0;
      widthold[i]=width[i];
      height[i]=-0.05*sin((i-1)*2*PI*3/(lattice_size_x-1))+0.03*gasdev(&idum);
      heightold[i]=height[i];
      flux[i]=0;h2[i]=0;w2[i]=0;
      deltaw[i]=0;deltah[i]=0;
      iup[i]=i+1;idown[i]=i-1;}
    iup[lattice_size_x]=1;idown[1]=lattice_size_x;
    timestep=1;
    elapsedtime=0.0;
    printed=0;
    while (elapsedtime<simulationlength)
     {printed=elapsedtime/outputinterval;
      maxdelw=0;
```

```
    for (i=1;i<=lattice_size_x;i++)
     flux[i]=((heightold[idown[i]]-heightold[i])/deltax+channelslope)/widthold[i];
    for (i=1;i<=lattice_size_x;i++)
     {del=0.5*timestep/deltax*(flux[idown[i]]-flux[i]);
      h2[i]=0.5*(heightold[idown[i]]+heightold[i])+del;
      w2[i]=widthold[i]-0.5*timestep*0.5*
       (flux[idown[i]]+flux[i]-2*channelslope/w0)/(h0-h2[i]);}
    for (i=1;i<=lattice_size_x;i++)
     flux[i]=((h2[i]-h2[iup[i]])/deltax+channelslope)/w2[i];
    for (i=1;i<=lattice_size_x;i++)
     {del=timestep/deltax*(flux[i]-flux[iup[i]]);
      height[i]+=del;
      width[i]-=0.5*timestep*(flux[iup[i]]+flux[i]-
       2*channelslope/w0)/(h0-heightold[i]);
      if (fabs(widthold[idown[i]]-widthold[iup[i]])/widthold[i]>maxdelw)
       maxdelw=fabs(widthold[idown[i]]-widthold[iup[i]])/widthold[i];
      if (width[i]<10) width[i]=10;
      if (width[i]>300) width[i]=300;}
    elapsedtime+=timestep;
    if ((maxdelw>0.5*deltax/timestep)||(timestep>deltax*deltax))
      {elapsedtime-=timestep;
       timestep/=2.0;
       for (i=1;i<=lattice_size_x;i++)
        {height[i]=heightold[i];
         width[i]=widthold[i];}}
     else
       if ((maxdelw<0.5*deltax*deltax/timestep/10)&&(timestep<deltax*deltax/10))
        timestep*=1.2;
    for (i=1;i<=lattice_size_x;i++)
      {heightold[i]=height[i];
       widthold[i]=width[i];}
    if ((int)(elapsedtime/outputinterval)>printed)
     {for (i=1;i<=lattice_size_x;i++)
       fprintf(fp1,"%d %f %f\n",i,height[i]/h0,width[i]/w0);
      printed=(int)(elapsedtime/outputinterval);}}
    fclose(fp1);
}
```

A6.3 | 1D model of spiral troughs on Mars

The following code integrates the 1D spiral trough model described in Section 7.7.

```
#define N 2
#define epssm 0.0001

float *xp,**yp,dxsav,a,eps;
int kmax,kount,*iup,*jup,*idown,*jdown;
```

```
void derivs(x,y,dydx)
float dydx[],x,y[];
{
        dydx[1]=y[2];
        dydx[2]=(y[2]*(1-y[2])*(y[2]-a)-y[1])/eps;
}
main()
{    FILE *fp1;
     int nbad,nok,t,lattice_size_x,i;
     float *sol,h1=0.1,hmin=0.0,x1,x2,*lap,deltax,fac;
     float *u,*v,timestep,kappa;


     fp1=fopen("spiralmodel1d","w");
     a=0.2;
     eps=0.05;
     deltax=0.2;
     timestep=deltax*deltax/5;
     kappa=timestep/(deltax*deltax);
     lattice_size_x=512;
     u=vector(1,lattice_size_x);
     v=vector(1,lattice_size_x);
     lap=vector(1,lattice_size_x);
     sol=vector(1,2);
     iup=ivector(1,lattice_size_x);
     idown=ivector(1,lattice_size_x);
     for (i=1;i<=lattice_size_x;i++)
      {iup[i]=i+1;
        idown[i]=i-1;}
     iup[lattice_size_x]=1;
     idown[1]=lattice_size_x;
     for (i=1;i<=lattice_size_x;i++)
      if ((i>lattice_size_x/2-5)&&(i<lattice_size_x/2-2))
       {u[i]=0.25;v[i]=0.0;}
      else
       {u[i]=0;v[i]=0;}
     /* plot initial condition first */
     for (i=1;i<=lattice_size_x;i++)
      fprintf(fp1,"%f %f %f\n",(i-lattice_size_x/2)/10.0,u[i],-v[i]);
     fac=timestep*kappa/deltax/deltax;
     x1=0;x2=timestep;
     for (t=1;t<=1000;t++)
      {for (i=1;i<=lattice_size_x;i++)
        lap[i]=0;
       for (i=1;i<=lattice_size_x;i++)
        {sol[1]=v[i];sol[2]=u[i];
         odeint(sol,N,x1,x2,epssm,h1,hmin,&nok,&nbad,derivs,rkqs);
         v[i]=sol[1];u[i]=sol[2];
         if (fabs(u[i])>0.00001) {lap[i]=-2*u[i];lap[iup[i]]=u[i];lap[idown[i]]=u[i];}}
```

```
     for (i=1;i<=lattice_size_x;i++)
       if (fabs(lap[i])>0) u[i]+=fac*lap[i];}
   for (i=1;i<=lattice_size_x;i++)
    fprintf(fp1,"%f %f %f\n",(i-lattice_size_x/2)/10.0,u[i],-v[i]);
   fclose(fp1);
}
```

Appendix 7

Codes for modeling stochastic processes

A7.1 Fractional-noise generation with Fourier-filtering method

The following code generates a 1D fractional noise (i.e. time series) using the Fourier-filtering method. The user is prompted for all of the information used to generate the noise. The code will produce noises with Gaussian or log-normal distributions depending on directions from the user. The random number seed idum generates a pseudo-random sequence of numbers. In order to generate distinct output, different values of idum must be used.

```
main()
{     float *precyear,beta,sum,sddevcomp,mean,sddev;
      int length,year,format;
      int idum;
      char outputfile[30];
      FILE *fp;

      printf("\nName of output file: ");
      scanf("%s",outputfile);
      printf("\nLength of desired noise: (factor of two, please) ");
      scanf("%d",&length);
      printf("\nGaussian or Log-normal dataset? (1 for gaussian 2 for l-n): ");
      scanf("%d",&format);
      printf("\nMean: (if log-normal this mean is that of the final,
       log transformed data) ");
      scanf("%f",&mean);
      printf("\nStandard Dev: ");
      scanf("%f",&sddev);
      printf("\nBeta: ");
      scanf("%f",&beta);
      printf("\nRandom number seed: (any negative integer) ");
      scanf("%d",&idum);
      fp=fopen(outputfile,"w");
      precyear=vector(1,2*length);
      /*we contruct a vector of length two times the input length so that we
      may cut off the ends to eliminate the periodicity introduced by filtering*/
```

```
for (year=1;year<=2*length;year++)
  precyear[year]=gasdev(&idum);
realft(precyear,length,1);
precyear[1]=0.0;
precyear[2]=precyear[2]/pow(0.5,beta/2.0);
for (year=3;year<=2*length;year++)
  precyear[year]=precyear[year]/pow((year/2)/(float)(2*length),beta/2.0);
realft(precyear,length,-1);
sum=0.0;
for (year=length/2;year<=length+length/2;year++)
   {sum+=precyear[year];}
sum=sum/length;
sddevcomp=0.0;
for (year=length/2;year<=length+length/2;year++)
  sddevcomp+=(precyear[year]-sum)*(precyear[year]-sum);
sddevcomp=sddevcomp/length;
/*below we rescale the amplitudes to restore the desired moments*/
for (year=length/2;year<=length+length/2;year++)
 {precyear[year]=(precyear[year]-sum)/sqrt(sddevcomp);
  precyear[year]=(precyear[year]*sddev)+mean;
  if (format==2) if (mean!=0.0) precyear[year]=
  sqrt(log(1+sddev*sddev/(mean*mean)))*precyear[year]+

  log(mean/sqrt(1+sddev*sddev/(mean*mean))));}
if (format==2) for (year=length/2;year<=length+length/2;year++)
 precyear[year]=exp(precyear[year]);
for (year=length/2;year<=length+length/2;year++)
 fprintf(fp,"%d %f\n",year-length/2,precyear[year]+(float)(year-length/2)/(length));
fclose(fp);
}
```

The following code generates a 2D Gaussian fractional noise (i.e. raster dataset) using the Fourier-filtering method. The user is prompted for all of the information used to generate the dataset.

```
main()
{    float fact,*precyear,betax,sum,mean,sddev,sddevcomp;
    int lengthx,lengthy;
    int *nn,i,j,idum,index;
    char outputfile[30];
    FILE *fp;

    printf("\nName of output file: ");
    scanf("%s",outputfile);
    printf("\nLength of desired noise: (factor of two, please) ");
    scanf("%d",&lengthx);
    lengthx=lengthx*2;
    nn=ivector(1,2);
    nn[1]=lengthx;
    nn[2]=lengthx;
    lengthy=lengthx;
```

```
printf("\nMean: ");
scanf("%f",&mean);
printf("\nStandard Dev: ");
scanf("%f",&sddev);
printf("\nBeta: ");
scanf("%f",&betax);
printf("\nRandom number seed: (any negative integer) ");
scanf("%d",&idum);
fp=fopen(outputfile,"w");
precyear=vector(1,2*lengthx*lengthy);
for (i=1;i<=lengthx;i++)
 for (j=1;j<=lengthy;j++)
  {precyear[2*(i-1)*lengthy+2*j-1]=gasdev(&idum);
   precyear[2*(i-1)*lengthy+2*j]=0.0;}
fourn(precyear,nn,2,1);
for (j=1;j<=lengthy;j++)
 {precyear[2*j-1]=0.0;
  precyear[2*j]=0.0;
  precyear[2*(j-1)*lengthy+1]=0.0;
  precyear[2*(j-1)*lengthy+2]=0.0;}
for (i=1;i<=lengthx/2;i++)
 for (j=1;j<=lengthy/2;j++)
  {fact=pow(sqrt(i*i+j*j)/(2*lengthx),(betax+1)/2.0);
   index=2*i*lengthy+2*j+1;
   precyear[index]=precyear[index]/(lengthx*lengthy*fact);
   precyear[index+1]=precyear[index+1]/(lengthx*lengthy*fact);
   index=2*i*lengthy+2*lengthy-2*j+1;
   precyear[index]=precyear[index]/(lengthx*lengthy*fact);
   precyear[index+1]=precyear[index+1]/(lengthx*lengthy*fact);
   index=2*lengthx*lengthy-2*(i-1)*lengthy-2*(lengthy-j)+1;
   precyear[index]=precyear[index]/(lengthx*lengthy*fact);
   precyear[index+1]=precyear[index+1]/(lengthx*lengthy*fact);
   index=2*lengthx*lengthy-2*lengthy-2*(i-1)*lengthy+2*(lengthy-j)+1;
   precyear[index]=precyear[index]/(lengthx*lengthy*fact);
   precyear[index+1]=precyear[index+1]/(lengthx*lengthy*fact);}
fourn(precyear,nn,2,-1);
sum=0.0;
for (i=1;i<=lengthx/2;i++)
 for (j=1;j<=lengthy/2;j++)
  sum+=precyear[2*(i-1)*lengthy+2*j-1];
sum=sum/(lengthx*lengthy);
sddevcomp=0.0;
for (i=1;i<=lengthx/2;i++)
 for (j=1;j<=lengthy/2;j++)
  sddevcomp+=(precyear[2*(i-1)*lengthy+2*j-1]-sum)*
   (precyear[2*(i-1)*lengthy+2*j-1]-sum);
sddevcomp=sqrt(sddevcomp/(lengthx*lengthy));
for (i=1;i<=lengthx/2;i++)
 for (j=1;j<=lengthy/2;j++)
```

```
        {index=2*(i-1)*lengthy+2*j-1;
         precyear[index]=(precyear[index]-mean)/sddevcomp;
         precyear[index]=(precyear[index]*sddev)+mean;
         fprintf(fp,"%f\n",precyear[index]);}
      fclose(fp);
}
```

A7.2 | Stochastic model of Pleistocene ice ages

The following code is the basis for the stochastic-resonance model of ice ages. The model outputs
four columns with the form: t (yr), T, v, and B_h.

```
#define maxoutputlength 10000000

main()
{    FILE *fp1;
     int n,t,t2,idum,duration,nsteps;
     float tau,per,drift,eta,*T,*B,Bh,*v,deltaT,timestep;

     fp1=fopen("iceagemodel","w");
     T=vector(1,maxoutputlength);
     B=vector(1,maxoutputlength);
     v=vector(1,maxoutputlength);
     eta=0.9;
     tau=75;
     drift=0.2;
     T[1]=0;B[1]=0;v[1]=0;
     idum=-7834;
     timestep=100;      /* yr */
     duration=3000000;  /* yr */
     nsteps=(int)(duration/timstep);
     for (t=1;t<=nsteps;t++)
      {if (T[t]<-1) deltaT=-T[t]-3.2*sqrt(-1-T[t])+eta*gasdev(&idum)+drift;
         else deltaT=-T[t]+eta*gasdev(&idum)+drift;
       Bh=0;
       for (t2=t-2*tau;t2<t;t2++)
        if (t2>1) Bh+=-(T[t2]-1)*exp(-fabs(t-t2)/tau)/tau;
       B[t+1]=Bh;
       if (T[t]<-1) v[t+1]=fabs(B[t+1]-B[t]); else v[t+1]=0;
       deltaT+=25*v[t+1];
       T[t+1]=T[t]+deltaT;
       fprintf(fp1,"%d %f %f %f\n",t*100.0,T[t],v[t+1],Bh);}
     fclose(fp1);
}
```

References

Ahnert, F., 1970, Functional relationships between denudation, relief, and uplift in large mid-latitude drainage basins, *American Journal of Science*, **208**, 243–63.

Allen, P. A. 1997, *Earth Surface Processes*, New York, Oxford University Press.

Amar, J. G. and Family, F., 1989, Numerical solution of a continuum equation for interface growth in $2+1$ dimensions, *Physical Review A*, **41**, 3399–402.

Anderson, R. S. and Humphrey, N. F., 1990, Interaction of weathering and transport processes in the evolution of arid landscapes, in Cross, T., ed., *Quantitative Dynamic Stratigraphy*, New York, Prentice Hall, pp. 349–61.

Anderson, R. S. and Bunas, K. L., 1993, Grain size segregation and stratigraphy in aeolian ripples modelled with a cellular automaton, *Nature*, **365**, 740–3.

Andrews, J. T., 1972, Glacier power, mass balances, velocities, and erosion potential, *Zeitschrift für Geomorphologie*, **13**, 1–17.

Andrews, D. J. and Bucknam, R. C., 1987, Fitting degradation of shoreline scarps by a nonlinear diffusion model, *Journal of Geophysical Research*, **92**, 12857–67.

Andrews, D. J. and Hanks, T. C., 1985. Scarp degraded by linear diffusion: inverse solution for age, *Journal of Geophysical Research*, **90**, 10193–208.

Arrowsmith, J. R., Rhodes, D. D., and Pollard, D. D., 1998, Morphologic dating of scarps formed by repeated slip events along the San Andreas Fault, Carrizo Plain, California, *Journal of Geophysical Research*, **103**, 10141–60.

Avouac, J.-P., 1993, Analysis of scarp profiles: Evaluation of errors in morphologic dating, *Journal of Geophysical Research*, **98**, 6745–54.

Avouac, J.-P. and Peltzer, G., 1993, Active tectonics in southern Xingiang, China: Analysis of terrace riser and normal fault scarp degradation along the Hotan-Qira fault system, *Journal of Geophysical Research*, **98**, 21773–807.

Bagnold, R. A., 1941, *The Physics of Blown Sand and Desert Dunes*, London, Chapman and Hall.

Baldwin, J. A., Whipple, K. X., and Tucker, G. E., 2003, Implications of the stream-power river incision model for the post-orogenic decay of topography, *Journal of Geophysical Research*, **108**, doi:10.1029/2001JB000550.

Bamber, J. L., Layberry, L. P., and Gogenini, S. P., 2001, A new ice thickness and bedrock dataset for the Greenland ice sheet 1: measurement, data reduction, and errors, *Journal of Geophysical Research*, **106**, 33773–80.

Bamber, J. L., Hardy, R. J., Huybrechts, P., and Joughin, I., 2000, A comparison of balance velocities, measured velocities and thermomechanically modelled velocities for the Greenland ice sheet, *Annals of Glaciology*, **30**, 211–16.

Barkley, D., 1991, A model for fast computer simulation of waves in excitable media, *Physica D*, **49**, 61–70.

Begin, Z. B., 1981, Stream curvature and bank erosion: A model based upon the momentum equation, *Journal of Geology*, **89**, 497–504.

Begin, Z. B., Meyer, D. F., and Schumm, S. A., 1981, Development of longitudinal profiles of alluvial channels in response to base-level lowering, *Earth Surface Processes and Landforms*, **6**, 49–68.

Benjamin, M. T., Johnson, N. M., and Naeser, C. W., 1987, Recent rapid uplift in the Bolivian Andes: Evidence from fission-track dating, *Geology*, **15**, 680–3.

Bercovici, D., 1994, A theoretical model of cooling viscous gravity currents with temperature-dependent viscosity, *Geophysical Research Letters*, **21**, 1177–80.

Boone, S. J. and Eyles, N., 2001, Geotechnical model for great plains hummocky moraine formed by till deformation below stagnant ice, *Geomorphology*, **38**, 109–24.

Boulton, G. S., 1974, Processes and patterns of subglacial erosion. In Coates, D. R., ed., *Glacial Geomorphology*, Binghamton, New York, State University of New York, pp. 41–87.

Bowden, G. J., Barker, P. R., Shestopal, V. O., and Twidell, J. W., 1983, The Weibull distribution and wind power statistics, *Wind Energy*, **7**, 85–98.

Braun, J., Zwartz, D., and Tomkin, J. H., 1999, A new surface processes model combining glacial and fluvial erosion, *Annals of Glaciology*, **28**, 282–90.

BSC (Bechtel SAIC Company), 2004a, *Atmospheric Dispersal and Deposition of Tephra from a Potential Volcanic Eruption at Yucca Mountain, Nevada*, Report ANL-MGR-GS-000002 REV 02, Las Vegas, Nevada, Bechtel SAIC Company.

BSC (Bechtel SAIC Company), 2004b, *Characterize Eruptive Processes at Yucca Mountain, Nevada*, Report ANL-MGR-GS-000002 REV 02, Las Vegas, Nevada, Bechtel SAIC Company.

Bucknam, R. C. and Anderson, R. E., 1979, Estimation of fault-scarp ages from a scarp-height-slope-angle relationship, *Geology*, **7**, 11–14.

Budd, W. F. and Warner, R. C., 1996, A computer scheme for rapid calculations of balance-flux distributions, *Annals of Glaciology*, **23**, 21–7.

Bull, W. B., 1979, Threshold of critical power in streams, *Geological Society of America Bulletin*, **90**, 453–64.

Bull, W. B., 1991, *Geomorphic Responses to Climatic Change*, New York, Oxford University Press.

Bull, W. B., 1996, Global climate change and active tectonics: Effective tools for teaching and research, *Geomorphology*, **16**, 217–32.

Bull, W. B., 1997, Discontinuous ephemeral streams, *Geomorphology*, **19**, 227–76.

Caine, N., 1980, The rainfall-intensity-duration control of shallow landslides and debris flows, *Geografiska Annaler*, **62**, 23–7.

Callander, R. A., 1978, River meandering, *Annual Review of Fluid Mechanics*, **10**, 129–58.

Carslaw, H. S. and Jaeger, J. C., 1959, *Conduction of Heat in Solids*, Oxford, Clarendon Press.

Carson, M. A. and Kirkby, M. J., 1972, *Hillslope Form and Process*, New York, Cambridge University Press.

Chadwick, O. A. and Davis, J. O., 1990, Soil-forming intervals caused by eolian sediment pulses in the Lahontan basin, northwestern Nevada, *Geology*, **18**, 243–6.

Chepil, W. S., 1956, Influence of moisture on erodibility of soil by wind, *Proceedings of the Soil Science Society of America*, **20**, 288–92.

Chase, C. G. and Wallace, T. C., 1988, Flexural isostasy and uplift of the Sierra Nevada of California, *Journal of Geophysical Research*, **93**, 2795–802.

Chorley, R. J., 1959, The shape of drumlins, *Journal of Glaciology*, **3**, 339–55.

Christensen, G. E. and Purcell, C., 1985, Correlation and age of Quaternary alluvial-fan sequences, Basin and Range province, southwestern United States. In Weide, D. L., ed., *Soils and Quaternary Geology of the Southwestern United States*, Geological Society of America Special Paper 203, Boulder, Colorado, pp. 115–22.

Clarke, C. D., Knight, J. K., and Gray, J. T., 2000, Geomorphological reconstruction of the Labrador Sector of the Laurentide Ice Sheet, *Quaternary Science Reviews*, **19**, 1343–66.

Clark, M. K., Maheo, G., Saleeby, J., and Farley, K. A., 2005, The non-equilibrium landscape of the southern Sierra Nevada, California, *GSA Today*, **15**, 4–10.

Cook, J. P. and Pelletier, J. D., 2007, Relief threshold for eolian transport across alluvial fans, *Journal of Geophysical Research*, **112**, doi:10.1029/2006JF000610.

Cooke, R. U. and Reeves, R. W., 1976, *Arroyos and Environmental Change in the American Southwest*, Oxford, Clarendon Press.

Costa-Cabral, M. C. and Burges, S. J., 1994, Digital elevation model networks (DEMON): A model of flow over hillslopes for computation of contributing and dispersal areas, *Water Resources Research*, **30**, 1681–92.

Coudert, L., Frappa, M., Viguier, C., and Arias, R., 1995, Tectonic subsidence and crustal flexure in the Neogene Chaco basin of Bolivia, *Tectonophysics*, **243**, 277–92.

Crowe, B. M. and Perry, F. V. 1990, Volcanic probability calculations for the Yucca Mountain Site: Estimation of volcanic rates, *Proceedings of the Topical Meeting on Nuclear Waste Isolation in the Unsaturated Zone*, FOCUS í89, September 17–21, 1989, Las Vegas, Nevada. La Grange Park, Illinois, American Nuclear Society, pp. 326–34.

CRWMS M&O (Civilian Radioactive Waste Management System Management and Operating Contractor), 1996, *Probabilistic Volcanic Hazard Analysis for Yucca Mountain, Nevada*, Report No. BA0000000-01717-2200-00082, Rev 0, Las Vegas, Nevada, CRWMS M&O.

Culling, W. E. H., 1960, Analytical theory of erosion, *Journal of Geology*, **68**, 336–44.

Culling, W. E. H., 1963, Soil creep and the development of hillside slopes, *Journal of Geology*, **71**, 127–61.

Czarnecki, J. B., 1997, *Geohydrology and Evapotranspiration at Franklin Lake playa, Inyo County, California*, US Geological Survey Water-Supply Paper 2377.

Dade, W. B. and Friend, P. F., 1998, Grain size, sediment transport regime, and channel slope in alluvial rivers, *Journal of Geology*, **106**, 661–75.

DeCelles, P. G. and DeCelles, P. C., 2001, Rates of shortening, propagation, underthrusting, and flexural wave migration in continental orogenic systems, *Geology*, **29**, 135–8.

Denton, G. H. and Hughes, T. J., 1981, *The Last Great Ice Sheets*, New York, John Wiley & Sons.

Dewers, T. and Ortoleva, P., 1994, Nonlinear dynamical aspects of deep basin hydrology: fluid compartment formation and episodic fluid release, *American Journal of Science*, **294**, 713–55.

Dowdeswell, J. A., Unwin, B., Nuttall, A. M., and Wingham, D. J., 1999, Velocity structure, flow instability and mass flux on a large Arctic ice cap from satellite radar interferometry, *Earth and Planetary Science Letters*, **167**, 131–40.

Dunne, T., 1978, Formation and controls of channel networks, *Progress in Physical Geography*, **4**, 211–39.

Dyke, A. S., Andrews, J. T., Clark, P. U., *et al.*, 2002, The Laurentide and Inuitian ice sheets during the Last Glacial Maximum, *Quaternary Science Revews*, **21**, 9–31.

Dyke, A. S. and Prest, V. K., 1987, Late Wisconsinan and Holocene history of the Laurentide ice sheet, *Geographie Physique et Quaternaire*, **41**, 237–64.

Entekhabi, D. and Rodriguez-Iturbe, I., 1994, Analytical framework for the characterization of the space-time variability of soil moisture, *Advances in Water Resources*, **17**, 35–46.

Erskine, R. H., Green, T. R., Ramirez, J. A., and MacDonald, L. H., 2006, Comparison of grid-based algorithms for computing upslope contributing area, *Water Resources Research*, **42**, doi:10.1029/2005WR004648.

Evensen, E. B., 1971, The relationship of macro to microfabrics of tills and the genesis of glacial landforms in Jefferson County, Wisconsin. In Goldthwait, R. P., ed., *Till, A Symposium*, Columbus, Ohio, Ohio State University Press, pp. 345–64.

Fagherazzi, S., Howard, A. D., and Wiberg, P. L., 2002, An implicit finite difference method for drainage basin evolution, *Water Resources Research*, **38**, doi:10.1029/2001WR000721.

Fairfield, J. and Laymarie, P., 1991, Drainage networks from grid digital elevation models, *Water Resources Research*, **27**, 709–17.

Fife, P. G., 1976, Pattern formation in reacting and diffusing systems, *Journal of Chemical Physics*, **64**, 554–64.

Fisher, D. A., 1993, If Martian ice caps flow: Ablation mechanisms and appearance, *Icarus*, **105**, 501–11.

Flemings, P. B. and Jordan, T. E., 1989, A synthetic stratigraphic model of foreland basin development, *Journal of Geophysical Research*, **94**, 3851–66.

Flint, R., 1971, *Glacial and Pleistocene Geology*, New York, John Wiley & Sons.

Folk, R. L., 1976, Rollers and ripples in sand, streams, and sky: Rhythmic alteration of transverse and longitudinal vortices in three orders, *Sedimentology*, **23**, 649–69.

Fowler, A. C., 2001, An instability mechanism for drumlin formation. In Maltman, A. J., *et al.*, eds., *Deformation of Glacial Materials*, London, Geological Society of London, pp. 307–19.

Freeman, G. T., 1999, Calculating catchment area with divergent flow based on a rectangular grid, *Computers & Geosciences*, **17**, 413–22.

Freeze, R. A. and Cherry, J. A., 1979, *Groundwater*, New York, Prentice Hall.

French, R. H., 1985, *Open-channel Hydraulics*, New York, McGraw-Hill.

French, R. H., 1987, *Hydraulic Processes on Alluvial Fans*, The Netherlands, Elsevier.

Furbish, D. J. and Fagherazzi, S., 2001, Stability of creeping soil and implications for hillslope evolution, *Water Resources Research*, **37**, 2607–18.

Gale, H. J., Humphreys, D. L. O., and Fisher, E. M. R., 1964, Weathering of Cesium-137 in soil, *Nature*, **201**, 257–61.

Gardner, W. R., 1958, Some steady-state solutions of the unsaturated moisture flow equations with application to evaporation from a water table, *Soil Science*, **85**, 228–32.

Gardner, W. R. and Fireman, M., 1958, Laboratory studies of evaporation from soil columns in the presence of a water table, *Soil Science*, **85**, 244–9.

Gasparini, N. M., Bras, R. L., and Whipple, K. X., 2006, Numerical modeling of non-steady-state river profile evolution using a sediment-flux-dependent incision model. In Willett, S., *et al.*, eds., *Tectonics, Climate, and Landscape Evolution*, Geological Society of America Special Publication, **398**, pp. 127–41.

Gershenfeld, N., 1999, *The Nature of Mathematical Modeling*, New York, Cambridge University Press.

Gile, L. H., Hawley, J. W., and Grossman, R. B., 1981, *Soils and Geomorphology in the Basin and Range area of Southern New Mexico: Guidebook to the Desert Project*, New Mexico Bureau of Mines and Mineral Resources Memoir 39.

Gillette, D. A., Adams, J. B., Endo, A., Smith, D., and Kihl, R., 1980, Threshold velocities for input of soil particles into the air by desert soils, *Journal of Geophysical Research*, **85**, 5621–30.

Gillette, D. A., Hardebeck, E., and Parker, J., 1997, Large-scale variability of wind erosion mass flux rates at Owens Lake 2: Role of roughness change, particle limitation, change of threshold friction velocity, and the Owen effect, *Journal of Geophysical Research*, **102**, 25989–98.

Gillette, D. A., Ono, D., and Richmond, K., 2003, A combined modeling and measurement technique for estimating windblown dust emissions at Owens (dry) Lake, California, *Journal of Geophysical Research*, **109**, doi:10.1029/2003JF000025.

Gillis, R. J., Horton, B. K., and Grove, M., 2006, Thermochronology, geochronology, and upper crustal structure of the Cordillera Real: Implications for Cenozoic exhumation of the central Andean plateau, *Tectonics*, **25**, doi:10.1029/200600TC6007.

Glen, J. W., 1955, The creep of polycrystalline ice, *Proceedings of the Royal Society of London A*, **228**, 519–38.

Gong, Y. and Christini, D. J., 2003, Antispiral waves in reaction-diffusion systems, *Physical Review Letters*, **90**, art. 088302.

Greeley, R. and Iversen, J. D., 1985, *Wind as a Geological Process on Earth, Mars, Venus, and Titan*, New York, Cambridge University Press.

Grove, J. M., 1972, The incidence of landslides, avalanches, and floods in eastern Norway during the little ice age, *Arctic and Alpine Research*, **4**, 131–8.

Gubbels, T., Isacks, B. L., and Farrar, E., 1993, High-level surfaces, plateau uplift, and foreland development, Bolivian central Andes, *Geology*, **21**, 695–8.

Hack, J. T., 1957, *Studies of Longitudinal Stream Profiles in Virginia and Maryland*, US Geological Survey Professional Paper 294B, Washington, DC, pp. 42–97.

Hack, J. T., 1979, *Rock Control and Tectonism: Their Importance in Shaping the Appalachian Highlands*, US Geological Survey Professional Paper 1126-B, Washington, DC.

Hairsine, P. B. and Rose, C. W., 1992, Modelling water erosion due to overland flow using physical principles: 1. Uniform flow, *Water Resources Research*, **28**, 237–43.

Hallet, B., 1979, A theoretical model of glacial abrasion, *Journal of Glaciology*, **23**, 39–50.

Hallet, B., 1990. Spatial self-organization in geomorphology: From periodic bedforms and patterned ground to scale-invariant topography, *Earth Science Reviews*, **29**, 57–75.

Hallet, B., 1996, Glacial quarrying: A simple theoretical model, *Annals of Glaciology*, **22**, 1–8.

Hanks, T. C., 2000, The age of scarplike landforms from diffusion equation analysis. In Noller, J. S., Sowers, J. M., and Lettis, W. R., eds., *Quaternary Geochronology, Methods and Applications*, Washington, DC, American Geophysical Union, pp. 313–38.

Hanks, T. C. and Andrews, D. J., 1989, Effect of far-field slope on morphologic dating of scarplike landforms, *Journal of Geophysical Research*, **94**, 565–73.

Hanks, T. C., Buckman, R. C., Lajoie, K. R., and Wallace, R. E., 1984, Modification of wave-cut and faulting-controlled landforms, *Journal of Geophysical Research*, **89**, 5771–90.

Harbor, J., 1992, Numerical modeling of the development of U-shaped valleys by glacial erosion, *Geological Society of America Bulletin*, **104**, 1364–75.

Harp, E. L. and Jibson, R. L., 1995, *Inventory of Landslides Triggered by the 1994 Northridge, California Earthquake*, US Geological Survey Open File Report, 95-213.

Hart, J. K., 1997, The relationship between drumlins and other forms of subglacial glaciotectonic deformation, *Quaternary Science Reviews*, **16**, 93–107.

Hasselmann, K., 1971, Stochastic climate modeling, Part I: Theory, *Tellus*, **6**, 473–85.

Hawkes, H. E., 1976, The downstream dilution of stream sediment anomalies, *Journal of Geochemical Exploration*, **6**, 345–58.

Heimsath, A. M., Dietrich, W. E., Nishiizumi, K., and Finkel, R. C., 1997, The soil production function and landscape equilibrium, *Nature*, **388**, 358–61.

He, Q. and Walling, D. E., 1997, The distribution of fallout ^{137}Cs and ^{210}Pb in undisturbed and cultivated soils, *Applied Radio Isotopes*, **48**, 677–90.

Heizler, M. T., Perry, F. V., Crowe, B. M., Peters, L., and Appelt, R., 1999, The age of the Lathrop Wells volcanic center: An ^{40}Ar/^{39}Ar dating investigation, *Journal of Geophysical Research*, **104**, 767–804.

Helgen, S. O. and Moore, J. N., 1996, Natural background determination and impact quantification in trace metal-contaminated river sediments, *Environmental Science and Technology*, **30**, 129–35.

Hetenyi, M., 1979, *Beams on Elastic Foundations*, Ann Arbor, Michigan, University of Michigan Press.

Hindmarsh, R. C. A., 1998, Drumlinization and drumlin-forming instabilities. Viscous till mechanisms, *Journal of Glaciology*, **44**, 293–314.

Hoffmann, D. J. and Rosen, J. M., 1987, On the prolonged lifetime of the El Chichon sulfuric acid aerosol cloud, *Journal of Geophysical Research*, **92**, 9825–30.

Hooke, R. L., 1972, Geomorphic evidence for late-Wisconsin and Holocene tectonic deformation, Death Valley, California, *Geological Society of America Bulletin*, **83**, 2073–98.

Horton, B. K. and DeCelles, P. G., 1997, The modern foreland basin adjacent to the central Andes, *Geology*, **25**, 895–8.

Horton, R. E., 1945, Erosional development of streams and their drainage basins: Hydrophysical approach to quantitative morphology, *Geological Society of America Bulletin*, **56**, 275–370.

House, P. K., Pearthree, P. A., and Vincent, K. R., 1991, Flow patterns, flow hydraulics, and flood-hazard implications of a recent extreme alluvial-fan flood in southern Arizona, *Geological Society of America Abstracts with Programs*, **23** (5), 121.

House, M. A., Wernicke, B. P., and Farley, K. A., 1998, Dating topography of the Sierra Nevada, California, using apatite (U-Th)/He ages, *Nature*, **396**, 66–9.

House, M. A., Wernicke, B. P., and Farley, K. A., 2001, Paleo-geomorphology of the Sierra Nevada, California, from (U-Th)/He ages in apatite, *American Journal of Science*, **301**, 77–102.

Hovius, N., Stark, C. P., and Allen, P. A., 1997, Sediment flux from a mountain belt derived by landslide mapping, *Geology*, **25**, 231–5.

Howard, A. D., 1978, Origin of the stepped topography of the Martian poles, *Icarus*, **84**, 2581–99.

Howard, A. D., 2000, The role of eolian processes in forming the surface features of the Martian polar layered deposits, *Icarus*, **144**, 267–88.

Howard, A. D., Cutts, J. A., and Blasius, K. R., 1982, Stratigraphic relationships within Martian polar cap deposits, *Icarus*, **50**, 161–215.

Howard, A. D. and Kerby, G., 1983, Channel changes in badlands, *Geological Society of America Bulletin*, **94**, 739–52.

Howard, A. D. and Knutson, T. R., 1984, Sufficient conditions for river meandering: A simulation approach, *Water Resources Research*, **20**, 1659–67.

Hsu, L. and Pelletier, J. D., 2004, Correlation and dating of Quaternary alluvial-fan surfaces using scarp diffusion, *Geomorphology*, **60**, 319–35.

Huang, C. H., 1999, On solutions of the diffusion-deposition equation for point sources in turbulent shear flow, *Journal of Applied Meteorology*, **38**, 250–4.

Huang, J. and Turcotte, D. L., 1989, Fractal mapping of digitized images: Applications to the topography of Arizona and comparisons with synthetic images, *Journal of Geophysical Research*, **94**, 7491–5.

Hulme, G., 1973, Turbulent lava flow and the formation of lunar sinuous rilles, *Modern Geology*, **4**, 107–17.

Hurst, H. E., Black, R. P., and Simaika, Y. M., 1965, *Long-Term Storage: An Experimental Study*, London, Constable.

Jackson, T. J., 1993, Washita '92 Data Sets: digital data, version 12/20.

Jackson, T. J. and Le Vine, D., 1996, Mapping surface soil moisture using an aircraft-based passive microwave instrument: Algorithm and example, *Journal of Hydrology*, **184**, 85–99.

James, L. A., 1989, Sustained storage and transport of hydraulic gold mining sediment in the Bear River, California, *Annals of the Association of American Geographers*, **79**, 570–92.

Jarzemba, M. S., 1997, Stochastic radionuclide distributions after a basaltic eruption for performance assessments of Yucca mountain, *Nuclear Technology*, **118**, 132–41.

Jarzemba, M. S., La Plante, P. A., and Poor, K. J., 1997, *ASHPLUME Version 1.0A Code for Contaminated Ash Dispersal and Deposition, Technical Description and Users Guide*, CNWRA Report 97-004, San Antonio, Texas, Rev. 1 Center for Nuclear Waste Regulatory Analyses.

Jibson, R. W., 1996, Use of landslides for paleoseismic analysis, *Engineering Geology*, **43**, 291–323.

Johnson, A. M., 1984, Debris flow. In Brunsden, D. and Prior, D. B., eds., *Slope Instability*, New York, John Wiley & Sons, pp. 257–361.

Johnson, K. A. and Sitar, N., 1990, Hydrologic conditions leading to debris-flow initiation, *Canadian Geotechnical Journal*, **27**, 789–801.

Jones, C. H., 1987, Is extension in Death Valley accommodated by thinning of the mantle lithosphere beneath the Sierra Nevada, California?, *Tectonics*, **6**, 449–73.

Kardar, M., Parisi, G., and Zhang, Y.-C., 1986, Dynamic scaling of growing interfaces, *Physical Review Letters*, **56**, 889–92.

Keefer, D. K., 1984, Landslides caused by earthquakes, *Geological Society of America Bulletin*, **95**, 406–21.

Kenyon, P. M. and Turcotte, D. L., 1985, Morphology of a delta prograding by bulk sediment transport, *Geological Society of America Bulletin*, **96**, 1457–65.

Kida, H., 1983, General circulation of air parcels and transport characteristics derived from a hemispheric GCM. Part 2: Very long-term motions of air parcels in the troposphere and stratosphere, *Journal of the Meteorological Society of Japan*, **61**, 510–22.

Kirby, E. and Whipple, K. X., 2001, Quantifying differential rock-uplift rates via stream-profile analysis, *Geology*, **29**, 415-18.

Kirchner, J. W., 1993, Statistical inevitability of Horton's Laws and the apparent randomness of stream channel networks, *Geology*, **21**, 591–4.

Kolb, E. J. and Tanaka, K. L., 2001, Geologic history of the polar regions of Mars based on Mars Global Surveyor data, *Icarus*, **154**, 22–39.

Komatsu, G. and Baker, V. R., 1994, Meander properties of Venusian channels, *Geology*, **22**, 67–70.

Komintz, M. and Pisias, N. G., 1979, Pleistocene climate: Deterministic or stochastic?, *Science*, **204**, 171–2.

Langbein, W. B. and Leopold, L. B., 1966, *River Meanders: Theory of Minimum Variance*, US Geological Survey Professional Paper 422-H, Washington, DC.

Leopold, L. B., Wolman, M. G., and Miller, J. P., 1964, *Fluvial Processes in Geomorphology*, San Francisco, Freeman & Company.

Leopold, L. B., Emmett, W. W., and Myrick, R. M., 1966, *Channel and Hill Slope Processes in a Semi-arid Area*, New Mexico, US Geological Survey Professional Paper 352-G, Reston, Virginia, pp. 193–253.

Livingstone, I., 1986, Geomorphological significance of wind flow patterns over a Namib linear dune. In Nickling, W. G., ed., *Aeolian Geomorphology*, Boston, Allen and Unwin, pp. 97–112.

Lliboutry, L., 1979, Local friction laws for glaciers; a critical review and new openings, *Journal of Glaciology*, **23**, 67–95.

Loughridge, M. S., 1986, Relief map of the earth's surface, *EOS, Transactions of the American Geophysical Union*, **67**, 121.

Lowry, A. R., Ribe, N. M., and Smith, R. B., 2000, Dynamic elevation of the Cordillera, western United States, *Journal of Geophysical Research*, **105**, 23371–90.

Ludwig, W. and Probst, J.-L., 1998, River sediment discharge to the oceans: Present-day controls and global budgets, *American Journal of Science*, **298**, 265–95.

Luke, J. C., 1972, Mathematical models for landform evolution, *Journal of Geophysical Research*, **77**, 2460–4.

Luke, J. C., 1974, Special solutions for nonlinear erosion problems, *Journal of Geophysical Research*, **79**, 4035–40.

MacGregor, K. R., Anderson, R. S., Anderson, S. P., and Waddington, E. D., 2000, Numerical simulations of glacial-valley longitudinal profile evolution, *Geology*, **28**, 1031–4.

Marcus, W. A., 1987, Copper dispersion in ephemeral stream sediments, *Earth Surface Processes and Landforms*, **12**, 217–28.

Marshall, S. J., Tarasov, L., Clarke, G. K. C., and Peltier, W. R., 2000, Glaciological reconstruction of the Laurentide Ice Sheet: Physical processes and modeling challenges, *Canadian Journal of Earth Science*, **37**, 769–93.

Marticorena, B. and Bergametti, G., 1995, Modeling the atmospheric dust cycle: 1. Design of a soil-derived dust emission scheme, *Journal of Geophysical Research*, **100**, 16415–30.

Marticorena, B., Bergametti, G., Gillette, D., and Belnap, J., 1997, Factors controlling threshold friction velocity in semiarid and arid areas of the United States, *Journal of Geophysical Research*, **102**, 23277–87.

Masek, J. G. and Turcotte, D. L., 1993, A diffusion-limited aggregation model for the evolution of drainage networks, *Earth and Planetary Science Letters*, **119**, 379–86.

Masoller, C., 2002, Noise-induced resonance in delayed feedback systems, *Physical Review Letters*, **88**, n. 034102.

Mattson, A. and Bruhn, R. L., 2001, Fault slip rates and initiation age based on diffusion equation modeling: Wasatch fault zone and eastern Great Basin, *Journal of Geophysical Research*, **106**, 13739–50.

McFadden, L. D., Eppes, M. C., Gillespie, A. R., and Hallet, B., 2005, Physical weathering in arid landscapes due to diurnal variation in the direction of solar heating, *Geological Society of America Bulletin*, **117**, 161–73.

McFadden, L. D., Ritter, J. B., and Wells, S. G., 1989, Use of multi-parameter relative-age estimates for age estimation and correlation of alluvial fan surfaces on a desert piedmont, Eastern Mojave Desert, California, *Quaternary Research*, **32**, 276–90.

McFadden, L. D., Wells, S. G., and Jercinovich, M. J., 1987, Influences of eolian and pedogenic processes on the origin and evolution of desert pavements, *Geology*, **15**, 504–8.

McKenzie, D., 1984, The generation and compaction of partially molten rock, *Journal of Petrology*, **25**, 713–65.

Menzies, J., 1979, The mechanics of drumlin formation with particular reference to the change in pore-water content of the till, *Journal of Glaciology*, **22**, 373–84.

Miller, J. R., 1997, The role of fluvial geomorphic processes in the dispersal of heavy metals from mine sites, *Journal of Geochemical Exploration*, **58**, 101–18.

Montgomery, D. R. and Brandon, M. T., 2002, Topographic controls on erosion rates in tectonically active mountain ranges, *Earth and Planetary Science Letters*, **201**, 481–9.

Montgomery, D. R. and Dietrich, W. E., 1999, Where do channels begin?, *Nature*, **336**, 232–4.

Mulch, A., Graham, S. A., and Chamberlain, C. P., 2006, Hydrogen isotopes in Eocene river gravels and paleoelevation of the Sierra Nevada, *Science*, **313**, 87–9.

Mullins, H. T. and Hinchey, E. J., 1989, Erosion and infill of New York State Finger Lakes: Implications for Laurentide ice sheet deglaciation, *Geology*, **17**, 622–5.

Murase, T. and McBirney, A. R., 1970, Viscosity of lunar lavas, *Science*, **167**, 1491–3.

Murray, A. B. and Paola, C., 1994, A cellular model of braided rivers, *Nature*, **371**, 54–6.

Murray, J. D., 1990, *Mathematical Biology*, New York, Springer-Verlag.

Nagumo, J., Yoshizawa, S., and Arimoto, S., 1965, Bistable transmission lines, *IEEE Transactions on Circuit Theory*, **12**, 400–12.

Nanson, G. C. and Hickin, E. J., 1983, Channel migration and incision on the Beatton River, *Journal of Hydraulic Engineering*, **109**, 327–37.

New York State Geological Survey, 1999, *Surficial Geology of the Finger Lakes Region*, Digital data available at www.nysm.nysed.gov/gis.html#region.

Newman, W. A. and Mickelson, D. M., 1994, Genesis of Boston Harbor drumlins, Massachusetts, *Sedimentary Geology*, **91**, 333–43.

Nicolis, C. and Nicolis, G., 1984, Is there a climatic attractor?, *Nature*, **311**, 529–32.

North, G., Cahalan, R., and Moeng, R., 1981, Energy balance climate models, *Reviews of Geophysics*, **19**, 90–121.

Nye, J. F., 1951, The flow of glaciers and ice-sheets as a problem in plasticity, *Proceedings of the Royal Society of London Series A*, **207**, 554–70.

Nye, J. F., 1952, The mechanics of glacier flow, *Journal of Glaciology*, **2**, 82–93.

O'Callaghan, J. F. and Mark, D. M., 1984, The extraction of drainage networks from digital elevation data, *Computer Vision Graphics and Image Processesing*, **28**, 323–44.

Oerlemans, J., 1980a, Continental ice sheets and the planetary radiation budget, *Quaternary Research*, **14**, 349–59.

Oerlemans, J., 1980b, Model experiments on the 100,000 year glacial cycle, *Nature*, **287**, 430–2.

Ohmori, H. and Hirano, M., 1988, Magnitude, frequency, and geomorphological significance of rocky mud flows, landcreep, and the collapse of steep slopes, *Zeitschrift für Geomorphologie N. F.*, **67**, 55–65.

Ohmori, H. and Sugai, T., 1995, Toward geomorphometric models for estimating landslide dynamics and forecasting landslide occurrence in Japanese mountains, *Zeitschrift für Geomorphologie N. F.*, **101**, 149–64.

Packard, F., 1974, The hydraulic geometry of a discontinuous ephemeral stream on a bajada near Tucson, Arizona, Ph.D. thesis, University of Arizona, Tucson, Arizona.

PAG (Pima Association of Governments), 2000, Digital DTM data, www.pagnet.org/RDC.

Paola, C., Heller, P. L., and Angevine, C., 1992, The large-scale dynamics of grain-size variation in alluvial basins, 1: Theory, *Basin Research*, **4**, 73–90.

Parker, G., 1975, Meandering of supraglacial melt streams, *Water Resources Research*, **11**, 551–2.

Payne, A. J., Huybrechts, P., Abe-Ouchi, A., *et al.*, 2000, Results from the EISMINT model intercomparison: The effects of thermomechanical coupling, *Journal of Glaciology*, **46**, 227–38.

Pearthree, P. A., Demsey, K. A., Onken, J., Vincent, K. R., and House, P. K., 1992, *Geomorphic Assessment of Flood-prone Areas on the Southern Piedmont of the Tortolita Mountains, Pima County, Arizona*, Arizona Geological Survey Open-File Report 91-11, scale 1:12,000 and 1:24,000.

Peckham, S. D., 1995, New results for self-similar trees with applications to river networks, *Water Resources Research*, **31**, 1023–9.

Pelletier, J. D., 1997, Analysis and modeling of the natural variability of climate, *Journal of Climate*, **10**, 1331–42.

Pelletier, J. D., 2003, Coherence resonance and ice ages, *Journal of Geophysical Research*, **108**, doi:10.1029/2002JD003120.

Pelletier, J. D., 2004a, Estimation of three-dimensional flexural-isostatic response to unloading: Rock uplift due to Late Cenozoic glacial erosion in the western US, *Geology*, **32**, 161–4.

Pelletier, J. D., 2004b, How do spiral troughs form on Mars?, *Geology*, **32**, 365–7.

Pelletier, J. D., 2004c, The influence of piedmont deposition on time scales of mountain-belt denudation, *Geophysical Research Letters*, **31**, doi:10.1019/2004GL020052.

Pelletier, J. D., 2004d, Persistent drainage migration in a numerical landform evolution model, *Geophysical Research Letters*, **31**, doi:10.1029/2004GL020802.

Pelletier, J. D., 2006, Sensitivity of playa windblown-dust emissions to climatic and anthropogenic change, *Journal of Arid Environments*, **66**, 62–75.

Pelletier, J. D., 2007a, A Cantor set model of eolian dust accumulation on desert alluvial fan terraces, *Geology*, **35**, 439–42.

Pelletier, J. D., 2007b, Erosion-rate determination from foreland basin geometry, *Geology*, 35, 5–8.

Pelletier, J. D., 2007c, Numerical modeling of the Cenozoic fluvial evolution of the southern Sierra

Nevada, California, *Earth and Planetary Science Letters*, **259**, 85–96.

Pelletier, J. D., Cline, M. L., and DeLong, S. B., 2007, Desert pavement dynamics: Numerical modeling and field-based calibration, *Earth Surface Processes and Landforms*, **32**, 1913–27.

Pelletier, J. D., Cline, M. L., DeLong, S. B., Harrington, C. D., and Keating, G. N., 2008, Dispersion of channel-sediment contaminants in fluvial systems: Application to tephra redistribution following a potential volcanic eruption at Yucca Mountain, *Geomorphology*, **94**, 226–46.

Pelletier, J. D. and Cook, J. P., 2005, Deposition of playa windblown dust over geologic time scales, *Geology*, **33**, 909–12.

Pelletier, J. D. and DeLong, S., 2004, Oscillations in arid alluvial channels, *Geology*, **32**, 713–16.

Pelletier, J. D., DeLong, S., Al Suwaidi, A. H., *et al.*, 2006, Evolution of the Bonneville shoreline scarp in west-central Utah: Comparison of scarp-analysis methods and implications for the diffusion model of hillslope evolution, *Geomorphology*, **74**, 257–70.

Pelletier, J. D., Harrington, C. D., Whitney, J. W., *et al.*, 2005a, Geomorphic control of radionuclide diffusion in desert soils. *Geophysical Research Letters*, **32**, doi:10:1029/2005GL024347.

Pelletier, J. D., Malamud, B. D., Blodgett, T., and Turcotte, D. L., 1997, Scale-invariance of soil moisture variability and its implications for the frequency-size distribution of landslides, *Engineering Geology*, **48**, 255–68.

Pelletier, J. D., Mayer, L., Pearthree, P. A., *et al.*, 2005b, An integrated approach to alluvial-fan flood hazard assessment with numerical modeling, field mapping, and remote sensing, *Geological Society of America Bulletin*, **117**, 1167–80.

Pelletier, J. D. and Turcotte, D. L., 1997, Long-range persistence in climatological and hydrological time series: Analysis, modeling, and application to drought hazard assessment, *Journal of Hydrology*, **203**, 198–208.

Pelletier, J. D. and Turcotte, D. L., 1999, Self-affine time series: II. Applications and models, *Advances in Geophysics*, **40**, 91–166.

Peltier, W. R., 1994, Ice-age paleotopography, *Science*, **265**, 198–201.

Petit, J. R., Jouzel, J., Raynaud, D., *et al.*, 1990, Paleoclimatological implications of the Vostok core dust record, Nature, **343**, 56–8.

Peixoto, J. P. and Oort, A. H., 1992, *Physics of Climate*, Washington, DC, American Institute of Physics.

Pinet, P. and Souriau, M., 1988, Continental erosion and large-scale relief, *Tectonics*, **7**, 563–82.

Poage, M. A. and Chamberlain, C. P., 2002, Stable isotope evidence for a Pre-Middle Miocene rain shadow in the western Basin and Range: Implications for the paleotopography of the Sierra Nevada, *Tectonics*, **21**, 1601–18.

Pollard, D., 1983, A coupled climate-ice sheet model applied to the Quaternary ice ages, *Journal of Geophysical Research*, **88**, 7705–18.

Pope, G. L., Rigas, P. D., and Smith, C. F., 1998, Statistical summaries of streamflow data and characteristics of drainage basins for selected streamflow-gaging stations in Arizona through water year 1996, US Geological Survey Water Resources Investigations Report 98-4225, Washington, DC.

Porter, S. C., Pierce, K. L., and Hamilton, T. D., 1983, Late Wisconsin mountain glaciation in the western United States. In Wright, H. E., Jr. and Porter, S. C., eds., *Late Quaternary Environments of the United States*, Vol. 1, Minneapolis, Minnesota, University of Minnesota Press, pp. 71–115.

Press, W. H., Flannery, B. P., Teukolsky, S. A., and Vetterling, W. T., 1992, *Numerical Recipes in C: The Art of Scientific Computing*, 2nd edition, New York, Cambridge University Press.

Press, W. H., Flannery, B. P., Teukolsky, S. A., and Vetterling, W. T., 2007, *Numerical Recipes in C: The Art of Scientific Computing*, 3rd edition, New York, Cambridge University Press.

Pye, K., 1987, *Aeolian Dust and Dust Deposits*, London, Academic Press.

Raupach, M. R., Gillette, D. A., and Leys, J. F., 1993, The effect of roughness elements on wind erosion threshold, *Journal of Geophysical Research*, **98**, 3023–9.

Rawls, W. J., Ahuja, L. R., Brakensiek, D. L., and Sirmohammadi, A., 1992, Infiltration and soil water movement. In Maidment, D. R., ed., *Handbook of Hydrology*, New York, McGraw-Hill, pp. 5.1–5.62.

Reeh, N., 1982, A plasticity-theory approach to the steady-state shape of a three-dimensional ice sheet, *Journal of Glaciology*, **28**, 431–55.

Reeh, N., 1984, Reconstruction of the ice covers of Greenland and the Canadian Arctic Islands by 3-dimensional, perfectly plastic ice-sheet modeling, *Annals of Glaciology*, **5**, 115–21.

Reheis, M. C., Slate, J. L., Throckmorton, C. K., *et al.*, 1996, Late-Quaternary sedimentation on the Leidy Creek fan, Nevada-California: Geomorphic responses to climate change, *Basin Research*, **12**, 279–99.

Reheis, M., 1999, *Extent of Pleistocene Lakes in the Western Great Basin*, US Geological Survey Miscellaneous Field Studies Map MF-2323, Denver, Colorado, US Geological Survey.

Reheis, M. C., 2006, A 16-year record of eolian dust in Southern Nevada and California, USA: Controls on dust generation and accumulation, *Journal of Arid Environments*, **67**, 487–520.

Reheis, M. C., Goodmacher, J. C., Harden, J. W., *et al.*, 1995, Quaternary soils and dust deposition in southern Nevada and California, *Geological Society of America Bulletin*, **107**, 1003–22.

Reneau, S. L., Drakos, P. G., Katzmann, D. L., *et al.*, 2004, Geomorphic controls on contaminant distribution along an ephemeral stream, *Earth Surface Processes and Landforms*, **29**, 1209–23.

Reynolds, R. L., Yount, J. C., Reheis, M., *et al.*, 2007, Dust emission from wet and dry playas in the Mojave Desert, USA, *Earth Surface Processes and Landforms*, **32**, 1811–27.

Rhoads, B. L. and Welford, M. R., 1991, Initiation of river meandering, *Progress in Physical Geography*, **15**, 127–56.

Richardson, W. R., 2002, Simplified model for assessing meander bend migration rates, *Journal of Hydraulic Engineering*, **128**, 1094–7.

Ridky, R. W. and Bindschadler, R. A., 1990, Reconstruction and dynamics of the Late Wisconsin "Ontario" ice dome in the Finger Lakes region, New York, *Geological Society of America Bulletin*, **102**, 1055–64.

Riebe, C. S., Kirchner, J. W., Granger, D. E., and Finkel, R. C., 2000, Erosional equilibrium and disequilibrium in the Sierra Nevada, California, *Geology*, **28**, 803–6.

Riebe, C. S., Kirchner, J. W., Granger, D. E., and Finkel, R. C., 2001, Minimal climatic control on erosion rates in the Sierra Nevada, California, *Geology*, **29**, 447–50.

Rodriguez-Iturbe, I., Vogel, G. K., Rigon, R., Entekhabi, D., and Rinaldo, A., 1995, On the spatial organization of soil moisture fields, *Geophysical Research Letters*, **22**, 2757–60.

Roering, J. J., Kirchner, J. W., and Dietrich, W. E., 1999, Evidence for nonlinear, diffusive sediment transport and implications for landscape morphology, *Water Resources Research*, **35**, 853–70.

Ruxton, B. P. and McDougall, I., 1967, Denudation rates in northeast Papua New Guinea from Potassium-Argon dating of lavas, *American Journal of Science*, **265**, 545–61.

Sadler, P. M., 1981, Sediment accumulation rates and the completeness of stratigraphic sections, *Journal of Geology*, **89**, 569–84.

Sadler, P. M. and Strauss, D. J., 1990, Estimation of completeness of stratigraphic sections using empirical data and theoretical models, *Journal of the Geological Society of London*, **147**, 471–85.

Safran, E. B., Bierman, P. R., Aalto, R., *et al.*, 2005, Erosion rates driven by channel network incision in the Bolivian Andes, *Earth Surface Processes and Landforms*, **30**, 1007–24.

Samet, J. M., Dominici, F., Curriero, F. C., Cousac, I., and Zeger, S. L., 2000, Fine particulate air pollution and mortality in 20 US cities, 1987–1994, *New England Journal of Medicine*, **343**, 1742–9.

Schoof, C., 2002, Mathematical models of glacier sliding and drumlin formation, D. Phil. Thesis, Oxford University, Oxford.

Schumm, S. A., 1977, *The Fluvial System*, New York, John Wiley & Sons.

Schumm, S. A. and Hadley, R. F., 1957, Arroyos and the semiarid cycle of erosion, *American Journal of Science*, **255**, 161–74.

Schumm, S. A. and Parker, R. S., 1973, Implications of complex response of drainage systems for Quaternary alluvial stratigraphy, *Nature*, **243**, 99–100.

Shao, Y. and Raupach, M. R., 1993, Effect of saltation bombardment on the entrainment of dust by wind, *Journal of Geophysical Research*, **98**, 12719–26.

Sharp, R. P., 1963, Wind ripples, *Journal of Geology*, **71**, 617–36.

Shepard, R. G., 1985, Regression analysis of river profiles, *Journal of Geology*, **93**, 377–84.

Sinha, S. K. and Parker, G., 1996, Causes of concavity in longitudinal profiles of rivers, *Water Resources Research*, **32**, 1417–28.

Sklar, L. and Dietrich, W. E., 2001, Sediment and rock strength controls on river incision into bedrock, *Geology*, **29**, 1087–90.

Sklar, L. and Dietrich, W. E., 2004, A mechanistic model for river incision into bedrock by saltating bedload, *Water Resources Research*, **40**, doi:10.1029/2003WR002496.

Slingerland, R. L. and Snow, R. S., 1987, Mathematical modeling of graded river profiles, *Journal of Geology*, **95**, 15–33.

Small, E. E., Anderson, R. S., Repka, J. L., and Finkel, R., 1997, Erosion rates of alpine bedrock summit surfaces deduced from [10]Be and [26]Al, *Earth and Planetary Science Letters*, **150**, 413–25.

Smalley, I. J. and Unwin, D. J., 1968, The formation and shape of drumlins and their distribution and orientation in drumlin fields, *Journal of Glaciology*, **7**, 377–90.

Smith, F. B., 1962, The problem of deposition in atmospheric diffusion of particulate matter, *Journal of Atmospheric Sciences*, **19**, 429–34.

Smith, R. B., 2003, Advection, diffusion, and deposition from distributed sources, *Boundary-Layer Meteorology*, **107**, 273–87.

Smolarkiewicz, P. K., 1983, A simple positive definite advection scheme with small implicit diffusion, *Monthly Weather Review*, **111**, 479–86.

Snyder, N., Whipple, K. X., Tucker, G., and Merritts, D., 2000, Landscape response to tectonic forcing: DEM analysis of stream profiles in the Mendocino triple junction region, northern California, *Geological Society of America Bulletin*, **112**, 1250–63.

Sonder, L. J. and Jones, C. H., 1999, Western United States extention: How the West was widened, *Annual Reviews of Earth and Planetary Sciences*, **27**, 417–62.

Spaulding, W. G., 1990, Vegetation dynamics during the last deglaciation, southeastern Great Basin, USA, *Quatenary Research*, **33**, 188–203.

Spiegelman, M., 1993, Flow in deformable porous media. Part 1: Simple analysis, *Journal of Fluid Mechanics*, **247**, 17–38.

Spiegelman, M. and McKenzie, D., 1987, Simple 2-D models for melt extraction at mid-ocean ridges and island arcs, *Earth and Planetary Science Letters*, **83**, 137–52.

Spiegelman, M., Keleman, P. B., and Aharonov, E., 2001, Causes and consequences of flow organization during melt transport: The reaction infiltration instability, *Journal of Geophysical Research*, **106**, 2061–78.

Squires, R. R. and Young, R. L. 1984, *Flood Potential of Fortymile Wash and its Principal Southwestern Tributaries, Nevada Test Site, Southern Nevada*, US Geological Survey Water Resources Investigations Report 83-4001, Carson City, Nevada, US Geological Survey.

Stanford, S. D. and Mickelson, D. M., 1985, Till fabric and deformational structures in drumlins near Waukesha, Wisconsin, USA, *Journal of Glaciology*, **31**, 220–8.

Stark, C. P. and Hovius, N., 2001, The characterization of landslide size distributions, *Geophysical Research Letters*, **28**, 1091–4.

Stock, G. M., Anderson, R. S., and Finkel, R. C., 2004, Pace of landscape evolution in the Sierra Nevada, California, revealed by cosmogenic dating of cave sediments, *Geology*, **32**, 193–6.

Stock, G. M., Anderson, R. S., and Finkel, R. C., 2005, Rates of erosion and topographic evolution of the Sierra Nevada, California, inferred from cosmogenic ^{26}Al and ^{10}Be concentrations, *Earth Surface Processes and Landforms*, **30**, 985–1006.

Stommel, H., 1965, *The Gulf Stream*, Berkeley, California, University of California Press.

Strudley, M. W., Murray, A. B., and Haff, P. K., 2006, Regolith-thickness instability and the formation of tors in arid environments, *Journal of Geophysical Research*, **111**, doi:10.1029/2005JF000405.

Summerfield, M. A. and Hulton, N. J., 1994, Natural controls on fluvial denudation rates in major world drainage basins, *Journal of Geophysical Research*, **99**, 13871–83.

Takle, E. S. and Brown, J. M., 1978, Note on use of Weibull statistics to characterize wind-speed data, *Journal of Applied Meteorology*, **17**, 556–9.

Tarboton, D. G., 1997, A new method for the determination of flow directions and upslope areas in grid digital elevation models, *Water Resources Research*, **33**, 309–19.

Terzaghi, K., 1962, Stability of steep slopes on hard unweathered rock, *Geotechnique*, **12**, 251–70.

Tokunaga, E., 1984, Ordering of divide segments and law of divide segment numbers, *Transactions of the Japanese Geomorphological Union*, **5**, 71–7.

Tomkin, J. H. and Braun, J., 2002, The influence of alpine glaciation on the relief of tectonically active mountain belts, *American Journal of Science*, **302**, 169–90.

Tsimring, L. S. and Pikovsky, A., 2001, Noise-induced dynamics in bistable systems with delay, *Physical Review Letters*, **87**, n. 250602.

Tucker, G. E. and Bras, R. L., 1998, Hillslope processes, drainage density, and landscape morphology, *Water Resources Research*, **34**, 2751–64.

Tulaczyk, S., Kamb, W. B., and Engelhardt, H. F., 2000, Basal mechanics of Ice Stream B, West Antarctica: 2, Undrained plastic bed model, *Journal of Geophysical Research*, **105**, 483–94.

Turcotte, D. L. and Schubert, G., 2002, *Geodynamics: Applications of Continuum Physics to Geological Problems*, 2nd edition, New York, Cambridge University Press.

United States Geological Survey, 2005, *Southwest Climate Impact Meteorological Stations (CLIM-MET)*, Digital data available at http://esp.cr.usgs.gov/info/sw/clim-met/climetdata.html.

Unruh, J. R., 1991, The uplift of the Sierra Nevada and implications for late Cenozoic epeirogeny in the western Cordillera, *Geological Society of America Bulletin*, **103**, 1395–404.

Ussami, N., Shiraiwa, S., and Dominguez, J. M., 1999, Basement reactivation in a sub-Andean foreland flexural bulge: The Pantanal wetland: SW Brazil, *Tectonics*, **18**, 25–39.

Valentine, G. A., Krier, D., Perry, F. V., and Heiken, G., 2005, Scoria cone construction mechanisms, *Geology*, **33**, 629–32.

Van Genuchten, M. Th., 1980, A closed-form equation for predicting the hydraulic conductivity of unsaturated soils, *Soil Science Society of America Journal*, **44**, 892–8.

van Kampen, N. G., 2001, *Stochastic Processes in Physics and Chemistry*, Amsterdam, Elsevier.

van Wees, J. D. and Cloetingh, S., 1994, A finite-difference technique to incorporate spatial variations in rigidity and planar faults into 3-D models for lithospheric flexure, *Geophysical Journal International*, **117**, 179–95.

Vialov, S. S., 1958, Regularities of glacial shields movement and the theory of plastic viscous flow, *International Association for Hydrological Sciences Publication*, **47**, 266–75.

von Engeln, O. D., 1961, *The Finger Lakes Region: Its Origin and Nature*, Ithaca, New York, Cornell University Press.

Voss, R. F. and Clarke, J., 1976, Flicker ($1/f$) noise: Equilibrium temperature and resistance fluctuations, *Physical Review B*, **13**, 556–73.

Wakabayashi, J. and Sawyer, T. L., 2001, Stream incision, tectonics, uplift, and evolution of topography of the Sierra Nevada, California, *Journal of Geology*, **109**, 539–62.

Warrick, A. W., 1988, Additional solutions for steady-state evaporation from a shallow water table, *Soil Science*, **146**, 63–6.

Watts, A. B., 2001, *Isostasy and Flexure of the Lithosphere*, New York, Cambridge University Press.

Watts, A. B., Lamb, S. H., Fairhead, J. D., and Dewey, J. F., 1995, Lithospheric flexure and bending of the central Andes, *Earth and Planetary Science Letters*, **134**, 9–21.

Webb, R. W., 1946, Geomorphology of the middle Kern River Basin, southern Sierra Nevada, California, *Geological Society of America Bulletin*, **57**, 355–82.

Weldon, R. J., 1986, The cause and timing of terrace formation in Cajon Creek, southern California. In Weldon, R. J., The late Cenozoic geology of Pajon Pass: Implications for tectonics and sedimentation along the San Andreas Fault, unpublished Ph.D. dissertation, California Institute of Technology, Pasadena, CA, pp. 99–190.

Wells, S. G., McFadden, L. D., Poths, J., and Olinger, C. T., 1995, Cosmogenic ^{3}He surface-exposure dating of stone pavements: Implications for landscape evolution in deserts, *Geology*, **23**, 613–16.

Werner, B. T., 1995, Eolian dunes: Computer simulations and attractor interpretation, *Geology*, **23**, 1107–10.

Western Regional Climate Center (Desert Research Institute), 2005, *Station Wind Rose: Amargosa Valley Nevada*, Digital data available at www.wrcc.dri.edu/cgi-bin/wea_windrose.pl?nvamar.

Whipple, K. X. and Tucker, G. E., 1999, Dynamics of the stream-power river incision model: Implications for height limits of mountain ranges, landscape response timescales, and research needs, *Journal of Geophysical Research*, **104**, 17661–74.

Whipple, K. X., Hancock, G. S., and Anderson, R. S., 2000, River incision into bedrock: Mechanics and relative efficacy of plucking, abrasion, and cavitation, *Geological Society of America Bulletin*, **112**, 490–503.

Whipple, K. X., Kirby, E., and Brocklehurst, S. H., 1999, Geomorphic limits to climate-induced increases in topographic relief, *Nature*, **401**, 39–43.

Whitney, J. W., Taylor, E. M., and Wesling, J. R., 2004, Quaternary stratigraphy and mapping in the Yucca Mountain area. In Keefer, W. R., *et al.*, eds., *Quaternary Paleoseismology and Stratigraphy of the Yucca Mountain Area*, Nevada, US Geological Survey Professional Paper 1689, Washington, DC, pp. 11–23.

Whitehouse, I. E. and Griffiths, G. A., 1983, Frequency and hazard of large rock avalanches in the central Southern Alps, New Zealand, *Geology*, **11**, 331–4.

Whittecar, G. R. and Mickelson, D. M., 1979, Composition, internal structures, and a hypothesis of formation for drumlins, Waukesha County, Wisconsin, USA, *Journal of Glaciology*, **22**, 357–70.

Wieczoreck, G. F., 1987, Effect of rainfall intensity and duration on debris flows in central Santa Cruz Mountains, California. In Costa, J. E. and Wieczorek, G. F., eds., *Debris Flows/Avalanches: Process, Recognition, and Mitigation: Reviews of Engineering Geology*, **7**, pp. 93–104.

Willett, S. D., Slingerland, R., and Hovius, N., 2001, Uplift, shortening, and steady-state topography in active mountain belts, *American Journal of Science*, **301**, 455–85.

Willgoose, G., Bras, R. L., and Rodriguez-Iturbe, I., 1991, A coupled channel network growth and hillslope evolution model: 1. Theory, *Water Resources Research*, **27**, 1671–84.

Wilson, I. G., 1972, Aeolian bedforms: Their development and origins, *Sedimentology*, **19**, 173–210.

Winfree, A. T., 1987, *When Time Breaks Down*, Princeton, New Jersey, Princeton University Press.

Wobus, C. W., Crosby, B. T., and Whipple, K. X., 2006, Hanging valleys in fluvial systems: Controls on occurrence and implications for landscape evolution, *Journal of Geophysical Research*, **111**, doi:10.1029/2005JF000406.

Young, M. H., McDonald, E. V., Caldwell, T. C., Benner, S. G., and Meadows, D. G., 2004, Hydraulic properties of a desert soil chronosequence, Mojave Desert, California, *Vadose Zone Journal*, **3**, 956–63.

Zelcs, V. and Driemanis, A., 1997, Morphology, internal structure, and genesis of the Burtnieks drumlin field, Northern Vidzeme, Latvia, *Sedimentary Geology*, **111**, 73–90.

Zhang, P., Molnar, P., and Downs, W. R., 2001, Increased sedimentation rates and grain sizes 2–4 Myr ago due to the influence of climate change on erosion rates, *Nature*, **410**, 891–7.

Index

avalanche, 244
calculatealongchannelslope, 239
computeflexure, 249
fillinpitsandflats, 235
fourn, 258
hillslopediffusioninit, 244
hillslopediffusion, 229
malloc.h, 222
math.h, 222
nr.h, 222
nrutil.h, 222
realft, 256
setupgridneighborsperiodic, 224
setupgridneighbors, 223
setupmatrices, 249
stdio.h, 222
stdlib.h, 222
3DEM, 223

advection equation
    analytic methods for, 88–90
    contrasting behavior to diffusion
        equation, 31, 87
    introduction to, 87
    method of characteristics, 88
    numerical methods for, 90–93
        Lax method, 91
        stability criteria, 91
        Two-step Lax–Wendroff method,
            91
        upwind differencing method,
            92
alluvial channels
    diffusion equation model of,
        8, 102
    introduction to, 4
alluvial fan
    cut and fill cycles on, 38
    introduction to, 9
    windblown sand transport across,
        17
Alternating Direction Implicit (ADI)
    method
    code for 2D implementation of
        diffusion equation, 229
    code for 2D implementation to
        solve flexure equation, 259
    introduction to, 63
    use of to solve the 2D flexure
        equation, 122

Amargosa Valley
    example of dust emission from
        playas, 13, 53
    Lathrop Wells cone
        age of, 79
        diffusive evolution of, 51
        study site for contaminant
            dispersion model, 79–83
        study site for flood hazard
            assessment, 73
        study site for potential fluvial
            system contamination, 75
        study site for radionuclide
            dispersion in soils, 46
Andes, 119
    best-fit flexural parameter of, 113
    Cordillera Real
        glacial erosion and exhumation
            of, 28
    example of 1D flexure solution,
        113
    example of 2D flexure solution,
        115
    example of curvature effects in 2D
        flexure, 116
    example of foreland basin model,
        119–120
    landslide distributions from, 209
    use as a paleo analog for western
        US, 1
Antarctic ice sheet
    introduction to, 20
    model reconstruction of, 138
Appalachian Mountains, 101
arid cycle of erosion, 169
arroyos, 169
ASHPLUME plume deposition model,
    83

balance velocity method, for ice
    velocities, 136, 149
basal sliding
    introduction to, 20
Basin and Range, 1–2
Beartooth Mountains
    cirques of, 27
bedload transport, 8, 171
bedrock channels
    abrasion, 4
    cavitation, 4

coupled evolution with alluvial
    channels
    models for, 101–103
    introduction to, 4
    plucking, 4
sediment flux driven model
    behavior of in block-uplift
        example, 94
    code for 2D implementation,
        248
    introduction to, 6, 94
    role in model of persistent
        mountain belt topography,
        105–107
sediment flux driven model
    application to the Sierra
        Nevada, 96–101
stream power model
    application to the Sierra
        Nevada, 96–101
    as a type of advection equation,
        87
    behavior of in block-uplift
        example, 94
    code for 2D implementation,
        243
    introduction to, 5
    role in model of persistent
        mountain belt topography,
        105–107
Bessel function
    code for numerical integration,
        232
    code for summing series solutions
        for volcanic cones, 233
    use of for modeling diffusive
        evolution of cinder cones, 49
boundary condition
    introduction to, 34
boundary conditions
    periodic vs. fixed, 223
Brooks Range
    glacier model reconstruction of,
        146

channel head
    definition of, 4
channel meandering, 164
    introduction to, 164
    model of, 165–166

channel oscillations
 code for implementation, 270
 model for, 169–173
 observations of, 169, 173
cinder cones
 use of diffusion equation for,
  49–51
cirques, 27
climate oscillations
 ice ages
  model for, 210–220
 role in modulating sediment
  production from drainage
  basins, 11
 role of in controlling dust
  emission, 194
coherence resonance
 introduction to, 211
 model for ice ages
  code implementing the model,
  277
compaction, *see* deformable porous
  media
complementary error function
 definition of, 37
conservation of mass
 role in the diffusion model of
  landform evolution, 9, 30, 170
contaminant dispersion in fluvial
  systems
 introduction to, 74
 previous work on, 75
 role of bed scour in, 77
cosmogenic isotopes
 erosion rates in Sierra Nevada, 93
 role in dating fan terraces, 10
 role in understanding hillslope
  evolution, 3
Crank–Nicholson method, 62

D∞ flow routing, 68
debris flow
 rheology of, 22
deformable porous media, 174
delivery ratio, sediment, 119
delta progradation
 diffusion equation model for,
  51–53
dendrochronology
 use of to infer channel bank
  migration rates, 166
desert pavement
 coevolution with dust deposition,
  15

introduction to, 15
 role in trapping windblown dust,
  15, 194
 role of parent-material texture in,
  15
deterministic models
 limitations of, 188
diffusion equation
 analytic methods for, 34–57
 as one part of a more complex
  model, 32
 model for hillslope evolution
  introduction to, 3
  requirements for applicability,
  30, 33
 nondimensionalization of, 37
 use in foreland basin modeling,
  116
 with stochastic noise, 191–193
Diffusion-Limited Aggregation (DLA)
 application to drainage network
  evolution, 198
diffusivity
 typical values for western US, 33
 units of, 30
discretization
 introduction to, 57
divides
 no-flux boundary condition, 2, 34
drainage density
 definition of, 4
 role in producing sediment-flux
  variations through time, 10
 role of in contaminant dispersion
  model, 77
 use of to distinguish channels
  from hillslopes in 2D
  drainage network evolution
  modeling, 96
droughts, 191
drumlins
 controls on geometry, 182
 introduction to, 23, 174
 role of sediment deformation in
  creating, 174
 stratigraphy of, 182
 use of to reconstruct ice flow
  directions, 23
dust cycle, 53, 194
 importance of, 14
 introduction to, 13
 open system nature of, 13
dust transport and deposition
 model of

application to Amargosa Valley,
  54–57
 introduction to, 14

elasticity, *see* flexure
entrainment
 introduction to, 14
eolian dunes
 barchans
  how they form, 19
 introduction to, 18
 longitudinal
  how they form, 19
 relationship to ripples and
  megadunes, 18
 role of grain size in, 18
 role of sand supply and
  wind-direction variability in,
  18
 spacing of, 168
 Werner's model, 166–169
  code for implementation, 267
  introduction to, 19
equilibrium line altitude (ELA)
 introduction to, 26
 late Miocene lowering of, 29
 localized erosion beneath, 27
 map of in the western US, 121
 role in model of cirque formation,
  27
erosion rates
 determined by cosmogenic
  isotopes, *see* cosmogenic
  isotopes
 relationships with elevation and
  relief, 101, 103
error function
 definition of, 37
 use of for modeling radionuclide
  concentrations in soil, 48
 use of in solutions for scarp
  evolution, 43
Eulerian vs. Lagrangian methods,
  90
excitable media, 185
explicit vs. implicit numerical
  methods, 58, 62
extension
 role in Basin and Range
  topography, 1

fault scarps
 application of the diffusion
  equation to, 42–45

Finger Lakes, 181
  drumlin field, 182
  introduction to, 23
  model for formation of,
      157–158
  model of ice flow over, 144
  observations of, 159
firn, 20
flexure
  coherence method
      use of to map elastic thickness,
          111
  elastic thickness
      definition of, 111
  equation
      introduction to, 111
  forebulge
      definition of, 112
  integral method for, 113
  introduction to, 109
  series solutions, 112–116
  solution methods in 1D, 111–113
  solution methods in 2D, 113–116
  wavelength
      comparison of values for rock
          vs. ice loading, 112
      role in controlling glacial
          erosion, 152
      typical range, 109
flood-envelope curve
  role of in contaminant dispersion
      model, 78
flow routing methods
  $\rho8$, 67
  $D\infty$, 68
  D8, 66
  DEMON, 68
  introduction to, 66
  multiple flow direction (MFD)
      routing method, see Multiple
      Flow Direction (MFD) routing
      method
  range of applications for, 66
  what is best?, 69
flux-conservative equations, 88
foreland basins
  modeling of, 116–120
Forward Time Centered Space (FTCS)
      method
  code for 1D implementation of
      nonlinear diffusion model,
      227
  code for 2D implementation of
      diffusion equation, 225

failure of for the advection
    equation, 90
  introduction to, 57
  stability criterion for, 59
  why use it?, 59
Fourier filtering method
  code for 1D generation of
      fractional noises, 274
  code for 1D implementation to
      solve flexure equation, 256
  code for 2D generation of
      fractional noises, 275
  code for 2D implementation to
      solve flexure equation, 258
  use of for solving the flexure
      equation, 112–116
  use of to generate fractional
      noises, 190
Fourier series method
  for solving the diffusion equation,
      35
  use of for linear equations only,
      36
fractal scaling
  of drainage networks, 196
  of dust accumulation, 195
  of landslide distributions, 205–210
  of soil moisture fields, 207
  of topographic transects, 207
fractional noises
  geoscience applications of, 190
  how to construct, 190
  introduction to, 190

Gaussian plume
  definition of, 15
glacial buzzsaw hypothesis, 28
glacial erosion
  Hallet's model for, 21, 151
  introduction to, 21
  modeling flexural response to
      unloading, 120–123
glacially-carved lakes
  frequency-size distribution of, 24
  introduction to, 24
  model for, 153
  near ice margins, 157
glaciers
  1D profile modeling of
      uniformly-sloping bed, 127
      wavy bed, 127
  difference from ice sheets, 25
  introduction to, 20
  role in sediment production, 12

Glen's Flow Law
  1D profile modeling of ice body
      subject to, 128
  introduction to, 22
Great Lakes
  as outliers in the frequency-size
      distribution of
      glacially-carved lakes, 25
  model for, 154
  role of flexure in controlling, 154
Greenland ice sheet
  model reconstruction of, 137
  use in code implementing the
      sandpile model, 265
Greenland ice sheet (LIS)
  introduction to, 19
groundwater flow
  as a type example of an
      advection-diffusion process,
      32

Hanaupah Canyon
  application of ADI method to
      modeling evolution of, 63
  example of terrace evolution in,
      38
  type example of a fluvial system,
      1–12
  use of for illustrating
      multiple-direction
      flow-routing methods, 68
hanging valleys, 26
hillslope processes
  bioturbation, 33
  creep, 33
  mass movements, 33
      model for hillslope evolution
          with, 60, 61
      models for, 205–210
  model for tephra transport at
      Yucca Mountain, 75
  rain splash, 33
  rilling, 33
  slope wash, 33
      model for hillslope evolution
          with, 33, 59
  those that are diffusive, 33
Horton's Laws
  introduction to, 196
  statistical inevitability of, 197
hummocky moraine, 178
Hurst phenomenon, 189
hydrologically-corrected DEM
  code for implementation, 235

ice sheets
  basal shear stress
    observations of, 132
  cold vs. warm-based, 21
ice-albedo feedback, 214
  role of in controlling bistability of
    the climate system, 215
implicit method
  computational advantage of,
    62
  introduction to, 62
  stability of, 62
isostasy
  Airy limit, 113
  in the evolution of the Sierra
    Nevada, 97
  introduction to, 109
  role in ice sheets
    introduction to, 23
    model of, 127, 139
  role in prolonging the mountain
    belt denudation, 109
  role in relief production, 28
  role in shaping Basin and Range
    topography, 1
  role in the formation of Great
    Lakes, 25
  role of in 2D bedrock drainage
    network modeling, 94
  time-dependent response of,
    111

Kardar–Parisi–Zhang (KPZ) equation,
    207
kei(r)
  series expansions for, 257
  use as a basis function for series
    solutions to the flexure
    equation, 114
Kelvin functions, see kei(r)
knickpoints
  introduction to, 6
  modeling of using advection
    equation, 90

Langevin equations, 191
Lathrop Wells cone, see Amargosa
    Valley
Laurentide ice sheet (LIS)
  impact on global climate, 215
  introduction to, 23
  model of, 138
lava channels
  meandering of, 164

lava flow
  2D radially-symmetric model
    code for implementation,
      263
  modeling of
    temperature-dependent
    viscous behavior of, 130–
      132
  rheology of, 22, 125
LIDAR, 67, 71
linear stability analysis, 165
  application to channel
    meandering, 164
  introduction to, 161
  of oscillating alluvial channels,
    171
lithology
  role in controlling bedrock
    channel erosion, 5, 103
load-accumulation feedback, 216, 218
load-advance feedback, 217
longitudinal profile
  analytic model for coupled
    bedrock-alluvial channels,
      103
  code implementing 1D coupled
    bedrock-alluvial model, 242
  influence on glacial erosion, 21
  model for steady-state bedrock
    channel, 5
Lorentzian spectrum, 212

Mars
  application of flow routing
    methods to, 69
  eolian dunes on, 168
  fluvial activity on, 12
  spiral troughs
    code for 1D model
      implementation, 271
    model of, 183–187
method of characteristics, see
    advection equations
Microsoft
  Excel, 223
  Visual C++, 222
Mohr–Coulomb criterion, 207
moisture
  role in controlling particle
    entrainment
    introduction to, 14
  role of in triggering mass
    movements, 205
Monte Carlo methods, 188

moraines
  application of the diffusion
    equation to, 40–42
mortality
  role of particulate matter in
    causing, 14
Multiple Flow Direction (MFD)
    routing method
  code for implementation, 236
  code for implementation in flood
    hazard analysis, 239
  introduction to, 67
  mathematical definition of, 67
  use of for improving USGS DEMs,
    71
  use of in contaminant dispersal
    model, 78
  use of in estimating flood hazards,
    72–74

New York State drumlin field, 181
Newton iteration, 130
non-Newtonian fluids
  analytic solutions for, 126–129
  introduction to, 22
  numerical solutions for, 129–132
  rheology of, 22
nonlinear advection equations, 90
nonlinear transport on hillslopes
  introduction to, 3
Numerical Recipes
  use with codes in this book,
    222

palimpsests, 25
perfectly plastic model
  application to climate modeling,
    216
  application to ice sheets and
    glaciers, 22
  application to thrust sheet
    mechanics, 147
  generalization of to
    threshold-sliding behavior,
      127
  introduction to, 22
  limitations of, 22
  use of in modeling 1D ice sheet
    profiles, 125
persistent mountain belt paradox
  introduction to, 101
  role of piedmont deposition in,
    101–103
piedmont, see alluvial fan

playa
    introduction to, 12
    role of hydrology in dust
        production
        introduction to, 13
    sand-dominated, 17
Pleistocene–Holocene transition
    role in triggering terrace
        formation, 10
pluvial shorelines
    application of the diffusion
        equation to, 42–45
    introduction to, 12
polar coordinates
    use of for solving diffusive
        evolution of cinder cones,
        49
positive feedback
    between channel width and slope
        in alluvial channels, 161
    between glaciers, climate, and
        topography, 28
    elevation-accumulation feedback
        in ice sheets, 20
    ice-albedo feedback, 213
    in hillslope evolution, 4
    role of in geomorphic instabilities,
        161
precision, single vs. double, 222

radiation balance
    role of in climate models, 213
radionuclides
    atmospheric nuclear testing as a
        man-made source, 46
    dispersion in soils
        diffusion equation model for,
        45–49
        processes of, 48
random walk
    application to dust accumulation
        on alluvial fan terraces,
        194–196
    bounded, 195
    introduction to, 193
Rayleigh–Taylor instability, 162–164
    analogy to drumlin formation,
        178
recirculation zones
    role in controlling eolian
        deposition, 18, 167
relief production in glaciated terrain
    introduction to, 28
RiverTools, 223

Rogen moraine, 179
root-finding techniques, 50

salt domes
    as an example of the
        Raleigh–Taylor instability,
        162
saltation
    characteristic distance of, 167
    introduction to, 13
sandpile method
    code for implementation, 264
    use of in 2D perfectly plastic
        model, 135
secondary flow, 164
Shreve magnitude, 197
Sierra Nevada
    knickpoint propagation
        introduction to, 6
        modeling of, 93–101
    non-equilibrium landscape of,
        93
similarity method
    application to solving diffusion
        equations
        introduction to, 36
        use in modeling deltaic
        sedimentation, 53
    use of in modeling terrace profile
        evolution, 40
sine function
    as functional form of late-stage
        terraces following base level
        fall, 40
sine-generated curve, 166
soil moisture
    variations in space and time
        modeling of, 206
        observations of, 206
solitary waves, 173
spring sapping
    role of in drainage network
        evolution, 198
Stefan–Boltzmann Law, 212
Stokes' Law, 175
Strahler ordering, 196
stratigraphic completeness, 196
stream function
    use of in fluid mechanics
        problems, 162, 175
sublimation
    role of in forming spiral troughs,
        184
surficial geologic mapping, 10

Taylor expansion
    use of in discretization, 57
terraces
    dust deposition on
        introduction to, 13
    implementation of 2D diffusion
        model, 226
    introduction to, 9
    mechanisms for formation, 10
    morphological age dating of, 39
    on Mars, 12
    relative age indicators of, 10
    role in controlling flood risk,
        10
    transient response to base level fall
        use of diffusion equation to
        model, 38–40, 61–62
thermal erosion, 166
thermochronology, 29
threshold of critical power, 171
Tibet
    archeological ruins of, 45
till fabric, 182
time series analysis
    autocorrelation function, 189
    introduction to, 188
    power spectrum, 189
    of Pleistocene climate, 212
Tokunaga side branching, 197
topographic inversion, 131
topographic steady state
    convergence of stream-power and
        sediment flux driven models
        in, 8
    in hillslopes, 3
    in ice sheet modeling, 129
    model for hillslope profile during
        transient approach to, 37
    model for hillslope profile in,
        34
    slope vs. area curve, 5
    use of in foreland basin modeling,
        117
topologically distinct channel
        networks (TDCNs), 197
transport capacity
    role in controlling bedrock vs.
        alluvial channel types, 4, 8
tridiagonal matrix
    use of in implicit numerical
        methods, 62
Tule Valley
    shoreline scarps in, 42
two-step Lax–Wendroff, 173

U-shaped valleys, 26
US Geological Survey
   DEMs
      problems with, 70
Uinta Mountains
   glacial erosion in, 27
UNIX cc compiler, 222
Ural Mountains, 101

V-shaped valley, 26
vegetation

fossil record of in packrat
   middens, 11
influence on weathering, 3,
   10
Vialov solution, 128
visualization, 223
Vostok ice core, 214, 219
   comparison with coherence
      resonance model of ice ages,
      220
   histogram of, 220

power spectrum of, 220

weathering front
   role in hillslope evolution, 3
Wind River Range
   glacial erosion in, 26
Wisconsin drumlin field, 180

Yucca Mountain
   volcanic hazard associated with,
      75